Plant Pathology and Plant Pathogens

Plant Pathology and Plant Pathogens

Fourth Edition

John A. Lucas

Rothamsted Research
Harpenden
Hertfordshire, UK

WILEY Blackwell

This fourth edition first published 2020
© 2020 John Wiley & Sons Ltd

First Edition 1977 by Blackwell Science Ltd
Second Edition 1982
Third Edition 1998
Reprinted 2002

Registered Office(s)
John Wiley & Sons, Inc., 111 River Street, Hoboken, NJ 07030, USA
John Wiley & Sons Ltd, The Atrium, Southern Gate, Chichester, West Sussex, PO19 8SQ, UK

Editorial Office
The Atrium, Southern Gate, Chichester, West Sussex, PO19 8SQ, UK

For details of our global editorial offices, customer services, and more information about Wiley products visit us at www.wiley.com.

Wiley also publishes its books in a variety of electronic formats and by print-on-demand. Some content that appears in standard print versions of this book may not be available in other formats.

Library of Congress Cataloging-in-Publication Data

Names: Lucas, J. A. (John Alexander), 1948– author.
Title: Plant pathology and plant pathogens / prof. John A. Lucas,
    Rothamsted Research, West Common, Harpenden, Hertfordshire, UK.
Description: Fourth edition. | Hoboken : Wiley-Blackwell, 2020. | Includes
    bibliographical references and index.
Identifiers: LCCN 2019051991 (print) | LCCN 2019051992 (ebook) | ISBN 9781118893869
    (paperback) | ISBN 9781118893821 (adobe pdf) | ISBN 9781118893852 (epub)
Subjects: LCSH: Plant diseases. | Phytopathogenic microorganisms–Control.
    | Plant-pathogen relationships.
Classification: LCC SB731 .D5 2020 (print) | LCC SB731 (ebook) | DDC
    632–dc23
LC record available at https://lccn.loc.gov/2019051991
LC ebook record available at https://lccn.loc.gov/2019051992

Cover Design: Wiley
Cover Image: Confocal microscopy image of a GFP-labelled strain of the cereal eyespot fungus *Oculimacula* colonizing a wheat coleoptile. Source: Bowyer et al. 2000.

Set in 9.5/12.5pt STIXTwoText by SPi Global, Pondicherry, India

Printed in the UK by Bell & Bain Ltd, Glasgow.

10 9 8 7 6 5 4 3 2 1

Contents

Preface

Plant pathology is an applied science concerned primarily with practical solutions to disease problems in agriculture, horticulture, and forestry. The study of plant disease and the development of methods for its control continue to be vital elements in the drive to improve crop productivity and ensure global food security. Projected world population growth and the impacts of climate change make this a priority. From a more strategic perspective, the experimental analysis of the interactions between plants and pathogens has become fertile territory for scientists interested in emerging areas of plant biology, including recognition and response systems, signal pathways, and innate plant immunity. The use of molecular genetic techniques has provided new insights into how pathogens cause disease and how plants defend themselves against attack. In turn, this new understanding is suggesting novel ways in which plant disease might be controlled.

This book provides an introduction to the main elements of plant pathology, including the applied aspects of disease identification, assessment and control, and fundamental studies of host–pathogen interactions. The aim is to place the subject not only in the context of agricultural science but also in the wider spheres of ecology, population biology, cell biology, and genetics. The impact of molecular biology and genomics on the diagnosis and analysis of plant disease is a recurrent theme.

A large number of people have helped me in the production of this new edition. First, I should acknowledge the invaluable help of many of my colleagues at Rothamsted Research who have generously provided advice, assistance, and materials for use in the book. They include Kim Hammond-Kosack, Jason Rudd, Kostya Kanyuka, Nicola Hawkins, Ana Machado, Bart Fraaije, Jon West, Stephanie Heard, Anastasia Sokolidi, Kevin King, John Jenkyn, Roger Plumb, James Bell, David Hughes, Mike Adams, Graham Shephard, and Sally Murdoch. Joe Helps advised on mathematical modeling as well as replotting data. The library staff, Tim Wales, Catherine Fernhead, and Chris Whitfield, helped with tracing publications and scanning figures. Special thanks are due to Lin Field and Sheila Bishop who supported me as a visiting worker in the Biointeractions and Crop Protection department. Numerous others in the plant pathology and crop protection community have also supplied images, articles, advice, and data for use in figures and tables, including Bruce McDonald, Mike Coffey, George Sundin, Pamela Gan, Richard O'Connell, Petra Boetink, Yaima Arocha-Rosete, Geert Haesaert, David Cooke, Paul Birch, Diane Saunders, Judith Turner, Moray Taylor, Sarah Holdgate, Nicola Spence, Ana Lopez, Hans Cools, Mike Davey, Jurriaan Ton, Stephen Rolfe, Paul Bowyer, Clive Brasier, Joan Webber, John Mansfield,

Lili Huang, Victoria Routledge, Lucy Carson-Taylor, Jason Pollock, Dierdre Haines, David Whattoff, Graham Matthews, Haydn Beddows, Dennis Gonsalves, and Malcolm Briggs. Thanks go to Holly Regan-Jones for her skilful copy editing and Dave Gardner who re-drew many of the Figures. Vimali Joseph provided vital support during the final production process. My appreciation should also be extended to previous colleagues at the University of Nottingham, especially Alison Daniels and the late Brian Case, who took many of the studio photos featured in this edition. Space does not allow me to acknowledge the many other colleagues who agreed to the use of published results.

Finally, thanks are also due to my daughter Maria Gabriella, whose grasp of digital technology greatly exceeds my own, and my wife Adélia, not only for her forbearance throughout the time taken to complete this project but also for her active contribution searching out all manner of useful information on the internet.

John A. Lucas
January 2020

List of Abbreviations

ABA	Abscisic acid
AUDPC	Area under disease progress curve
BCA	Biological control agent
CDA	Controlled droplet application
CDPK	Calcium-dependent protein kinase
CK	Cytokinin
CWDE	Cell wall-degrading enzyme
DAMP	Damage-associated molecular pattern
DSS	Decision support system
EHM	Extrahaustorial membrane
ELISA	Enzyme-linked immunosorbent assay
ETI	Effector-triggered immunity
GA	Gibberellic acid
GLA	Green leaf area
GPS	Global positioning system
GWAS	Genome-wide association studies
HGT	Horizontal gene transfer
HIGS	Host-induced gene silencing
HLB	Huanglongbing disease of citrus, also known as citrus greening
HR	Hypersensitive response
HRGP	Hydroxyproline-rich glycoprotein
Hrp	Hypersensitive response and pathogenicity
HST	Host-specific toxin
IAA	Indole acetic acid
IPPC	International Plant Protection Convention
ISR	Induced systemic resistance
JA	Jasmonic acid
LAMP	Loop-mediated isothermal amplification assay
LGT	Lateral gene transfer
LPS	Lipopolysaccharide
LYD	Lethal yellowing disease of coconuts
MAMP	Microbe-associated molecular pattern
MAPK	Mitogen-activated protein kinase

NDVI	Normalized difference vegetation index
NGS	Next-generation sequencing, also known as high-throughput sequencing
NLR	Family of receptor proteins characterized by nucleotide-binding site and leucine-rich repeat domains
NO	Nitric oxide
NRPS	Nonribosomal peptide synthase
PAL	Phenylalanine ammonia lyase
PAMP	Pathogen-associated molecular pattern
PCD	Programmed cell death
PCR	Polymerase chain reaction
PG	Polygalacturonase
PGIP	Polygalacturonase-inhibiting protein
PKS	Polyketide synthase
PRA	Pest risk analysis
PRR	Pattern recognition receptor
PTI	PAMP-triggered immunity
QTL	Quantitative trait locus
REMI	Restriction enzyme-mediated integration
RIP	Ribosome-inactivating protein
RLK	Receptor-like kinase
RLP	Receptor-like protein
ROS	Reactive oxygen species
RT-PCR	Real-time polymerase chain reaction
RUE	Radiation use efficiency
SA	Salicylic acid
SAR	Systemic acquired resistance
SIGS	Spray-induced gene silencing
SNP	Single nucleotide polymorphism
T3SS	Type 3 secretion system
UAV	Unmanned aerial vehicle
WRKY	(pronounced *worky*) Family of transcription factors containing a WRKY amino acid domain

About the Companion Website

This book is accompanied by a companion website:

The URL is **www.wiley.com/go/Lucas_PlantPathology4**

The website includes:

Figures from the book

Part I

Plant Disease

We see our cattle fall and our plants wither without being able to render them assistance, lacking as we do understanding of their condition.

(J.C. Fabricius, 1745–1808)

The health of green plants is of vital importance to everyone, although few people may realize it. As the primary producers in the ecosystem, green plants provide the energy and carbon skeletons upon which almost all other organisms depend. The growth and productivity of plants determine the food supply of animal populations, including the human population. Factors affecting plant productivity, including disease, therefore affect the quantity, quality, and availability of staple foods throughout the world. Nowadays crop failure, due to adverse climate, pests, weeds, or diseases, is rare in developed agriculture, and instead there are surpluses of some foods. Nevertheless, disease still takes a toll, and much time, effort, and money are spent on protecting crops from harmful agents. In developing countries, the consequences of plant disease may be more serious, and crop failure can damage local or national economies, and lead directly to famine and hardship. Improvements in the diagnosis and management of plant disease are a priority in such instances. Furthermore, the pressures on plant productivity are increasing. The area of cultivated land available per person on the planet declined from around 0.4 ha in the 1960s to less than 0.3 by the year 2000 (FAO data), and as the human population continues to multiply the area will further decrease.

As well as supplying staple foods, plants provide many other vital commodities such as timber, fibers, oils, spices, and drugs. The use of plants as alternative renewable sources of energy and chemical feedstocks is becoming more and more important, as other resources such as fossil fuels are depleted and the need to mitigate climate change becomes a priority. Finally, the quality of the natural environment, from wilderness areas to urban parks, sports fields, and gardens, also depends to a large extent on the health of plants.

Healthy plants provide a series of benefits for the farmer, food chain, and the environment (Table 1). The yield and quality of crop products are ensured, and healthy plants are more efficient at using precious resources such as water and nutrients. In doing so, they

Plant Pathology and Plant Pathogens, Fourth Edition. John A. Lucas.
© 2020 John Wiley & Sons Ltd. Published 2020 by John Wiley & Sons Ltd.
Companion website: www.wiley.com/go/Lucas_PlantPathology4

Table 1 Benefits of healthy plants

- Greater yield
- Superior quality
- More competitive with weeds
- Easier harvesting
- Fewer residual nutrients, reduced pollution
- Improved control of soil erosion
- Lower carbon footprint

also prevent losses of nitrogen and other nutrients to the wider environment, and reduce pollution of rivers and ground water, which in many areas is a potential problem for drinking supplies. Vigorous, healthy plants are more competitive with weeds, and are easier to harvest than crops that are stunted or collapsed. Plant root systems play an important role in reducing soil erosion, and thereby help to conserve another precious resource. Finally, given the mounting concerns about climate change, it should be noted that healthy crops have a lower carbon footprint than diseased crops, due to their greater productivity and more efficient use of inputs per area of cultivated land.

The science of **plant pathology** is the study of all aspects of disease in plants, including causal agents, their diagnosis, physiological effects, population dynamics and control. It is a science of synthesis, using data and techniques from fields as diverse as agriculture, microbiology, meteorology, engineering, genetics, genomics, and biochemistry. But first and foremost, plant pathology is an applied science, concerned with practical solutions to the problem of plant disease. Part of the appeal of the subject is to be found in this mixture of pure and applied aspects of biology.

The scope of plant pathology is difficult to define. On a practical level, any shortcoming in the performance of a crop is a problem for the plant pathologist. In the field, he or she may well be regarded in the same way as the family doctor – expected to provide advice on all aspects of plant health! A distinction is often drawn between **disease** caused by infectious agents and **disorders** due to noninfectious agents such as mineral deficiency, chemical pollutants, or adverse climatic factors. The main emphasis of this book is on disease caused by plant pathogenic microorganisms such as fungi, oomycetes, bacteria, and viruses. Under favorable conditions, these pathogens can multiply and spread rapidly through plant populations to cause destructive disease **epidemics**. Many of the principles discussed apply equally well, however, to other damaging agents such as insect pests and nematodes.

A fundamental concept in plant pathology is the **disease triangle** (Figure 1) which shows that disease results from an interaction between the host plant, the pathogen, and the environment. This can be enlarged to include a further component, the host–pathogen complex (Figure 1), which is not simply the sum of the two partners, as the properties of each are changed by the presence of the other.

A comprehensive analysis of plant disease must take all four components into account. Obviously, one needs to be familiar with the characteristics of the host and the pathogen in isolation. The successful establishment of a pathogen in its host gives rise to the host–pathogen complex. Unraveling the dynamic sequence of events during infection, the

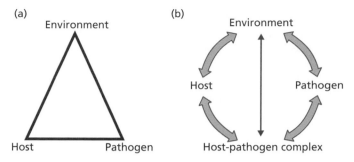

Figure 1 The disease triangle (a) incorporating the host–pathogen complex (b).

molecular "cross-talk" taking place between the partners, is one of the most challenging problems in experimental biology. In addition, the effects of the environment on each of other components must be understood. This includes not only physical and chemical factors but also macro- and microbiological agents. The two-way arrow between the host and the environment in Figure 1b should not be overlooked, as populations of crop plants often have important effects on their surrounding microclimate. For example, the relative humidity within a crop canopy is higher than that outside and this will favor the development of some microorganisms. Plants attacked by pests and pathogens often produce volatile compounds that can act as signals sensed by neighboring plants. Effects of pathogens on the environment are more subtle but may be significant; some fungi, for instance, produce the volatile hormone ethylene, which can in turn affect the development of adjacent host plants.

This book is intended to provide an outline of the main elements of modern plant pathology. The approach is designed to achieve a balance between laboratory and field aspects of the subject, and to place the phenomenon of plant disease in a wider biological context. Research in plant pathology can be broadly divided into **tactical** and **strategic** aspects. The former is concerned with providing solutions to disease problems by identifying causal agents and evaluating the most cost-effective options for their control. The latter is longer term and aims to understand fundamental aspects of plant disease such as pathogen ecology, population biology, host–pathogen interactions, and plant immunity. This knowledge can then be applied to devise improved methods of disease control.

The book is divided into three parts. The first focuses on the problem of plant disease, the causal agents and their significance, disease diagnosis, and the development of epidemics in plant populations. This account highlights the influence of environmental factors on the multiplication and spread of pathogens, and the use of climate data in disease prediction. The second part deals with host–pathogen interactions: how pathogens gain entry to the host, how their growth and development in the plant lead to disease symptoms, and how the plant responds. The outcome of any host–pathogen confrontation depends on a dynamic interplay between factors determining microbial pathogenicity and the active mechanisms of plant immunity. This interaction ultimately determines host–pathogen specificity, whereby any pathogen is able to cause disease in only a restricted range of host plants. Plant and molecular biologists are studying host–pathogen systems as experimental models to probe the mechanisms of gene expression and regulation, and to reveal details of the evolutionary "arms race" taking place between plants and pathogens. Part III deals with the practical business of disease control, often described as **crop protection**. This covers the management of disease by means of chemicals, breeding for resistance, and alternative biological approaches. Finally, the combined use of cultural practices and all these other measures to provide sustainable, integrated systems for disease control is described.

A comprehensive treatment of individual diseases and the methods used in their control is beyond the scope of a text of this length. For the sake of brevity, specific pathogens or the diseases they cause are often mentioned without further explanation. This approach may be likened to that adopted in many ecology texts, in which the reader is expected to be familiar with most of the higher plants or animals discussed therein. There is, however, an appendix listing all the pathogens and diseases mentioned in the book, together with brief details which will enable the reader to obtain further information about particular diseases. More detail concerning specific aspects of pathology may be obtained by consulting the recommended further reading.

Further Reading

Books

Agrios, G. (2005). *Plant Pathology*, 5e. Amsterdam: Elsevier Academic Press.

Schumann, G.L. and D'Arcy, C.J. (2009). *Essential Plant Pathology*, 2e. St Paul, Minnesota: American Phytopathological Society Press.

Strange, R.N. (2003). *Introduction to Plant Pathology*. Chichester: Wiley.

Reviews and Papers

Foley, J.A., Ramankutty, N., Brauman, K.A. et al. (2011). Solutions for a cultivated planet. *Nature* 478: 337–342.

Godfray, H.C.J., Beddington, J.R., Crute, I.R. et al. (2010). Food security: the challenge of feeding 9 billion people. *Science* 327: 812–818.

Savary, S., Bragaglio, S., Willocquet, L. et al. (2017). Crop health and its global impacts on the components of food security. *Food Security* 9: 311–327. https://doi.org/10.1007/s12571-017-659-1.

Scientific Journals

Many scientific journals contain reviews and research papers relevant to plant pathology. One especially useful source is the Annual Review of Phytopathology. Others include:
Advances in Botanical Research
Annals of Applied Biology
Crop Protection
European Journal of Plant Pathology
Fungal Genetics and Biology
Journal of Phytopathology
Molecular Plant–Microbe Interactions

Molecular Plant Pathology
Mycological Research
New Phytologist
Pest Management Science
Plant Disease
The Plant Cell
The Plant Journal
Physiological and Molecular Plant Pathology
Phytopathology
Plant Pathology
PLOS Pathogens
American Phytopathological Society: www.apsnet.org
British Society of Plant Pathology: www.bspp.org.uk
European Foundation for Plant Pathology: www.efpp.net
Review of Plant Pathology, an abstracts database of plant pathology research: www.cabi.org/publishing-products/online-information-resources/review-of-plant-pathology

1

The Diseased Plant

Since it is not known whether plants feel pain or discomfort, and since, in any case, plants do not speak or otherwise communicate to us, it is difficult to pinpoint exactly when a plant is diseased.

(George N. Agrios, 1936–2010)

The significance of disease in plants varies depending upon biological, agricultural, and socioeconomic factors. At one extreme, disease may be so severe that the farmer is faced with total crop failure, and the need for control measures is immediately obvious. In other cases, it may be difficult to define disease symptoms, the cause of the problem is not initially clear, and any benefits obtained from control measures are not easy to predict. This chapter discusses the nature of disease and surveys the range of pathogens, pests, and other agents which adversely affect plants. The impact of disease, both in natural plant communities and in agriculture, forestry and horticulture, is then considered.

Concepts of Disease

To fully understand the nature of disease, one must first identify the processes occurring during the growth and development of the healthy plant. Such an analysis may be done at three levels:

- the sequence of events comprising the normal plant life cycle
- the physiological processes involved in plant growth and development
- the metabolic pathways and molecular reactions underlying these processes.

Seed germination, maturation of vegetative structures, the initiation of reproduction, and the formation and dispersal of fruits and seeds are all critical phases of the life cycle at which disease may occur (Figure 1.1). At each stage in this developmental sequence, the integration of several physiological processes is essential for the continued development of the plant. Cell division and differentiation, the fixation and utilization of energy (photosynthesis and biosynthesis), transport of water and nutrients (transpiration and translocation), and storage of reserve compounds are all necessities for growth. Each of these functions involves a complex series of molecular events which comprise the overall metabolism of

Plant Pathology and Plant Pathogens, Fourth Edition. John A. Lucas.
© 2020 John Wiley & Sons Ltd. Published 2020 by John Wiley & Sons Ltd.
Companion website: www.wiley.com/go/Lucas_PlantPathology4

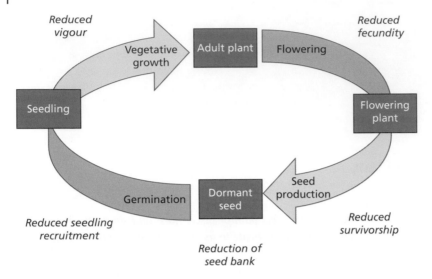

Figure 1.1 A plant life cycle and some effects of disease.

the plant. The nature and regulation of metabolism are themselves determined by the genetic make-up of the plant, interacting with the environment in which it is growing.

Disease may disrupt the activities of the plant at one or more of these levels. Some disorders involve subtle alterations in metabolism which do not affect the successful completion of the life cycle. Certain diseases caused by viruses have only slight effects on the growth of the plant; in such cases, it may be difficult even to recognize the existence of a disease problem. For instance, potato virus X was known as potato healthy virus until virus-free seed potatoes became widely available. Comparisons with infected plants then showed the virus to be capable of causing a 5–10% loss in yield. Other more destructive diseases may interfere with numerous molecular, cellular, and physiological processes and lead to premature death of the plant.

While everyone is familiar with the idea of disease, in practice there may be difficulties in drawing a precise distinction between healthy and diseased plants. No single definition of disease has found universal acceptance; the most widely used involves some reference to the "normal" plant, for instance "a condition where the normal functions are disturbed and harmed" (Holliday 1989). However, there is no consensus as to the exact extent of deviations from this norm which may constitute the diseased state. The problem of defining normality, in terms of the processes outlined above, is further complicated by the variation inherent in all plant populations. Such variation is particularly common in natural populations, especially where hybrids occur, but even within apparently uniform populations of crop plants, there may be differences between individuals. Such differences either have a genetic basis or are due to environmental factors operating during the growth of the crop. If, for instance, one sows seed of an old cereal variety alongside that of a modern, improved cultivar, one will observe major differences between the two crops. In particular, the modern cultivar will be shorter, form much larger seed heads and heavier grains, and the final yield will be greater. The difference in this case is due to intensive selection and genetic improvement rather than to any disease in the old variety, but this example highlights the importance of understanding the initial potential of the plant before accurate estimates of disease can be obtained.

Damage or Disease?

It can be argued that short-term harmful effects on plants, such as injury due to grazing, do not constitute disease. Indeed, some plants, such as the grasses, are well adapted to regular grazing and respond with increased growth if so affected. In cases where damage is sustained over a longer period of time, such as progressive destruction of roots by migratory nematodes or distortion of aerial shoots by exposure to persistent herbicides, the outcome is clearly within the scope of pathology. However, these fine distinctions are of limited use in arriving at a working definition of disease. Such a definition will depend in part on the situation in which it is intended to be used. For example, the biochemist may well be concerned with a malfunction involving a single enzyme and hence view disease as a specific metabolic lesion, whereas the farmer is normally only interested in changes which affect the overall performance of the crop and reduce its value.

Although at present, definitions of disease lack precision, it may ultimately be possible to describe all malfunctions in terms of biochemical changes. To date, this has been achieved in only a few exceptional cases, notably in diseases caused by fungi which produce host-specific toxins, where all the symptoms are due to a single toxic compound acting at a specific target site (see Chapter 8).

Symptoms of Disease

A doctor diagnoses illness in a patient by looking for visible or measurable signs that the body is not functioning normally. Such signs are known as **symptoms** and they may occur singly or in characteristic combinations and sequences. For example, someone suffering from influenza may have a sore throat, fever, and muscular aches and pains. Such a group of symptoms occurring together and in a regular sequence is termed a disease **syndrome**. For many diseases, the occurrence of a particular combination of symptoms is sufficient to arrive at an accurate diagnosis. Alternatively, symptoms may be common to a wide variety of diseases (for instance, fever is a generalized response to both infection and certain types of injury). In such cases, detailed microbiological and biochemical analyses will be necessary to detect other diagnostic symptoms.

Similar considerations apply to the diagnosis of disease in plants. Just as with doctors and human disease, plant pathologists must be aware of the range of visual disease symptoms, the organs affected (Figure 1.2) and what these suggest as the cause of the problem.

The major symptoms of disease in plants are listed in Table 1.1 on the basis of the functions affected. This approach is used because it directs attention to the underlying nature of the disorder. For instance, the presence of galls or other cancerous growths immediately suggests some malfunction in the control of cell division; this in turn implicates a hormonal imbalance and/or genetic change in host cells. It should be realized that this classification of symptoms is to some extent arbitrary and nonspecific. Permanent wilting provides

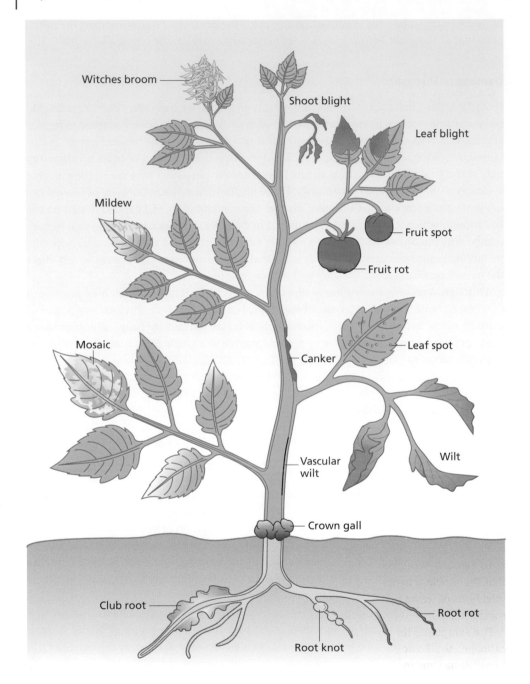

Figure 1.2 Some disease symptoms caused by pathogens infecting different plant organs.

Table 1.1 Symptoms of disease in plants

Symptom	Function affected	Examples
Stunting	General development	Take-all of cereals, barley yellow dwarf, Napier grass stunt (phytoplasma)
Necrosis (cell death)	General	Whole plant – damping off of seedlings
		Leaf tissues – potato late blight, botrytis gray mold of vegetables and ornamentals
		Storage tissues – *Erwinia* and *Dickeya* rot of potatoes and vegetables
		Woody tissues – apple canker, fireblight, chestnut blight
Chlorosis	Photosynthesis	Cereal rusts, beet mild yellowing virus, halo blight of bean, cassava mosaic diseases
Wilting	Water relations	Panama disease of bananas, *Verticillium* wilt of tomatoes, bacterial wilt of cucurbits
Hypertrophy	Growth regulation	Club root of brassicas, maize smut, peach leaf curl
Hyperplasia	Growth regulation	Crown gall, cocoa witches broom, peach leaf curl
Leaf abscission	Growth regulation	Leaf blight of rubber, coffee rust, black spot of roses
Etiolation	Growth regulation	Bakanae disease of rice
Inhibition of flowering and fruiting	Reproduction	Choke of grasses
		Ergot of grasses, cereal smut diseases
Abnormal coloration	Pigment synthesis	Grapevine leaf roll virus, citrus greening, tulip breaking virus

a useful example. Although this symptom suggests that something is interfering with the uptake and transport of water, the symptom itself tells us little about the actual site or cause of the interference. The problem could be due either to a blockage in the vascular system, as in vascular wilt diseases, or to a general destruction of root tissues. It is also possible that the problem has little to do with water uptake or transport; in some diseases, such as infections of leaves by rust fungi, wilting is a sign of excessive water loss due to increased transpiration.

The symptoms listed in Table 1.1 will also interact in numerous ways. In club root of cabbage, the basic symptoms are hypertrophy (abnormal enlargement of cells) and hyperplasia (uncontrolled cell proliferation) in root tissue (Figure 1.3), but the first visible symptom is often wilting of the aerial parts of the plant. Any disruption of normal root development inevitably affects other functions such as water and nutrient transport. In view of the highly integrated nature of life processes, it is hardly surprising that attempts to define symptoms often lack precision.

The relative importance of any symptom will vary, depending not only upon its duration and severity but also on the habit or life form of the plant affected. Hence,

(a)

(b)

Figure 1.3 Club root disease of brassicas. (a) Primary infection causes distortion of root hairs, which contain plasmodia of the pathogen *Plasmodiophora brassicae*. Bar = 50 µm (b). Secondary infection of the main root leads to division and enlargement of cortical cells to produce the typical "clubbed" root symptom.

(a)   (b)

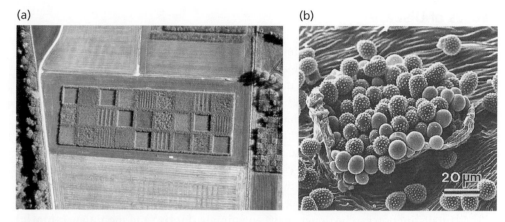

Figure 1.4 Rust of willows caused by *Melampsora* species. (a) Aerial view of experimental trial of willow clones in summer. Plots which appear empty have lost their leaves due to severe rust infection. *Source:* Courtesy of D.J. Royle. (b) Scanning electron micrograph of rust pustule on willow leaf showing spiny urediospores of the fungus. *Source:* Spiers and Hopcroft (1996).

necrosis in the stem of an herbaceous seedling will probably lead to the death of the whole plant, while necrotic lesions (known as cankers) in the stem of a woody perennial may only result in the loss of a twig or branch. If, however, such a lesion girdles the trunk of a tree, then translocation will be disrupted to the extent that the plant will die.

Pathogens which actually kill plants are the exception. More commonly, disease symptoms indicate an impairment of the efficiency of plant physiological and metabolic processes (Table 1.1). Some symptoms, such as local changes in pigmentation, may be trivial in terms of overall plant performance. Often, the most important consideration is the stage in the life cycle at which symptoms first appear. Severe chlorosis or even necrosis of the first-formed leaves of a cereal may have little effect upon the final yield, as these leaves senesce naturally during crop growth, and most of the photosynthetic products required for grain filling are provided by the top three leaves and ear tissues. Accelerated abscission of leaves is unlikely to be a problem in annual herbaceous plants but in perennials, it may exert a severe drain on the food reserves of the plant. For instance, Figure 1.4 shows defoliation of some willow clones due to infection by the rust fungus *Melampsora*. Loss of photosynthetic tissue reduces the biomass produced by the crop. A similar symptom can be seen in coffee bushes affected by another rust, *Hemileia*, or in rubber trees affected by the leaf blight fungus *Microcyclus ulei*. In both these evergreen crops, early leaf fall is often followed by the production of a second flush of leaves. If these are also prematurely lost due to further infection then the plant loses vigor and may eventually die.

While visual symptoms are still routinely used to diagnose diseases and disorders in crops growing in the field, in recent years a range of molecular assay techniques have become available to directly detect the agents causing the symptoms. Such molecular diagnostics are discussed in more detail in Chapter 4.

Causes of Disease

Any agent capable of adversely affecting green plants may be regarded as lying within the scope of plant pathology. The principal agents involved in plant disease are shown in Figure 1.5. Partial or total crop failure may be due to one or more agents. Where more than one agent is responsible, each may act independently, or they may interact. In the latter instance, there may be **synergism**, that is, two or more agents acting in combination to cause symptoms that are more severe than those produced by either agent alone. Synergism has been demonstrated to occur with several combinations of viruses. For example, tobacco mosaic virus and potato virus X each cause relatively mild mottling symptoms in tomato. But if by chance they both occur together in the same host, then severe necrosis develops and this may even result in the death of the plant.

A useful distinction can be drawn between animate (**biotic**) and inanimate (**abiotic**) causes of disease (Figure 1.5). Many of the animate agents, including the microbial **pathogens**, the parasitic angiosperms, and some of the animal pests, are infectious. Due to their capacity for growth, reproduction, and dispersal, these agents spread from one host plant to another. Under particularly favorable conditions, they may be dispersed rapidly over wide areas and even entire continents.

Pests

Among the animals exploiting plants are many pests which cause damage to roots, leaves, shoots, flowers, and fruits. Usually these pests, which include insects such as aphids and leafhoppers, and some nematodes, spend relatively brief periods on individual plants

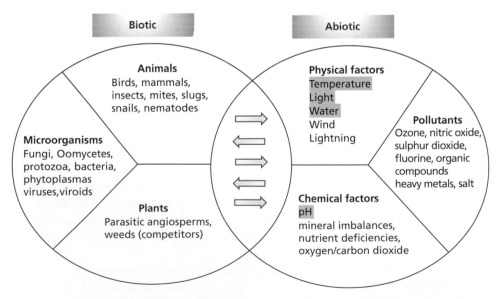

Figure 1.5 Agents responsible for plant disease, disorders, and damage. Highlighted factors are damaging when extremes occur.

before moving on to explore new food supplies. Other pests, such as leaf miners, gall-forming sawflies, and endoparasitic nematodes, spend their entire life cycle, or a major part of it, on one plant. Pest attack may result simply in a drain on host nutrients or, alternatively, in extensive destruction of tissues. Aphids and whiteflies on leaves and stems extract sap from the phloem with almost clinical efficiency. Many caterpillars are simply small herbivores; nevertheless, they can consume large areas of the leaf lamina.

Other pests cause more complex host responses or symptoms. Developing gall wasp larvae induce the formation of morphologically characteristic and often pigmented galls on leaves, while nematodes such as *Meloidogyne* spp. cause swellings, termed "knots," on the roots of tomatoes and potatoes. When these root-knot nematodes and the related endoparasitic cyst nematodes penetrate root tissues, host cells adjacent to the vascular system become enlarged and provide a specialized feeding site where nutrients are transferred to the sedentary worm. Such pests often show highly specialized adaptions to their respective hosts and, conversely, the plants mount defense reactions in response to attack which are similar to those induced by pathogenic microorganisms. Nematodes are of particular importance in the tropics where they damage numerous crop species, but some are also serious pests on temperate crops; for instance, cyst nematodes are the number 1 pest problem on potatoes in the UK and have infested around two-thirds of the land on which the crop is grown.

Larger animals such as birds or mammals can also be destructive pests. Winter grazing by rabbits can seriously reduce the final yield of autumn-sown crops such as wheat and oilseed rape. In Europe, pigeons also cause damage to oilseed rape, while in parts of Africa flocks of seed-eating finches, such as *Quelea*, are a major threat to crops of sorghum and millet.

Parasitic Plants and Weeds

Higher plants may cause disorders of or damage to other plants, either by acting directly as parasites diverting nutrients and water or as vigorous competitors or antagonists within mixed populations. Parasitic angiosperms are rare enough to be curiosities in many cool temperate countries, but elsewhere they are nuisances or economically important parasites (Table 1.2, Figure 1.6). The dwarf mistletoes, *Arceuthobium*, can kill or deform pines and other conifers, and even minor attacks reduce the quality of timber by causing the production of numerous large knots and irregularly grained, spongy wood. These parasites spread their sticky seeds by an explosive dispersal mechanism, leading to patches or foci of infestation within a plantation. By contrast, root parasites such as *Orobanche* and witchweed, *Striga*, produce numerous tiny seeds which lie dormant in the soil. The seeds are triggered to germinate by a stimulant from host roots. The parasite then attaches itself to the root by means of a specialized organ and diverts water and nutrients, leading to wilting, chlorosis, and stunting of the host. These parasites are difficult to control due to the large number of seeds they produce (in the case of *Striga*, as many as 200 000 per plant) and the long periods over which they remain viable.

On a world scale the most important angiosperm parasite is *Striga hermonthica* which attacks cereals such as maize, sorghum, millet, and rice. In many of the agricultural areas where it is most prevalent, for example sub-Saharan Africa, there are insufficient resources

Table 1.2 Angiosperms parasitic on other higher plants

Family, common name, genus	Geographic area	Crops attacked
Convolvulaceae Dodder (*Cuscuta*)	Europe, North America	Alfalfa, clover, potatoes, sugar beet
Lauraceae Dodder (*Cassytha*)	Tropics and subtropics	Citrus trees
Loranthaceae Dwarf mistletoe (*Arceuthobium*)	Worldwide	Gymnosperms
American true mistletoe (*Phoradendron*)	North America	Angiosperm trees
European true mistletoe (*Viscum*)	Europe	Angiosperm trees, especially apple
Orobanchaceae Broom rape (*Orobanche*)	Europe	Tobacco, sunflower, beans
Scrophulariaceae Witchweed (*Striga*)	Africa, Asia, Australia, North America	Maize, sorghum, rice, cowpea

to support expensive control measures and infested land may eventually be abandoned. Recently, cultivation systems have been developed that reduce *Striga* infestation by inter-cropping the cereal host with a different crop, usually a legume, that suppresses infection by the parasite (see later in this chapter). *Orobanche* is a significant problem in sunflower, tobacco, tomato, and especially faba bean, with a substantial proportion of the crop area in the Mediterranean region affected.

The deleterious effects of other higher plants are due to competition for space, light, water, and nutrients. Species which are vigorous competitors with crop plants are usually described as **weeds**. As well as affecting crop development, weeds may interfere with harvesting and their seeds can contaminate grain samples. They may also be important as alternative hosts for pests or pathogens which can subsequently spread to crops. In addition to direct competitive effects, some plants produce chemicals which inhibit the growth of neighboring plants. This phenomenon, analogous to microbial antibiosis, is known as **allelopathy**. Plant roots release a diverse range of chemicals which can act as potential inhibitors or defense compounds, but it is difficult to determine the extent to which these interactions operate in nature. Allelopathy is believed to influence plant succession and distribution in natural communities, and may also have significant effects in agricultural systems. The chemicals involved are of interest both as potential herbicides and as signal molecules affecting the growth and behavior of other organisms. The suppression of *Striga* by some legumes, described earlier, has been shown to be due to a combination of compounds that stimulate "suicidal germination" in the absence of the host and inhibitors that interfere with infection of roots.

Figure 1.6 Poplar tree infested by European mistletoe *Viscum album*. *Source:* Photo provided by John Lucas.

Abiotic Agents

Green plants, in common with all other living organisms, only flourish within a relatively narrow range of environmental conditions. Inside the plant, individual cells are able to exert control over their internal environment and thereby maintain conditions suitable for normal metabolism. However, the extent to which living cells can withstand alterations in the external environment is limited. Fluctuations in environmental conditions outside an acceptable range are therefore harmful and may result in irreversible damage. Green plants, unlike animals, are particularly susceptible to the effects of inanimate agents because they are sedentary and so are unable to escape from local changes in the environment. Plants also lack the sophisticated homeostatic mechanisms possessed by higher animals.

Many abiotic disease agents are, under other circumstances, normal components of the environment. The harmful effects of physical factors are associated with the incidence of extreme conditions. Light, for instance, while essential for green plants, may in excess cause a type of necrosis termed "scorch" on susceptible aerial parts of the plant. Low temperatures often result in frost damage. Plants differ greatly in their sensitivity to frost and typical symptoms include morphological deformation or death of part or all of the plant. Some plantation crops grown in subtropical regions, such as citrus and coffee, are especially vulnerable. For instance, frosts in southern Brazil in 1975 affected most of the coffee-growing area with production losses estimated at more than 60%. Many of these physical effects are relatively unsubtle and the symptoms associated with them are nonspecific. Drought is also an important cause of loss, due to reduced yield or even complete crop failure. Recently, prolonged dry spells in the USA, Australia, and Europe have highlighted this threat to crop productivity, and increased concerns about the impact of global climate disruption.

While the current models of climate change predict varying future scenarios, they all agree that greater fluctuations in temperature and rainfall are likely to occur, with consequent annual variations in crop yield. Rising temperatures may pose a threat not only due to drought. Episodes of higher than usual temperatures can disrupt important developmental processes, such as fertilization in cereals and fruit set in tomatoes. In the longer term, such effects may lead to changes in the areas where certain crops can be reliably grown. The frequency of extreme weather events is also predicted to increase. High winds can have catastrophic effects on plants; perennial plantation crops such as bananas have been destroyed by hurricanes in the West Indies, Cuba, and Central America. Hailstorms are particularly damaging to soft fruit crops and grapevines. There are other, less common environmental hazards. The massive eruption of Mount St Helens in Washington State in 1980 deposited volcanic ash over a wide area with a variety of effects on agriculture. Ash on plant leaves reduced photosynthesis by as much as 90%, some crops such as alfalfa actually lodged (collapsed) under the weight of ash, but eventual crop losses were less than expected, at around 7% of the total.

Chemical deficiencies or imbalances often result in distinctive symptoms, which may be diagnostic in the case of deficiencies of essential cations. For example, magnesium deficiency in swedes is associated with an abnormal purplish pigmentation in interveinal leaf areas, whereas boron deficiency in the same crop causes brown-heart symptoms in the storage root. Such deficiency diseases are commonplace, especially in the intensive cropping systems of present-day agriculture. In the UK, recent reductions in atmospheric sulfur, and subsequent deposition by rainfall, are now leading to deficiencies occurring in sulfur-demanding crops such as oilseed rape. Sulfur deficiency can affect crop quality as well as quantity, for instance by reducing the bread-making quality of wheat flour.

Excess amounts of certain mineral ions may be equally harmful, due to their effects on the availability or uptake of other essential ions. When insufficient iron is taken up, plants become chlorotic. Such a shortage may be due to inhibition of iron uptake by high levels of calcium or manganese in the soil, rather than any absolute shortage. Imbalances of soil nitrogen, phosphorus, and potassium result in the development of plant tissues which are particularly prone to infection by microorganisms or damage by other agents. Aluminum is the most abundant metallic element in soil, and at acid pH can become soluble and toxic to plant growth. This is an important constraint to crop production in some tropical soils. The problem can be rectified by raising the pH of soil by the addition of lime. Liming has been used successfully to open up new areas for crop production, notably the Brazilian cerrado (a type of savannah), which is now a major production region for soybeans.

A common difficulty in diagnosing disorders caused by chemical agents is the similarity between the symptoms they produce and those due to infection by microorganisms. Foliar symptoms in barley resulting from a deficiency of manganese resemble those caused by the leaf blotch fungus *Rhynchosporium*. Symptoms of other deficiency diseases bear a striking resemblance to those caused by viruses. Recently, molecular work on the cellular systems responsible for the uptake of specific nutrient ions has identified some of the transporter proteins involved. The genes encoding these transporters are in many cases regulated by levels of the appropriate nutrients. In the long term, it may therefore be possible to engineer components of such transport pathways to recognize a deficiency or excess of ions by producing a reporter chemical which is visible in the plant. These so-called "smart plants" would be sown at intervals within a crop and hence provide an early warning of nutrient imbalance.

Pollutants are substances which are either unnatural components of the environment, such as polychlorinated biphenyls (PCBs) and dichlorodiphenyltrichloroethane (DDT), or naturally occurring substances present in abnormal concentrations, such as ozone, sodium chloride and the acid mist which forms when oxides of sulfur and nitrogen dissolve in atmospheric moisture. Photochemical smog is increasingly common in urban areas and results from the interaction of waste gases, especially from automobiles, particulates, and sunlight. High concentrations of certain chemicals may be a normal feature of some habitats, but problems arise when human activities redistribute these substances. High salt concentrations, which are nontoxic to salt marsh plants, can severely injure inland species, as in the case of roadside communities which are damaged by splash following the application of sodium chloride to roads to prevent ice formation. Irrigation of desert soils can also lead to an accumulation of salts, a process described as **salination**.

The influence of many of these compounds on higher plants is now well known and the symptoms induced include abnormal growth due to meristem damage, chlorosis, and necrosis. Some species of plants are especially sensitive to particular pollutants. However, within a species the response of different cultivars may vary considerably. Plant species or genotypes which are tolerant of high levels of pollutants, and which can accumulate them from soil, are of value in reclaiming contaminated sites, a process known as **bioremediation**.

This discussion has only considered the direct effects of biological, physical, and chemical agents acting independently. In reality, all these agents interact with each other in a more or less complex manner. For instance, infection of the stem base of many crop plants predisposes them to collapse in wind or heavy rain; such lodging then results in problems at harvest. It is difficult to distinguish between the damaging effects of each of these factors. Other interactions are even more complex. The widespread defoliation and death of trees observed in some industrialized countries, a condition known as **forest decline**, is believed to be due to aerial pollution. However, there is dispute over the relative importance of different atmospheric pollutants and acid rain as contributory factors. It has even been claimed that the premature death of some trees is a normal part of the forest cycle. Most likely, several factors interact to affect tree health, including direct toxic effects, soil acidity, release of toxic ions such as aluminum, and indirect effects on root function, including inhibition of beneficial mycorrhizal fungi and enhanced activity of minor root pathogens. Forest decline provides a good example of a complex disease syndrome, and also illustrates the difficulty of reaching a conclusive diagnosis when several interacting factors are involved.

Significance of Disease

Disease in Natural Plant Communities

It is often assumed that disease outbreaks are less frequent and less severe in wild plant populations than in crops. This is because there are several important differences between natural and agricultural plant communities (Table 1.3). Wild species are more diverse, both genetically and in the age structure of the population. Hence individual plants will differ in their relative susceptibility to infection. Natural populations also tend to be spatially dispersed as part of a mixed plant community, thereby reducing opportunities for the spread of disease. A recent survey of grassland plots differing in the number of plant species present found that the number of groups of fungal pathogens increased with diversity of the plant community, but that the severity of infection on individual plants decreased. This supports the idea that disease is commonplace in natural plant populations but that mixed communities are less prone to severe disease outbreaks. Often, nutrients are added to crops as fertilizers, and in some cases this can lead to increased susceptibility to infection. Finally, it is likely that plants in natural communities have co-evolved with their pathogens over long periods of time, leading to some kind of host–pathogen equilibrium. Analysis of wild populations of the genetic model plant *Arabidopsis thaliana* found very high levels of polymorphism in genes determining recognition of pathogens, supporting the idea that reciprocal selection between host and pathogen over time has maintained a natural diversity for disease resistance. Modern, intensive agricultural systems and the deployment of crops in new areas have disturbed this balance.

In natural ecosystems, disease is one of the many factors which regulate populations (Figure 1.1) and hence determine the spectrum of species which are successful in any habitat. Pathogens affect the reproduction and longevity of plants and hence act as agents of natural selection. Figure 1.7 shows the effect of rust infection on the survival and reproductive capacity of the annual weed groundsel; infected plants produced fewer mature flowers, and hence seeds, and died earlier than uninfected plants. Repeated over several

Table 1.3 Natural and agricultural plant communities compared

	Natural	Agricultural
Genotypes	Diverse	Uniform
Age structure	Mixed	Uniform
Distribution	Dispersed	Crowded
Nutrient status	Often low	Usually high
Co-evolution span	Long	Short

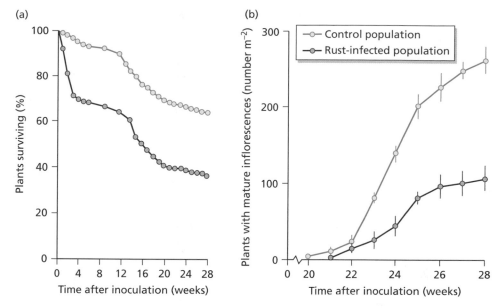

Figure 1.7 Effect of infection by the rust fungus *Puccinia lagenophorae* on survival and reproduction of the annual weed groundsel (*Senecio vulgaris*). (a) Percentage of the original population surviving in control and rust-infected populations. (b) Number of mature inflorescences formed by control and rust-infected plants. *Source:* After Paul and Ayres (1986a,b).

generations, such disease would therefore have a significant effect on population size. There is a practical spin-off from this observation as some pathogenic fungi may be useful as agents to control weeds, known as **mycoherbicides**. Experiments in which foliar fungal pathogens were allowed to infect or were excluded from a grassland community by fungicide treatment showed that natural levels of disease reduced the overall biomass of the community, but actually increased biodiversity by reducing the abundance of the most dominant grass species. In particular, disease will tend to limit the spread of species to less favorable geographic regions or habitats, as the impact of pathogens will be greater on plants growing under suboptimal conditions.

Disease may also accelerate change within established plant communities. For example, forest trees may be killed, thereby opening the canopy and allowing regeneration to proceed. The global epidemic of Dutch elm disease (Table 1.5) completely altered the structure and species composition of woodlands where elm was one of the dominant trees. An even more dramatic example of a disease outbreak in natural vegetation occurred in Australia where the root-rot pathogen *Phytophthora cinnamomi* has killed large areas of native forest (Table 1.5). The unusual feature of this epidemic was that many different species of trees and shrubs were affected by the same pathogen.

One explanation for the severity of such invasive diseases is that the pathogen responsible was almost undoubtedly introduced from another continent, and hence spread through a host population not previously exposed to infection. The term **new-encounter disease** has been proposed to describe such an epidemic arising from contact between a previously separate host and pathogen.

Disease in Agriculture, Horticulture, and Forestry

In agricultural ecosystems, disease is one of the factors influencing crop yield and quality. Farmers, foresters, and horticulturalists are, by and large, interested only in those changes in crop performance which influence cash return per hectare. The ideal situation, in which pathogens are avoided, excluded or eliminated, is therefore a theoretical rather than practical goal. On the farm, other priorities may prevail; choice of crop or cultivar is usually based on likely profitability, rather than resistance to pests or pathogens. Unless a financial return is guaranteed, control measures may be ignored or reduced in scale. As a consequence, the significance of disease, as perceived by the grower, will depend to a large extent on the market value of the crop. Inputs of chemicals or other actions designed to reduce disease are only justified when the likely impact on yield or quality will outweigh the cost of the measure. Even a possible bonus, such as restricted carry-over of the pathogen to the following season, may not provide sufficient incentive for any financial outlay.

Other parties with an interest in crop diseases are government advisory or extension pathologists, consultants, and representatives of the agrochemical industry. The relative resistance of new crop varieties to pathogens and the efficacy of commercial formulations of pesticides are assessed by advisory scientists under field conditions; recommendations for use may be based on these field trials. Independent consultants offer growers an overall package of advice for crop management, part of which concerns disease. Agrochemical companies provide information on the performance of their crop protection products, whilst government agencies regulating use of chemicals in the field issue guidelines and can impose restrictions on pesticide application. Nowadays, such advice includes strategies to reduce the risk of resistance developing to different pesticide classes. In recent years, more stringent legislation on the registration and use of agrochemicals, especially in Europe, has led to withdrawal of many crop protection products, and also affected the availability of new pesticides. Hence decisions on disease and pest control are influenced by many factors and often involve compromises driven by economic considerations or the regulatory system.

Crop Yield and Quality

The relationship between the amount of disease and loss of income is complex due to the many possible interactions between symptoms of disease and the final determinants of crop yield and quality. Figure 1.8 defines a number of yield levels for a hypothetical crop. The theoretical maximum yield is a value based on predictions from crop physiology; under field conditions, this yield level is not a practical possibility and hence there is some unavoidable loss. Attainable yield indicates the maximum level to be expected under optimum conditions in the field. With optimum inputs of fertilizer, water, and pesticides, this is the best yield the farmer can realistically hope for. The difference between this value and the actual yield (also described as farmer yield) obtained from the crop can be defined as an avoidable loss. In practice, attainable yield is not a realistic goal for most crops for simple economic reasons. To increase yield to this level requires so many inputs that the cost is greater than the eventual return at the end of the season. Instead, we can define a slightly lower threshold, the economic yield, which represents the break-even point at which the

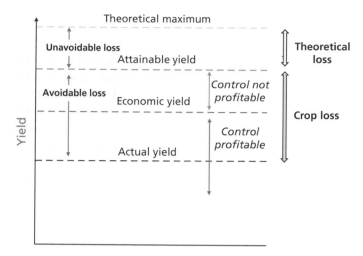

Figure 1.8 Relationship between yield levels and crop loss, indicating economic benefits of control. *Source:* Zadoks and Schien (1979).

input cost is balanced by the extra productivity of the crop. Any shortfall below this level is an avoidable loss which justifies the expense of a control measure.

For some crops, especially high-value fruits, vines, vegetables or ornamental plants, the quality of the product is as important as the yield. Under these circumstances, very little disease is tolerated, as any damage or blemish may have a disproportionate effect on crop value. Not surprisingly, the most intensive disease and pest control regimes available are used for such crops.

The Impact of Disease

The impact of plant disease depends on agricultural, biological, socioeconomic, and historical factors. In developed agricultural systems, impacts are usually measured in terms of reduced crop yield and quality, and overall effects on farm profitability. In less developed systems, the consequences of disease can be much more serious, affecting food security, regional and national economies, and social stability. Invasive pests and pathogens can also destroy native trees and alter natural ecosystems, as well as impacting on biodiversity (Table 1.4).

Some examples of these impacts and the pathogens responsible are given in Table 1.5. Prior to understanding of the germ theory of disease, and discovery of chemicals and other means of controlling epidemics, disease outbreaks could devastate crops and consequently cause famine and social disruption. It is believed that a plant disease, most likely a virus, contributed to the collapse of the ancient Mayan civilization in the ninth century. But the most notorious case is the potato late blight outbreak of 1845–1846, caused by the oomycete pathogen *Phytophthora infestans*, that spread rapidly across Europe and had particularly disastrous consequences in Ireland, where many communities were dependent on potatoes as their sole source of food. Around 1 million people died of starvation and countless others were displaced, many emigrating to the USA. Fortunately, with diversification

Table 1.4 Some impacts of plant disease

Developed agriculture

- Reduced crop yield
- Reduced crop quality
- Compromised product safety, e.g., mycotoxin contamination
- Reduced profitability

Developing agriculture

- Food security – malnutrition and famine
- Impact on communities or national economies
- Social instability

The natural environment

- Loss of key species or natural communities
- Damage to landscapes and leisure amenities

Table 1.5 Some examples of the impacts of specific plant diseases

Type of impact	Disease	Causal agent	Country/region affected
Famine	Late blight of potato	*Phytophthora infestans*	Europe 1845–1846
	Brown spot of rice	*Helminthosporium oryzae*	India 1942–1943
	Failure of maize crop	Maize mosaic virus?	Guatemala, ninth-century Mayan civilization
	Cassava mosaic disease	Cassava mosaic Gemini viruses	East Africa 1980s to present
Economic	Coffee rust	*Hemiliea vastatrix*	Sri Lanka 1870, now worldwide
	Cocoa swollen shoot	Cocoa swollen shoot virus	Ghana/Nigeria 1930–present
	Citrus canker	*Xanthomonas axonopodis* pv. *citri*	Florida 1912, 1986, 1995–present
Agricultural	Southern corn leaf blight	*Bipolaris maydis*	USA 1970
	Asian soybean rust	*Phakopsora pachyrhizi*	Asia 1900s, Africa 1995, Brazil 2001, USA 2004
	Black stem rust	*Puccinia graminis* f.sp. *tritici*	USA 1900s, new race Ug99 in Africa 1999, now Middle East and potentially Asia
Ecological	Dutch elm disease	*Ophiostoma novo-ulmi*	Northern hemisphere 1930, 1970–present
	Jarrah dieback	*Phytophthora cinnamomi*	Western Australia 1920–present
	Sudden oak death	*Phytophthora ramorum*	California 1995, UK 2002
	Ash dieback	*Hymenoscyphus fraxineus*	Poland 1990s, western Europe, UK 2012

of food sources, and improved crop protection, this scenario is now much less likely to be repeated in developed countries, but in subsistence agriculture is still a constant threat.

A more recent example is the spread of cassava mosaic disease (CMD) in Africa. A severe outbreak emerged in Uganda in the 1980s, with crop losses as high as 80–90%, and cultivation of this vital food crop was abandoned in some areas. CMD, that is now known to be caused by a complex of related Gemini viruses spread by whitefly vectors, has since invaded other countries in sub-Saharan Africa where it continues to affect food security. Recently, a different virus, cassava brown streak, has spread to East Africa to pose a further threat to this vital staple crop.

Other diseases have had serious economic impacts, such as coffee rust that devastated the industry in Sri Lanka and has now spread worldwide. More recent examples of global pandemics caused by rust fungi include Asian soybean rust, that has now spread to the major producing areas in Brazil and the USA, and black stem rust of wheat, a new variant of which (Ug99) emerged in Africa and is now spreading east, threatening wheat production areas in Asia. Soybean producers in the Americas now have to factor in the cost of fungicide treatments, while there are concerns that Ug99 might decrease wheat production in affected countries.

A further consequence of disease is the impact of measures taken to control pathogens that threaten export markets. Attempts to eradicate the destructive cocoa swollen shoot virus (CSSV) from Ghana by means of statutory removal of infected and surrounding trees not only entailed the most costly eradication campaign ever attempted, but also lead to political unrest. In the USA and South America, spread of the bacterial disease citrus canker has only been contained by burning huge numbers of infected citrus trees and nursery stock, combined with vigilant quarantine measures. In Florida, this included removing asymptomatic citrus trees from private gardens within affected areas, which inevitably brought plant health authorities into conflict with home owners. Outbreaks of crop diseases can also impact on agricultural practices and policy. The major epidemic of southern corn leaf blight in the United States in 1970 (see Chapter 5, Figure 5.1) raised doubts about the wisdom of achieving genetic uniformity in modern cereal crops and forced a reassessment of the breeding methods employed in the production of new cultivars.

Plant pathogens can also have major impacts on both natural and managed forests. Currently, sudden oak death is spreading in the western USA and Europe, while ash dieback has recently invaded the UK. With the expansion of international trade in plants and plant products, the frequency of invasions by exotic pests and pathogens is increasing, with serious implications for many native plant species (see Chapter 5).

Quantifying Losses Due to Disease

Even in areas of high agricultural efficiency, losses due to pathogens, pests, and weeds make constant inroads into production, and hence profits. Such estimates are notoriously difficult to compile but suggest that as much as one-third of total production is being lost. Table 1.6 shows estimated worldwide losses for six major crops due to weeds, pests, pathogens, and virus diseases during 2001–2003. Furthermore, comparison with data from earlier surveys (Figure 1.9) suggests that the situation has not improved in recent years, despite advances in the science and practice of crop protection. Some of the changes taking place

Table 1.6 Estimated losses (%) due to weeds, pests, and diseases in six major crops worldwide in 2001–2003

Crop	Weeds	Pests	Pathogens	Viruses	Total
Wheat	7.7	7.9	10.2	2.4	28.2
Rice	10.2	15.1	10.8	1.4	37.4
Maize	10.5	9.6	8.5	2.7	31.2
Potatoes	8.3	10.9	14.5	6.6	40.3
Soybeans	7.5	8.8	8.9	1.2	26.3
Cotton	8.6	12.3	7.2	0.7	28.8

Source: Oerke (2006).

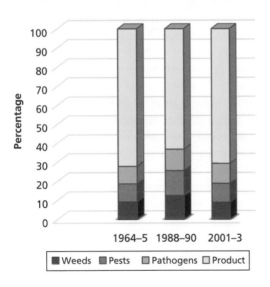

Figure 1.9 Comparison of proportions of total production of major food and cash crops lost to pathogens, pests and weeds estimated in 1964–1965 (by Cramer, 1967), and 1988–1990 and 2001–2003 (by Oerke et al. 1994; Oerke 2006). *Note:* Crops included in the calculation are not identical in the three surveys.

in modern agriculture may in fact have increased vulnerability to disease. The demand for improved agricultural productivity has led to large areas being planted with high-yielding, genetically identical cultivars. Similarly, the increasing cost of labour and the trend toward mechanization, which involves major capital expenditure, have also contributed to a reduction in the diversity of crop types planted. As a result, long-established systems of crop rotation have been discontinued in many areas.

A second aspect of modern agriculture which has undoubtedly aggravated disease problems is increased world trade in crop plants and plant products. Following the gradual shift from self-sufficiency, at a community and in many instances also at a national level, and the increasing relative affluence of some countries, large-scale transport of plant material and food produce over long distances has become commonplace. One consequence of this globalization of plant trade is the increased risk of introducing invasive pests and pathogens (see discussion of new-encounter diseases earlier). In addition, there is increasing long-term storage of produce, which in turn brings further pathological

Table 1.7 Estimates of postharvest losses likely to occur in the absence of effective disease control measures

	Commodity	Country of origin	Potential loss (%)
Loss during low-temperature storage	Apples	England, USA	2–50
	Carrots	England, USA	6–38
	Citrus fruits	Italy, USA	3–52
Loss during transport and marketing	Apples	USA	29
	Citrus fruits	USA	0–25
	Lettuce	USA	10–15
	Peaches	USA	15–24
	Strawberries	USA	25–35

Source: Data from Eckert (1977).

problems. Losses due to postharvest pests and diseases such as storage rots have tended to be underrated. It has been estimated that a significant proportion of all tropical produce may be destroyed, by a variety of agents, before it reaches the consumer. Even in developed countries, the scale of potential losses during long-term storage, transport, and marketing can be surprisingly high (Table 1.7). Overall, such losses are due to a range of factors, including not only biotic agents but also waste during processing, handling and by consumers. In developed countries, extensive refrigeration systems, more effective handling of produce and other control measures have now reduced the impact of postharvest pests and diseases. Nonetheless, more recent estimates of food losses indicate that substantial amounts of some commodities still go to waste. For instance, the combined loss of grain products, fresh fruit, and vegetables during retail and by consumers in the USA is between 20% and 40%. Consumer intolerance of substandard produce has placed a greater emphasis on food quality so, for instance, blemished or misshapen fruit and vegetables will often be rejected. Hence, pests or diseases causing superficial damage have assumed greater importance.

Disease is a Dynamic Phenomenon

The intensification of agriculture through plant breeding, widespread use of fertilizers, pesticides and plant growth regulators, with larger fields and shorter rotations, has brought in its wake new and sometimes severe disease problems. Changes in the types of crops grown, or in crop management practices, almost invariably lead to new challenges. Irrigation has opened up whole regions to agriculture, but the same water which brings life to the crop can also nourish and spread its microbial enemies. The extension of crops into new geographical areas has exposed them to novel disease agents; for instance, tropical plant species such as cocoa and cassava, which originated in South America, are grown extensively in Africa where they have succumbed to virus diseases spreading from native species. While breeding programs and chemical control measures have won notable victories in the campaign against plant disease, the situation is never static.

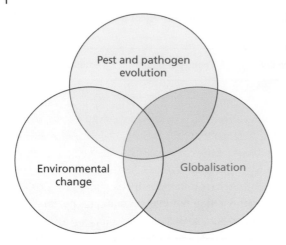

Figure 1.10 A convergence of forces increasing the threat of plant disease. *Source:* Lucas (2017b).

If anything, in recent years the threat posed by pests and pathogens has increased, rather than decreased (Figure 1.10).

The age we are living in has been described as the "Anthropocene," as no part of the planet is now unaffected by human activity. Environmental change has taken place on a massive scale, altering ecosystems, modifying the distribution of species, and reducing biodiversity. The globalization of trade and travel has redistributed crops and plant products and inadvertently introduced their enemies into new regions. There is now conclusive evidence that the global climate is changing as a result of human actions. Altogether, the pace of such change is accelerating the adaptation and evolution of biological systems, including the pathogens causing animal and plant diseases. There is a need therefore for constant vigilance to ensure that the plants we grow remain healthy and productive.

Further Reading

Books

Ainsworth, G.C. (1981). *An Introduction to the History of Plant Pathology*. Cambridge: Cambridge University Press.

Buczacki, S. and Harris, K. (2014). *Pests, Diseases and Disorders of Garden Plants*, 4e. London: Harper Collins.

Chakraborty, U. and Chakraborty, V. (eds.) (2015). *Abiotic Stresses in Crop Plants*. Wallingford: CABI.

Holliday, P. (1989). *A Dictionary of Plant Pathology*. Cambridge: Cambridge University Press.

Ingram, D.S. and Robertson, N. (1999). *Plant Disease: A Natural History*. London: Collins.

Perry, R.N. and Moens, M. (eds.) (2013). *Plant Nematology*, 2e. Wallingford: CABI.

Press, M.C. and Graves, J.D. (eds.) (1995). *Parasitic Plants*. London: Chapman & Hall.

Putnam, R.J. (ed.) (1989). *Mammals as Pests*. London: Chapman & Hall.

Radosevich, S.R., Holt, J.S., and Ghersa, C.M. (2007). *Ecology of Weeds and Invasive Plants: Relationship to Agriculture and Natural Resource Management*. Chichester: Wiley.

Robinson, J.B.D. (ed.) (1987). *Diagnosis of Mineral Disorders in Plants*, vol. 1–3. London: HMSO.

Schumann, G.L. (1991). *Plant Diseases: Their Biology and Social Impact*. St Paul, Minnesota: APS Press.

Smith, I.M., Dunez, J., Phillips, D.H. et al. (eds.) (1988). *European Handbook of Plant Diseases*. Oxford: Blackwell Scientific Publications.

Van Embden, H.F. and Harrington, R. (eds.) (2017). *Aphids as Crop Pests*, 2e. Wallingford: CABI.

Zimdahl, R.L. (2018). *Fundamentals of Weed Science*, 5e. San Diego, California: Academic Press.

Reviews and Papers

Anderson, P.K., Cunningham, A.A., Patel, N.G. et al. (2004). Emerging infectious diseases of plants: pathogen pollution, climate change and agrotechnology drivers. *Trends in Ecology & Evolution* 19 (10): 535–544.

Bos, L. and Parlevliet, J.E. (1995). Concepts and terminology on plant/pest relationships: toward consensus in plant pathology and crop protection. *Annual Review of Phytopathology* 33: 269–102.

Fears, R., Aro, E.-M., Pais, M.S., and ter Meulen, V. (2014). How should we tackle the global risks to plant health? *Trends in Plant Science* 19 (4): 206–208.

Fisher, M.C., Henk, D.A., Briggs, C.J. et al. (2012). Emerging fungal threats to animal, plant and ecosystem health. *Nature* 484 (7393): 186–194.

Hodges, R.J., Buzby, J.C., and Bennett, B. (2011). Foresight project on global food and farming futures. Postharvest losses and waste in developed and less developed countries: opportunities to improve resource use. *Journal of Agricultural Science* 149: 37–45.

Jarosz, A.M. and Davelos, A.L. (1995). Effects of disease in wild plant populations and the evolution of pathogen aggressiveness. *New Phytologist* 129: 371–387.

Oerke, E.C. (2006). Crop losses to pests. *Journal of Agricultural Science* 144: 31–43.

Rottstock, T., Joshi, J., Kummer, V., and Fischer, M. (2014). Higher plant diversity promotes higher diversity of fungal pathogens, while it decreases pathogen infection per plant. *Ecology* 95 (7): 1907–1917.

Savary, S., Bregaglio, S., Willocquet, L. et al. (2017). Crop health and its global impacts on the components of food security. *Food Security* 9: 311–327.

Vurro, M., Bonciani, B., and Vannacci, G. (2010). Emerging infectious diseases of crop plants in developing countries: impact on agriculture and socio-economic consequences. *Food Security* 2: 113–132.

2

The Microbial Pathogens

It was first necessary to determine if characteristic elements occurred in diseased parts of the body, which do not belong to the characteristics of the body, and which have not arisen from body characteristics.

(Robert Koch, 1843–1910)

Heterotrophic microorganisms, unlike autotrophs, are entirely dependent upon an external supply of organic carbon compounds. The ultimate source of most carbon compounds is green plants, but there are a variety of routes by which microbes can obtain these nutrients.

A large number of microorganisms are decomposers. These organisms utilize substrates in dead tissues and their activities eventually lead to the disappearance of plant and animal remains. Such decomposers play a key role in the ecosystem by releasing nutrients which would otherwise remain locked up in plant litter. Some microbes have, in addition, an ability to parasitize living plants; if, during invasion of the plant, they kill host cells this ensures a supply of dead tissues on which they can continue to grow. Other microorganisms are only able to obtain nutrients from living host cells, and establish more balanced relationships which may be of mutual benefit. The effects of microbes on plants therefore vary from severe damage and even death, to diversion of nutrients, to associations in which both partners gain some advantage. Hence, heterotrophic microorganisms are involved in a variety of ways in the movement of fixed carbon between different trophic levels in the ecosystem.

A comprehensive analysis of plant disease caused by microorganisms requires several different types of information. First, the causal agents must be identified. However, the usual criteria employed for distinguishing between microbial species are of limited value when dealing with microorganisms isolated from plants. Different isolates of the same species may vary widely in their ability to cause disease. It is important to understand the genetic basis of such variation, and the corresponding variations in the plant's response. Second, the nature of the host–parasite relationship needs to be considered; the biology of infection, sources of nutrients, the basis of damage to the host, and the effects of the environment. The diversity of relationships is enormous, but identifying some common features is helpful in providing basic guidelines for the control of contrasting types of pathogens.

Plant Pathology and Plant Pathogens, Fourth Edition. John A. Lucas.
© 2020 John Wiley & Sons Ltd. Published 2020 by John Wiley & Sons Ltd.
Companion website: www.wiley.com/go/Lucas_PlantPathology4

Pathogens and Pathogenesis

Considerable confusion surrounds the terms **pathogen** and **parasite**. While they are generally used to describe microbial disease agents, in particular the fungi, bacteria, and viruses, the distinction between the two terms has often been overlooked. They are not synonymous; a parasite is an organism having a particular type of nutritional relationship with a host, while the term pathogen refers to the ability of an organism to cause disease. They may be defined as follows.

- *Parasite*: an organism or virus living in intimate association with another living organism (host) from which it derives some or all of its nutrients, while conferring no benefit in return.
- *Pathogen*: an organism or virus able to cause disease in a particular host.

The allied term **pathogenesis** describes the complete process of disease development in the host, from initial infection to production of symptoms.

At first sight, the distinction between a parasite and a pathogen might appear subtle; indeed, in many cases the parasitic activities of an organism automatically lead to it being a pathogen as well. The diversion of nutrients from the host will cause some metabolic stress which will normally be expressed as disease. However, in other host–microorganism associations this stress may be offset by the microbe contributing nutrients in return. This is the case with root nodules of legumes, where the bacterium *Rhizobium* obtains carbohydrates from the host but also fixes atmospheric nitrogen, some of which the host subsequently utilizes. Mycorrhizal fungi infect plant roots but actually stimulate growth by assisting the uptake of scarce nutrients, especially phosphates, from the soil. The definition of a parasite given above takes account of situations such as these.

Where the invading microbe confers some beneficial effect, the term **symbiosis** has been used. As originally conceived, symbiosis (literally = living together) referred to any intimate or close association between organisms, irrespective of benefit or harm, and was subdivided as shown in Figure 2.1.

The advantage of this scheme is that it can accommodate relationships where the balance may shift from mutual benefit, termed **mutualism**, to injurious effects on one partner.

If one considers the terms parasite and pathogen from the reverse viewpoint, in other words the ability to cause disease, the difference becomes more obvious. While all parasites

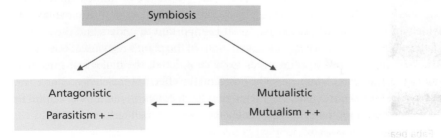

Figure 2.1   Symbiotic relationships: + positive effects on partner; – negative effects on partner.

are potentially pathogenic due to their diversion of host nutrients, many of the characteristic symptoms of disease cannot be explained on the basis of nutritional stress alone. The growth and development of a pathogen in its host, along with the response of the host to the presence of an alien organism, involve other interactions which have little to do with nutrition. Many of the more injurious effects of pathogens may be traced to toxic chemicals whose production may be incidental to their parasitic way of life (see Chapter 8). Looked at in this way, the statement "a good parasite is a poor pathogen" may appear to be justified. Any organism which is dependent upon another organism for its supply of nutrients might be expected to restrict its pathogenic effects to a minimum.

Biotrophs and Necrotrophs

Although there is an enormous variety of pathogens, an important distinction can be made between those which rapidly kill all or part of their host and others which co-exist with host tissues for an extended period without inflicting severe damage. The former category, referred to as **necrotrophs**, are often opportunist pathogens which invade wounds and juvenile or debilitated plant tissues. They grow intercellularly, producing cytolytic factors and then utilize the dead host tissues as a resource. The ability to attack a living host distinguishes these organisms from the **saprotrophs** which subsist exclusively on organic debris. In contrast, **biotrophs** do not kill their host immediately. They are, in fact, dependent upon viable host tissue to complete their development. Extreme biotrophy resembles mutualism in that it is difficult to discern any marked pathogenic effects.

Figure 2.2 shows two contrasting diseases of faba beans, both caused by fungi. In the first, chocolate spot caused by *Botrytis fabae*, the main symptom is dark necrotic lesions in which host cells have died. As the disease progresses the lesions expand and eventually coalesce to destroy the whole leaf. In the second, bean rust caused by *Uromyces viciae-fabae*, the leaves are covered with pustules producing rust-colored spores, but the tissues around the pustules are still green and alive The rust penetrates living cells and absorbs soluble nutrients that are then transported back into the center of the pustules to fuel spore production. Over time, this process will impact on the physiology and development of the host plant, but tissues are not directly destroyed by the pathogen.

Figure 2.2 Faba bean leaves infected by *Botrytis fabae* (*left*) showing necrotic lesions, and bean rust, *Uromyces viciae-fabae* (*right*) with pustules producing rust-colored spores.

Table 2.1 Main characteristics of necrotrophic and biotrophic pathogens.

Necrotrophs	Biotrophs
Morphological and biochemical features	
Host penetration via wounds or natural openings	Host penetration direct or via natural openings
Few special parasitic structures formed	Special parasitic structures, e.g., haustoria, typically formed
Host cells rapidly killed	Living host–pathogen interface
Toxins and cytolytic enzymes produced	Few or no toxins or cytolytic enzymes produced
Nutrients obtained from breakdown of host substrates	Nutrients diverted from host cells
Destruction of host resistance	Suppression or evasion of host resistance
Ecological features	
Wide host range	Narrow host range
Able to grow saprophytically away from host	Unable to grow away from host
Attack juvenile, debilitated or senescing tissues	Attack healthy hosts at all stages of development

The contrasting features of necrotrophs and biotrophs are summarized in Table 2.1. Biotrophs, in keeping with their more specialized parasitism, usually attack only a limited range of hosts. Biotrophic fungi like the rusts form differentiated infection structures including modified intracellular hyphae termed **haustoria** (see Chapter 6). Generally, these fungi do not produce large quantities of extracellular enzymes or toxins; during their co-evolution with the host, synthesis of hydrolytic enzymes may have been repressed or limited to localized sites where host cells are penetrated. Eventually, the ability to elaborate such enzymes may have been lost altogether.

An alternative view of these different lifestyles is that necrotrophs may have evolved from biotrophs through an increasing ability to produce enzymes capable of degrading complex substrates. This theory proposes that the first fungi were dependent on living plants, but gradually evolved independence by developing enzyme systems able to deal with polymeric carbon sources in plant litter. Such schemes can therefore be extended to include free-living saprotrophs but in the absence of any adequate fossil record, both versions are speculative. The advent of techniques for analyzing genome structure and molecular phylogeny may, however, provide fresh evidence to support or refute such evolutionary models.

A human analogy for these contrasting types of parasite has been proposed, as follows: necrotrophs are "thugs" while biotrophs are "con artists," reflecting their more devious way of obtaining resources from the host plant. However, the original idea that necrotrophs are unsophisticated pathogens is now being revised as we learn more about their strategies for invading plants and overcoming host defense. It turns out that some have the capacity to hijack host pathways, leading to programmed cell death, thereby releasing nutrients for their own use.

The impression may have been given that biotrophy and necrotrophy represent absolute categories; in reality, there is a continuous gradation between the two types of pathogen. At one extreme are the viruses, which can replicate only within living cells, and fungal biotrophs, such as the rusts and powdery mildews. At the other extreme are necrotrophs, such as the damping-off fungi and soft-rot bacteria. In between, one encounters pathogens with intermediate characteristics. For instance, the potato late blight pathogen *Phytophthora infestans* exhibits a high degree of host specificity and other biotrophic features such as haustoria, but it also causes relatively rapid necrosis of invaded tissues. Many pathogens pass through both a biotrophic and a necrotrophic phase during their life cycle. Plant pathogenic bacteria such as *Pseudomonas syringae* initially proliferate in intercellular spaces in leaves or fruits without apparent damage to host cells, but water-soaked lesions which become necrotic then appear. The apple scab fungus *Venturia inaequalis* grows beneath the cuticle of host leaves for several days without causing obvious necrosis, but as the lesions age, the host tissues are eventually killed and the typical scabs develop (Figure 2.3). Some species of the anthracnose fungus, *Colletrotrichum*, penetrate directly into host cells which remain alive for several days (see Chapter 6, Figure 6.9); subsequently, necrotic, spreading lesions are formed. The term **hemibiotroph** has been used to describe such behavior. The factors responsible for this switch from a balanced mode of parasitism to rapid killing of host cells have in many cases not yet been identified.

In nature, necrotrophs may grow on both living and dead host tissues. Pathogens such as *Pythium* and *Rhizoctonia* may be found growing actively in soil or on subterranean or aerial plant surfaces in competition with the natural microflora. In the absence of a suitable host, they may successfully complete their life cycle by utilizing dead organic resources. The ability of biotrophs to compete for dead organic matter is very limited or even nonexistent. These differences in patterns of natural occurrence of pathogens are reflected in their growth on laboratory culture media. Most necrotrophs are nutritionally undemanding; they grow well on a wide range of simple media. Biotrophs, on the other

 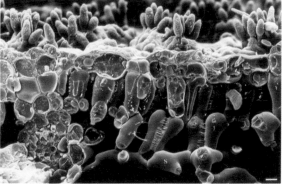

Figure 2.3 Apple scab disease caused by *Venturia inaequalis*. (*Left*) Scab lesions on fruit. (*Right*) Scanning electron micrograph of apple leaf fractured through a scab lesion, showing sporulation of the fungus on the surface, and intact, uncolonized host tissues beneath. Bar = 10 μm. *Source:* Courtesy of Alison Daniels.

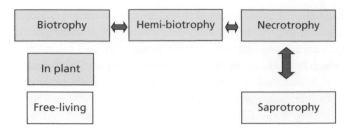

Figure 2.4 Nutritional modes in heterotrophic microorganisms.

hand, have traditionally been regarded as fastidious organisms and in extreme cases cannot be grown on any known culture media.

Distinctions based on the criterion of culturability are used to divide pathogens into two nutritional types: **facultative** and **obligate** parasites. A further refinement of this scheme distinguishes pathogens which are able to grow relatively well in pure culture, but which in nature are unable to compete with nonparasitic microbes. Such parasites are termed **ecologically obligate** in contrast to **biochemically obligate** organisms which are unable to grow apart from the living host either *in vivo* or *in vitro*. The basis of obligate parasitism remains largely unresolved; such microorganisms may be unable to synthesize essential metabolites and therefore have to obtain them from the host, lack particular nutrient uptake mechanisms, or may require developmental cues that are only provided in the presence of the host plant.

Figure 2.4 summarizes these different relationships and modes of nutrition in heterotrophic microorganisms.

Pathogen Classification

The classification of pathogenic microorganisms is based initially on the same morphological, physiological, and molecular criteria as other groups. However, conventional taxonomy does not accommodate all the characteristics of importance in pathology. Thus, different isolates of a pathogen which may appear identical in morphology and cultural characters can differ in pathogenicity and in the range of host species attacked. The same problem also occurs in medical microbiology. For instance, the common gut bacterium *Escherichia coli* is normally a harmless species resident in the human intestine, but certain isolates of this species can infect the gut, causing gastroenteritis and severe illness. The differences between the pathogenic isolates and normal *E. coli* are relatively minor and are coded for by a few genes often carried on extrachromosomal plasmids. Similar subtleties are common with plant pathogens. In some cases, differences in pathogenic behavior may be due to only a single gene. Differences in host range may be sufficient to define particular groups, or **pathotypes**, adapted to particular host species. In fungi, where such host specialization is clear, it may be possible to recognize **form species**. For instance, the black stem rust fungus *Puccinia graminis* occurs on various grasses including wheat (*P. graminis* f.sp. *tritici*) and barley (*P. graminis* f.sp. *hordei*). With plant pathogenic bacteria, particular **pathovars** adapted to different host plants may also be distinguished.

The classification of plant pathogenic viruses presents particular problems as many have very wide host ranges, infecting different plant species, genera, and even families. Nevertheless, different strains occur which vary in important properties such as the relative severity of disease they cause or frequency of transmission by different insect vectors. Such variation needs to be accommodated in any scheme for classifying viruses responsible for disease in plants.

Koch's Postulates

To determine with certainty that a particular microorganism is the cause of a disease rather than some incidental contaminant, it is necessary to critically examine its relationship with the host. This dilemma was first recognized in studies of pathogens of humans and other animals. In 1876, Robert Koch provided the first experimental proof of disease causation by applying a set of rules which have since come to be known as Koch's postulates. Koch considered that these rules must be satisfied before any microorganism can be regarded as a pathogen. The rules involve five steps outlined below.

1) The suspected pathogen must be consistently associated with the same symptoms.
2) The organism should be isolated into culture, away from the host. This precludes the possibility that the disease may be due to malignant tissues or other disorders of the host itself.
3) The organism should then be reinoculated into a healthy host.
4) Symptoms should then develop which are identical to those observed in the original outbreak of disease.
5) The causal agent should be reisolated from the test host into pure culture and be shown to be identical to the microorganism initially isolated.

An actual example of the use of Koch's rules is shown in Figure 2.5. An apparently new disease of orange trees, with symptoms of chlorosis, stunting, and dieback of branches, was reported in South America. Leaves from affected trees were surface sterilized and plated onto a nutrient medium. Colonies of a small, gram-negative bacterium were obtained. Suspensions of the bacterium were then injected into healthy citrus saplings, and after a period of incubation, some of these artificially inoculated trees developed symptoms very similar to those seen in the original infected tree. The same small bacterium was reisolated from these trees.

This procedure completed Koch's postulates and showed that the new disease, named citrus variegated chlorosis, was due to a bacterium. In reality, a lot more work, including light and electron microscopy and the use of specific antisera, was required to actually identify the agent as a new strain of the xylem-inhabiting pathogen *Xylella fastidiosa*. A few years later, in 2000, *X. fastidiosa* became the first cellular plant pathogen to have its complete genome sequenced.

Procedures for the detection and diagnosis of specific pathogens are described in more detail in Chapter 4.

This example shows that even today, Koch's rules are still relevant, although they cannot be rigidly applied in their original form to all pathogens. The most important exceptions

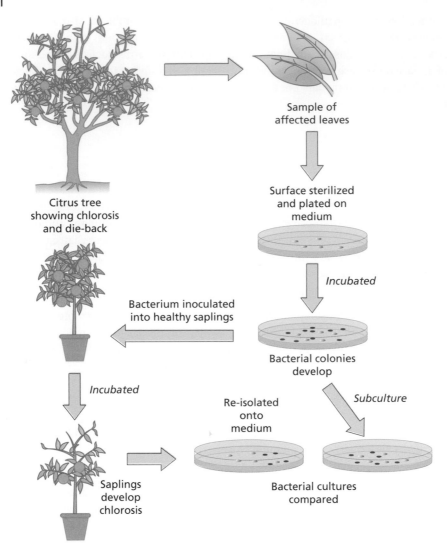

Figure 2.5 Use of Koch's postulates to establish the etiology of a new disease of citrus caused by the bacterium *Xylella fastidiosa*. *Source:* Based on Hartung et al. (1994).

in plant pathology are when the pathogen cannot be grown in artificial culture, for example the viruses and some biotrophic fungi. The problem of isolating viruses from their host plants is generally overcome by using indicator plants. These are alternative hosts which develop symptoms which are specific for a particular virus. Healthy specimens of the original host may then be reinoculated. In addition, electron microscopy of plant sap or of purified crystalline samples of the virus, coupled with serological techniques, may be employed to investigate the type(s) of virus present at each step of the procedure. There are also a number of new and powerful methods for detecting nucleic acid sequences specific to particular pathogens.

The application of Koch's rules to nonculturable fungal pathogens presents fewer problems because these agents produce spores. Such propagules can be removed from the host and then used in reinoculation experiments. In many instances, spore morphology is also a valuable aid to identification of the inoculated and reisolated pathogens.

Further difficulties in satisfying these postulates may be experienced in cases where symptoms result from mixed infections or when dealing with previously undescribed disease agents. For instance, few pathologists would have predicted the existence of the viroids, which scarcely conform to our preconceptions of a successful parasite (see Chapter 3).

Host Resistance and Pathogen Virulence

All crops are exposed to a wide variety of potentially pathogenic microorganisms present in soil, water, and the surrounding atmosphere. Yet most plants remain healthy most of the time. Consequently, the majority of pathogens are unable to infect the plants with which they come into contact. Even where a specific pathogen can attack a particular host species, there are marked variations in the extent to which individual plants succumb to disease. These differences are paralleled by variation in the pathogen population, reflecting differences in the genetic constitution of both the host and the pathogen. The ability of the pathogen to cause disease, and the host to respond to invasion, has been shown to be determined by specific genes. This discovery has important implications both in the analysis of disease and in its control.

Resistance and Susceptibility

When a microorganism makes contact with a plant, it may be able to penetrate the potential host or it may be completely excluded. Following penetration, development of the pathogen may be halted by a host response or, alternatively, growth continues within the host tissues.

Describing the interaction between a microbial pathogen and its host presents problems as the outcome needs to be defined in terms of both partners (Figure 2.6). At one extreme is the situation where a microorganism is incapable of causing disease in the host under any conditions, and so is described as a nonpathogen of that host. Likewise, plants able to

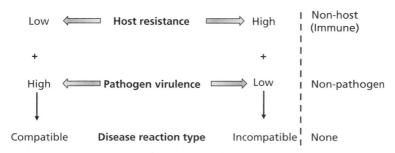

Figure 2.6 Relationships between host, pathogen, and disease reaction.

completely prevent penetration by a microbial agent are nonhosts, and are considered to be immune to that organism. The majority of interactions between microbes and green plants are likely to be of this type. It may, however, be difficult to establish whether a plant is immune to a particular pathogen, as the absence of visible symptoms does not automatically imply a failure to penetrate. Nowadays, the term **immunity** is mainly used in the context of innate plant defense, or should only be applied in cases where precise descriptions of the microbe–plant interactions are available.

In some instances, (see Chapter 9), pathogens penetrate their host only to be immediately prevented from further colonization by the death of the first living cells they enter. Such restricted development indicates that the host is **resistant** to the pathogen. Resistance, unlike immunity, is not an all-or-nothing property of the host. In practice, there may be a whole range of responses, varying from high resistance, where no visible symptoms are manifest, to low resistance, where the host succumbs completely to disease. Between these two extremes, resistance is described by a number of adjectives which, though imprecise, are of practical use in distinguishing between host reaction types.

Alternatively, differing degrees of pathogen development may be described in terms of host **susceptibility** (Figure 2.6). For each degree of resistance, there is a corresponding level of susceptibility. For instance, high resistance is equivalent to low susceptibility. Complete susceptibility is of considerable biological interest, as it appears to constitute the exception to the general rule that plants exhibit a degree of resistance. However, in this book these complementary descriptions of host responses are mainly considered in terms of resistance. This approach has significant practical advantages in that it emphasizes the character which is selected for by plant breeders (see Chapter 12).

The terms resistance and susceptibility describe conditions of the host. However, just as the host may vary in its ability to resist infection, so pathogens differ in their ability to invade and cause disease. Those microorganisms which cannot under normal circumstances induce disease in a host are regarded as nonpathogenic with respect to that host. Others which are able to penetrate but which have insignificant effects on the host may be termed **avirulent**; where the effects are more drastic, they are described as possessing some degree of **virulence**.

It should be noted that there are still some problems with this terminology. In clinical microbiology, a distinction was originally drawn between nonpathogenic and pathogenic microorganisms. Pathogenicity was considered to be an absolute property. Virulence was used to describe differences in the extent to which different strains of a pathogen caused disease. These clear-cut distinctions are valid for some pathogenic species, but it is now appreciated that the situation is more complex. Many microorganisms have the ability to acquire or lose traits which have a major effect on disease reaction type.

Ultimately, the phenomenon of pathogenicity needs to be analyzed in genetic terms, and the differences between species, strains, and pathotypes described in terms of the molecular interactions determining the disease phenotype. Avirulence, for instance, has a more precise meaning associated with the presence in the pathogen of a specific gene encoding a product which can be recognized by the plant, hence triggering resistance (see later in this chapter and Chapter 9). Hence some authors prefer to use the term **aggressiveness** to describe differences between pathogen strains in the amount of disease they cause in the host. Aggressiveness can only be measured under carefully controlled conditions where

Table 2.2   Some parameters used to measure aggressiveness

- Length of latent period (time from inoculation to production of new inoculum)
- Rate of multiplication of the pathogen in host tissues (used for bacteria and viruses)
- Rate of lesion expansion
- Number of lesions produced per amount of initial inoculum
- Eventual lesion size or extent of host tissue infected
- Number of spores or cells produced per unit area of host tissue

---

each pathogen strain is inoculated onto a defined host genotype in the same environment, allowing quantification of a range of different parameters (Table 2.2). Change in the aggressiveness of a pathogen is sometimes an important factor in the emergence of new invasive strains or increases in the severity of disease epidemics. For instance, the recent spread of yellow rust (*Puccinia striiformis*) to new regions has been linked to the appearance of strains with a more rapid disease cycle, greater production of spores, and the ability to cause epidemics under warmer conditions than those favored by previous strains.

Genetic Control of Resistance and Virulence

In common with all other biological characteristics, host resistance and pathogen virulence are genetically determined. However, these two properties can only be assessed in the presence of the other partner. In the majority of cases, host resistance or pathogen virulence are not obviously correlated with other phenotypic characters. Features of the pathogen, such as rapid and extensive growth or the production of cell wall-degrading enzymes, may or may not be related to virulence. Assessments of resistance and virulence are therefore based on disease reaction types. An interaction where symptoms are clearly expressed is described as a **compatible** disease reaction as opposed to an **incompatible** reaction where symptoms do not develop and the effect on the plant is minimal (Figure 2.6).

Host resistance is controlled by one or a few genes whose individual effects may be easily detected, or by a multiplicity of genes, each of which contributes only a small fraction of the property as a whole. The practical implications of this are described in more detail in Chapter 12. In a few instances, disease reaction type has been shown to be controlled by factors inherited through the host's cytoplasm. The best-known example of such cytoplasmic inheritance involves the reaction of maize to the leaf blight fungus *Bipolaris maydis*. In the past, the production of hybrid maize has involved the laborious task of detasselling by hand to avoid self-pollination occurring. The discovery of a cytoplasmically inherited mitochondrial factor for male sterility (*Cms*), which meant that cross-pollination was essential, removed the need for this operation. Because of this, cultivars possessing *Cms* came to predominate throughout the USA. Unfortunately, *Cms* was also correlated with susceptibility to a particular strain (race T) of *B. maydis*. As a result, the occurrence in 1970 of favorable conditions for the development of the pathogen resulted in a disastrous epidemic (see Chapter 5, Figure 5.1). In any breeding program, the possibility that the cytoplasm may be important in disease resistance must therefore be considered.

Although pathogen virulence, host resistance, and disease reaction types are genetically determined, the environment may modify the expression of any of these characters. For example, the $Sr6$ gene conferring resistance in wheat to black stem rust is effective at 20 °C, but is inoperative at 25 °C.

Gene-for-Gene Theory

From an evolutionary viewpoint, it is predictable that genetic systems determining virulence in the pathogen will be paralleled by genes conferring resistance in the host. This is because any mutation to virulence in a pathogen population will be countered by the selection of hosts able to resist this more aggressive pathogen. Evolutionary biologists describe such a dynamic process of complementary changes as an "arms race." Thus, in an ideal world we might envisage a perpetual stalemate, with host and pathogen populations being closely matched in resistance and virulence. Hence over a period of several years disease would be neither completely absent nor would it become rampant.

Support for these ideas has been obtained by field observations and experimental studies on several plant pathosystems. The planting of crop cultivars containing a limited number of specific resistance genes will tend to select for pathogen strains possessing complementary virulence. Strains of a pathogen which differ in specific virulence genes are usually called **races**. The races present in any one season or in a particular locality may be identified by their behavior on a selection of cultivars carrying different combinations of genes conferring resistance. A hypothetical interaction between several races of a pathogen and a differential set of host cultivars is shown in Table 2.3.

Further analysis of such differential interactions has shown that there is a precise genetic relationship between the two partners. The pioneering study was done by Harold Flor, who analyzed the genetics of host resistance and pathogen virulence using flax rust, *Melampsora lini*, as a model. Flor showed that for each host gene conferring resistance, there is a complementary gene in the pathogen determining virulence. This finding has become widely known as the **gene-for-gene theory** of host–pathogen interactions. On the basis of Flor's theory, the possible interactions between a pair of alleles governing

Table 2.3 Hypothetical interaction between four host cultivars and four pathogen races

Host	Pathogen			
	Race 1	Race 2	Race 3	Race 4
Cultivar 1	+	−	−	−
Cultivar 2	−	+	−	−
Cultivar 3	−	−	+	−
Cultivar 4	−	−	−	+

+ = compatible disease reaction (host susceptible, pathogen virulent).
− = incompatible disease reaction (host resistant, pathogen avirulent).

resistance in a plant and the corresponding pair determining virulence in the pathogen can be shown as a quadratic check (Figure 2.7). In this scheme, a resistant reaction occurs only where an allele for resistance in the host plant interacts with an allele for avirulence in the pathogen.

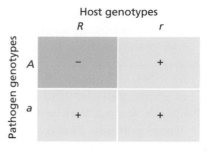

More recent analysis of mechanisms of pathogenicity in some necrotrophic fungi has shown that host susceptibility can in some cases be determined by the interaction of a pathogen toxin with a plant receptor causing sensitivity to this toxin. In such cases, susceptibility is the dominant trait, and the interaction has been described as an "inverse gene-for-gene" model (see Chapter 10, Figure 10.10).

Figure 2.7 The quadratic check, showing interactions between alleles of a host resistance gene and a pathogen gene for avirulence. Resistance (*R*) and avirulence (*A*) are usually dominant.

The gene-for-gene theory has important practical implications. The sequential introduction of new host cultivars which differ in resistance genes has been accompanied by corresponding changes in pathogen populations, whereby new races have successively come to predominate. The implications of this will be discussed further in Chapter 12. The gene-for-gene theory is also an important starting point for molecular models of host–pathogen specificity. Where single genes determine the outcome of a particular interaction, identification of the gene products involved should clarify how host–pathogen recognition occurs. Progress toward this goal is described in Chapter 10.

Further Reading

Books

Nash, A., Dalziel, R., and Fitzgerald, J. (2015). *Mims Pathogenesis of Infectious Disease*. Washington, DC: Academic Press.

Smith, S.E. and Read, D.J. (2010). *Mycorrhizal Symbiosis*, 3e. Washington, DC: Academic Press.

Articles

Andrivon, D. (1993). Nomenclature for pathogenicity and virulence: the need for precision. *Phytopathology* 83: 889–890. (and subsequent correspondence. See Phytopathology 85: 518–519, 1995).

Cooke, R.C. and Whipps, J.M. (1980). The evolution of modes of nutrition in fungi parasitic on terrestrial plants. *Biological Reviews* 55: 341–362.

Crute, I.R. (1994). Gene-for-gene recognition in plant-pathogen interactions. *Philosophical Transactions of the Royal Society of London. Series B, Biological Sciences* 346: 345–349.

Delaye, L., García-Guzmán, G., and Heil, M. (2013). Endophytes versus biotrophic and necrotrophic pathogens – are fungal lifestyles evolutionarily stable traits? *Fungal Diversity* 60 (1): 125–135. https://doi.org/10.1007/s13225-013-0240-y.

Hartung, J.S., Beretta, J., Brinasky, R.H. et al. (1994). Citrus variegated chlorosis bacterium: axenic culture, pathogenicity, and serological relationships with other strains of *Xylella fastidiosa*. *Phytopathology* 84: 591–597.

Kabbage, M., Yarden, O., and Dickman, M.B. (2015). Pathogenic attributes of *Sclerotinia sclerotiorum*: switching from a biotrophic to necrotrophic lifestyle. *Plant Science* 233: 53–60. https://doi.org/10.1016/j.plantsci.2014.12.018.

Pariaud, B., Ravigné, V., Halkett, F. et al. (2009). Aggressiveness and its role in the adaptation of plant pathogens. *Plant Pathology* 58: 409–424.

Simpson, A.J.G., Reinach, F.C., Arruda, P. et al. (2000). The genome sequence of the plant pathogen *Xylella fastidiosa*. *Nature* 406 (6792): 151–157.

3

Pathogen Biology

Scarcely any part of the organised world is free from the attack of parasites, a provision which is clearly one amongst many ordered by the Creator to maintain that balance amongst living beings ...

(Reverend M.J. Berkeley, 1803–1889)

By virtue of their lifestyle, parasites are faced with a number of problems which, though not unique, are nevertheless more acute than those confronting free-living organisms. Dependence on a host which is often dispersed in space and time means that parasites must possess effective mechanisms for transmission and survival between hosts. Furthermore, their habitat is a living organism. Each host has the ability to mount an active response to infection in order to ward off potential invaders. The host population is variable and can undergo changes from one generation to the next, so that the habitat is never static. However, the genetic flexibility of microorganisms, combined with their capacity for prolific reproduction, ensures that they maintain their status as successful parasites. Hence, microbial parasites are able to exploit many of the niches afforded by the diversity of green plants in natural communities.

The groups of microorganisms with which this book is concerned, the fungi, oomycetes, bacteria, phytoplasmas, viruses, and viroids, have certain features in common as plant pathogens. For instance, they are all capable of rapid reproduction or replication which results in the formation of numerous infective spores, cells or particles. Collectively, these may be described as propagules. Each group has, however, a distinctive and characteristic vegetative morphology and mode of growth (Figure 3.1). Such differences have important implications as regards the behavior of these different groups as plant pathogens.

What is the relative significance of these groups of microorganisms in pathology? Most of the well-known infectious diseases of humans and other animals are caused by bacteria and viruses. By contrast, plants are more commonly affected by fungi and viruses. This difference in the importance of particular groups of microorganisms can be explained partly on the basis of several differences between plants and animals as habitats for microbial growth (Table 3.1). Bacteria generally prefer warm, alkaline conditions with high nitrogen levels. Being unicellular, spread within the host is enhanced by a circulatory system. Plant-infecting bacteria often exploit vascular tissues to spread from initial sites of invasion.

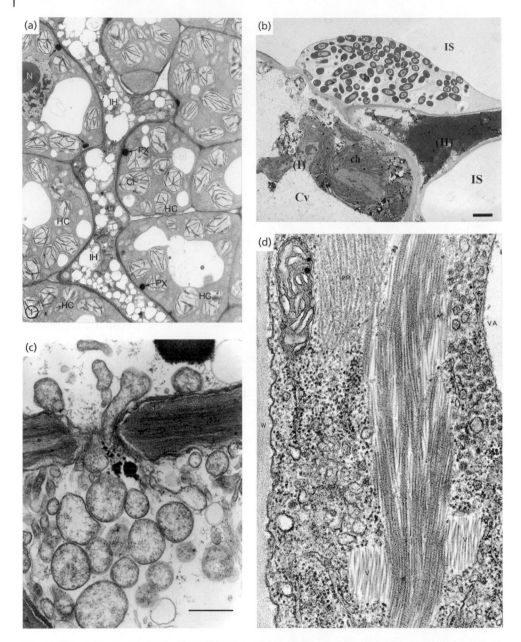

Figure 3.1   (a) Intercellular hypha (IH) of the oomycete *Peronospora viciae* in shoot tissue of pea plant showing host cells (HC) with chloroplasts (Cl). Dark deposits (Px) are sites at which haustoria will penetrate host cells. (x 3000) (b) Colony of the bacterium *Pseudomonas syringae* in intercellular space (IS) of *Arabidopsis* leaf. Adjacent host cells are disorganized (I) or have completely collapsed (II). Ch, chloroplast; Cv, cell vacuole. Bar = 2 µm (c) Phytoplasma cells within phloem sieve element showing variable morphology and cells passing through sieve plate. Bar = 0.5 µm (d) Crystalline aggregates of rod-shaped tobacco mosaic virus (TMV) particles in phloem parenchyma cell of tobacco (×30 000). *Source:* Photographs from Hickey and Coffey (1977); Soylu et al. (2005); Courtesy of Yaima Arocha; Esau (1968).

Table 3.1 Higher plants and animals compared as hosts

Plants	Animals
Anatomical features	
Rigid cell wall	No cell wall
No circulatory system	Circulatory system
Internal environment	
Acid pH	Alkaline pH
High C : N ratio	Low C : N ratio
No temperature regulation	Temperature regulation

Fungi are better equipped to overcome structural barriers such as the cell wall, and their filamentous growth habit allows them to penetrate and extend through plant tissues. It has been suggested that animals and plants have contrasting strategies for dealing with stress, with the well-developed sensory and movement capacity of animals allowing them to evade stresses, while plants are sedentary and must possess greater physiological tolerance and other means of defense. It is now appreciated, however, that both animals and plants have highly developed immune systems, albeit based on contrasting cellular and molecular mechanisms (see Chapter 9), so any successful pathogen must be able to counter or evade such defenses to infect the host. This is a common theme in both plant and animal pathology, based on the co-evolution of microbial pathogens with their hosts over long periods.

These generalizations, while of some value in explaining the relative importance of different groups as pathogens, can break down due to changes in the environment or properties of the interacting partners. Recently, for instance, invasive fungal diseases have become more common in human hosts, due primarily to the increased prevalence of immunocompromised individuals in the population. People with defective immunity are more vulnerable to opportunistic infection by fungi, including species that were not previously recorded as human pathogens.

Fungi as Plant Pathogens

Vegetative Growth of Fungi

Most plant-pathogenic fungi form **hyphae** (singular = hypha), which are filamentous cells that extend by apical growth and an ordered system of branching. The network of hyphae which results from such growth is called a **mycelium**, and the interconnected hyphal network derived from one propagule is termed a **colony**.

The apical mode of growth of most fungi is the key to the success of these organisms both as saprotrophs and parasites. Unlike unicellular organisms, filamentous fungi are able to extend through soil, plant litter or living tissues. As a hypha grows through the substrate, it secretes extracellular enzymes which digest complex molecules. The products of this process are then absorbed by the hypha. As the nutrients become exhausted, the hypha simply grows on to explore a new area. Not all filamentous fungi have the same enzymatic capabilities. For example, many of the necrotrophic pathogens which attack leaves or fruits are noted for their production of cell wall-degrading enzymes that can macerate tissues. Heart-rot pathogens of trees can secrete a more extensive repertoire of enzymes, including ligninases which enable them to utilize the complex lignocellulose constituents of wood. In contrast, biotrophic fungi usually produce fewer hydrolytic enzymes active against cell walls, and comparisons between the complete genomes of fungi with these contrasting lifestyles has confirmed that biotrophs generally have fewer genes encoding such enzymes.

The relationship between growing hyphal tips and the older, first-formed parts of the colony varies at different stages in the life cycle of a fungus. Three different patterns of behavior may be identified. In some fungi, as hyphae extend at their apices the older portions of the colony become moribund and may autolyse, or be destroyed by bacteria and grazing animals. Sporulation in these fungi occurs at or near the advancing margin of the colony. Following disintegration of the first-formed hyphae, the older parts of the lesion are quickly invaded by secondary organisms.

The hyphae of other fungi are longer-lived. They can act as transport systems from the older parts of the colony to the hyphal tips, and play an important role in exploratory growth in natural substrates and in finding new hosts. Unless the fungus is able to obtain nutrients by competing with saprotrophic microorganisms, it must fuel both its growth across inhospitable terrain and the subsequent infection processes needed to establish itself on a new host, by transporting nutrients from its established food base.

Hyphae growing on the plant surface or in soil face problems such as desiccation and competition with the associated microflora and fauna. They may therefore exhibit adaptations which improve their efficiency and success in this role. The **runner hyphae** of the take-all fungus *Gaeumannomyces graminis*, which facilitate rapid and extensive growth along the surface of roots, have thicker, darker-colored walls than hyphae involved in penetration and colonization of the root tissues. Other fungi produce cord-like structures known as **mycelial strands** consisting of a number of hyphae aggregated together (Figure 3.2). Some tree pathogens, such as *Armillaria mellea* and *Fomes lignosus*, form more highly differentiated strands, known as **rhizomorphs**, with a thickened and pigmented outer rind and internal hyphae that act as a highly efficient conducting system for water and nutrients. Rhizomorphs can grow 5–6 times faster than normal vegetative hyphae, and can extend over relatively long distances across the forest floor.

The third type of colonial behavior is exhibited by the rust fungi and other biotrophic pathogens. In these, the whole colony remains functional for a relatively long period and transport of nutrients takes place from the colony margin to the center, where there is usually some continuing activity such as sporulation. Rust fungi develop these integrated colonies within photosynthetic tissues and the continued activity of their colonies depends on these tissues remaining viable.

Individual hyphae are frequently modified in form to accomplish particular functions, and this is especially common among pathogenic species. These fungi have in many cases evolved specialized hyphal structures to aid adhesion, penetration, and colonization of the host. Appressoria, infection hyphae, and haustoria (Figure 3.3) are all examples of such hyphal adaptations (see Chapter 6).

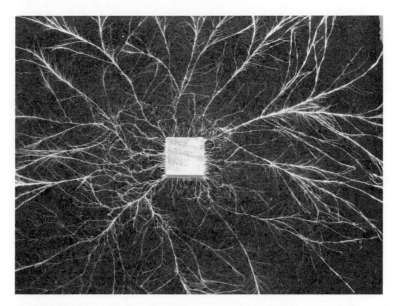

Figure 3.2 Mycelial strands formed by a basidiomycete fungus. The fungus has colonized a wood block and strands are radiating out across soil in search of another food source. Similar exploratory behavior is seen with some fungal pathogens of trees. *Source:* Photo by John Lucas, courtesy of Lynne Boddy.

(a)

(b)

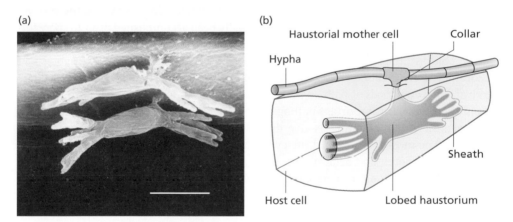

Figure 3.3 Specialized parasitic structures, known as haustoria, formed by the powdery mildew fungus *Blumeria graminis*. (a) Scanning electron micrograph of a fractured epidermal cell showing two branched haustoria with finger-like extensions inside the host cell. Scale bar = 10 µm. *Source:* Honegger (1985). (b) Diagrammatic interpretation showing the fungal hypha on the leaf surface and a haustorium within the epidermal cell. *Source:* After Bracker (1968).

Reproduction in Fungi

The fungi exhibit a very wide range of reproductive mechanisms. Asexual reproduction seems to be of prime importance in increasing the number of individuals. Epidemic spread of pathogens usually depends on the production of countless asexual spores. The Mastigomycotina and the Zygomycotina form asexual spores in **sporangia**. In some species, these are motile, when they are called **zoospores**, but in others they are nonmotile. The asexual spores of the Ascomycotina and Basidiomycotina, which are all nonmotile, are called **conidia**.

Sexual reproduction fulfills the dual role of increasing variation in the population and assisting survival through unfavorable periods. It occurs regularly in the life histories of many fungal pathogens, such as *A. mellea*, *Claviceps purpurea*, and *Zygoseptoria tritici*. In some species, however, it is apparently an exceptional or very rare event. In fact, one group of fungi, the Fungi Imperfecti (or Deuteromycotina), is defined on the basis of the absence or infrequency of sexual reproduction. *Penicillium* on citrus fruits, *Botrytis* on strawberries and *Cladosporium fulvum* on tomatoes are all examples of pathogens which seem to have relegated sexual reproduction to a minor role in their life cycles. With some imperfect fungi, however, the existence of a sexual stage may have been overlooked. The cereal eyespot pathogens, *Oculimacula* spp., were thought to reproduce only by conidia but subsequently a sexual stage was discovered on straw in several different countries. Even when sexual reproduction is a regular feature of a life cycle, it usually only occurs under a more limited range of host and environmental conditions than permit asexual reproduction.

The nature of sexual reproduction varies considerably among the major groups of fungi. Several types of sexual spore can be identified, with **zygospores**, **ascospores**, and **basidiospores** being formed by the Zygomycota, Ascomycota, and Basidiomycota respectively. Some sexual spores germinate soon after dispersal, as in the case of *Puccinia* basidiospores on barberry leaves and *Claviceps* ascospores on grass stigmas. In these fungi, the survival function ascribed to sexual reproduction has usually been accomplished prior to spore formation, by the perennation of structures which allow the fungus to overwinter or oversummer. In contrast, the sexual spores formed by the lower fungi, and some ascospores and basidiospores, are themselves able to withstand environmental stresses and thus facilitate survival between periods of pathological activity.

Some economically important pathogens, notably the rusts, have extremely complex life cycles in which as many as five different types of spore are formed in regular sequence. The most complex rust life cycles involve urediospores, teliospores, basidiospores, pycniospores, and aeciospores. Epidemic spread is essentially by urediospores, which are asexual conidia, whereas variation and survival are ensured by the other four types of spore.

Oomycetes as Plant Pathogens

A second group of filamentous microorganisms that includes a series of major plant pathogens is the oomycetes. These were originally classified as fungi, due to their growth habit and formation of spores. Further study has shown several significant differences between oomycetes and true fungi, including cell wall composition (oomycetes lack chitin, that occurs in fungal cell walls) and cell ploidy (most fungi are haploid, while oomycetes are diploid or polyploid). Molecular phylogeny, the study of evolutionary history based on gene sequences, has confirmed that the oomycetes are in fact a distinct kingdom, separate from plants and fungi, known as the Stramenopila.

Table 3.2 Some important oomycete plant pathogens

Genus/species	Host	Disease
Phytophthora infestans	Potato, tomato	Late blight
P. ramorum	Oak and other woody species	Sudden oak death
P. cinnamomi	Numerous woody perennials	Root rot
Pythium ultimum	Many seedlings	Damping-off
Albugo candida	Cruciferae, e.g., brassicas	White rust
Bremia lactucae	Lettuce	Downy mildew
Hyaloperonospora spp.	Cruciferae, e.g., *Arabidopsis*	Downy mildew
Peronosclerospora sorghi	Sorghum	Downy mildew
Pseudoperonospora cubensis	Cucurbitaceae, e.g., cucumber	Downy mildew
Sclerospora graminicola	Pearl millet	Downy mildew

Culturable; Nonculturable.

Pathogenic oomycetes cause economically important diseases of fish and many crop plants (Table 3.2). They include some of the most destructive plant pathogen genera known, such as *Pythium*, responsible for damping-off of seedlings, *Phytophthora*, translating as "plant destroyer," and the downy mildews, that include *Plasmopara viticola* on vines, *Peronospora tabacina* (tobacco blue mold), and several genera important as pathogens of tropical cereal crops. The biology of oomycete pathogens is in some ways analogous to true fungi, in that they are dispersed as spores, form infection structures such as appressoria and haustoria, and demonstrate apical growth through tissues (Figure 3.1). The asexual spores can be motile zoospores propelled by flagella or wind-borne sporangia; in some cases both types of spore can occur, for instance in *Phytophthora*, depending on environmental conditions. Sporangiospores can germinate and penetrate the host plant directly, or release zoospores that swim, encyst, and then penetrate. The sexual oospores, formed after crossing of two mating types, are thick-walled and can remain dormant for long periods as a means of survival.

The similarities between pathogenic oomycetes and true fungi appear to be a case of convergent evolution, where parallel solutions have emerged that enable them to gain entry to the host, interfere with plant immunity, and divert nutrients for their own use.

Bacteria as Plant Pathogens

Given the large number of bacterial genera, relatively few have been recorded as plant pathogens (Table 3.3). Some of the most important plant pathogenic species, for example *Pseudomonas syringae* and *Xanthomanas campestris*, are, however, subdivided into numerous pathovars distinguished by specialization on different hosts. The taxonomy of some of these is under review as studies of DNA homology have revealed large variations in relatedness between pathovars. The identity of other bacteria has been resolved only

Table 3.3 Main genera of plant-pathogenic bacteria and some example species and pathovars

Genus	Species	Pathovar or subspecies	Disease
Gram negative			
Agrobacterium	*tumefaciens*		Crown gall (wide host range)
Dickeya	*dadantii, solani*		Potato and other hosts
Erwinia	*amylovora*		Fireblight (apple, pear, other rosaceous hosts)
Pseudomonas	*syringae*	pv. *glycinea*	Bacterial blight of soybean
		pv. *phaseolicola*	Halo blight of beans
		pv. *tomato*	Bacterial speck of tomato
Pseudomonas	*savastanoi*	pv. *savastanoi*	Olive knot
Ralstonia	*solanacearum*		Brown rot of potato
			Bacterial wilt of tomato and tobacco
			Moko disease of banana
Xanthomonas	*campestris*	pv. *campestris*	Black rot of crucifers
		pv. *vesicatoria*	Bacterial spot of pepper and tomato
Xanthomonas	*axonopodis*	pv. *citri*	Citrus canker
Xanthomonas	*oryzae*	pv. *oryzae*	Bacterial blight of rice
Xylella	*fastidiosa*		Pierce's disease of grapevine
Candidatus Liberibacter	*asiaticus*		Citrus huanglongbing (citrus greening)
Gram positive			
Clavibacter	*michiganense*	ssp. *insidiosus*	Bacterial wilt of alfalfa
		ssp. *sepodonicus*	Potato ring rot
Streptomyces	*scabies*		Potato common scab

recently; small rickettsia-like organisms were first discovered in the vascular tissues of grapevines in 1973, and were later successfully cultured and shown to be a distinct species, *Xylella fastidiosa*. The host range of these xylem-limited species appears to be very wide. More recently, the causal agent of the devastating disease citrus huanglongbing (HLB; citrus greening) was shown to be a nonculturable gram-negative bacterium restricted to phloem tissues (Table 3.3).

The majority of plant-pathogenic bacteria are unicellular (Figure 3.4), with cell division by binary fission. The major exceptions to this generalization are the bacterial pathogens classified as Actinomycetes. *Streptomyces*, for example, forms a rudimentary branching mycelium composed of relatively narrow, septate filaments. Sometimes, individual bacterial cells aggregate to form substantial colonies, as in crown gall or in the cankers resulting from fireblight (*Erwinia amylovora*) infection. In other cases, cells spread throughout an entire organ or physiological system. Soft-rot bacteria in potatoes spread indiscriminately through tuber tissues, while vascular wilt bacteria in various hosts are widely dispersed

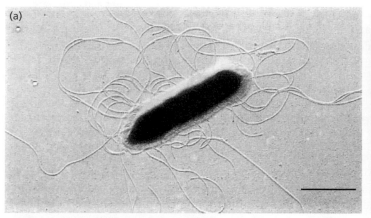

Figure 3.4 Bacterial cell structure. (a) Electron micrograph of the bacterium *Erwinia amylovora*, causal agent of fireblight disease, showing rod-shaped cell with flagella. Bar = 1 μm. *Source:* Courtesy of R.N. Goodman. (b) Scanning electron micrograph of cell of *Spiroplasma poulsonii* isolated from insect hemolymph. Bar = 1 μm. *Source:* Ramond et al. (2016).

within the xylem. Growth of single-celled bacteria often involves the secretion of extracellular enzymes, polysaccharides, and other products. However, most plant-pathogenic bacteria have only a limited ability to degrade cell wall polymers, such as cellulose and lignin. Instead, pectolytic enzymes digest substances in the middle lamella which provide the bacteria with nutrients and allow them to spread between the newly separated cells.

While unicellular bacteria cannot form a highly integrated system like the fungal mycelium, they are nonetheless able to coordinate their behavior via a form of cell-to-cell communication called **quorum sensing** (QS). When the population reaches a certain cell density, signal molecules act as autoinducers, switching on genes regulating a range of processes, including formation of biofilms on surfaces and secretion of factors involved in pathogenicity, defense, and survival. The most common QS signals are low molecular weight compounds known as N-acyl-homoserine lactones. QS is a mechanism enabling bacteria to act in a coordinated manner in natural environments, and also when invading a host.

Unlike fungi, bacteria cannot produce modified cells which facilitate entry into a host plant or extraction of nutrients from host cells. However, many plant-pathogenic bacteria possess flagella (Figure 3.4) and are therefore motile during some phases of their life cycle and capable of moving along nutrient gradients and toward host signal molecules. This may be important in habitats such as the soil surrounding plant roots, or on the surface of leaves. They also possess mechanisms for attachment to, and penetration of, host cells. Pathogenic gram-negative bacteria have evolved a complex surface structure, the Type III secretion system, that injects proteins into host cells, which then interfere with defense pathways (see Chapter 8).

Bacteria can undergo a type of sexual recombination, but less frequently than the fungi. They lack a clearly defined nucleus and instead have a circular chromosome, as well as extrachromosomal DNA in the form of plasmids. DNA passes from one cell, the donor, to another, the recipient, by transduction, transformation or conjugation. Transduction, which is mediated by bacteriophages, is possibly the most common method occurring in nature. Plasmids often carry genes involved in pathogenicity or resistance to antibiotics, and these can also be exchanged between cells. Pathogenic bacteria therefore possess effective mechanisms promoting genetic variation and the ability to colonize and survive within the living host.

Mycoplasma-Like Organisms as Plant Pathogens

A number of diseases once thought to be of viral etiology are in fact caused by a group of very small prokaryotic agents known as mycoplasma-like organisms (MLOs) or mollicutes. In particular, several yellow diseases, characterized by extensive chlorosis and a gradual decline in the host, are associated with the presence of large numbers of these organisms in the phloem (Figure 3.1). Despite their superficial similarity in electron micrographs of diseased tissue, it is now clear that this group includes at least two types of agent: the **phytoplasmas** and the **spiroplasmas**. Both types may be regarded as bacteria which have lost the ability to form a rigid cell wall; they are pleomorphic and can therefore translocate in phloem tissues by passing through the sieve plates between cells. Mollicutes have a dual-kingdom host range as many can replicate in insects as well as in plants. They possess very small genomes encoding fewer proteins than typical bacteria, lacking some key metabolic pathways, hence their dependence on another, living host. It seems likely that under natural conditions, they cannot survive apart from their host organisms. To date, no phytoplasma has been cultured away from its host, while spiroplasmas can be cultured on artificial media in which they typically assume a motile, helical form (Figure 3.4).

More than 100 diseases in a variety of crops are now known to be caused by phytoplasmas. These include coconut lethal yellowing (see Chapter 4, Figure 4.3), which has virtually destroyed the industry in the Caribbean, and rice yellow dwarf. Phytoplasma disease symptoms often feature major effects on plant development, such as stunting, deformed leaves and flowers, and excessive branching, known as witch's broom. Spiroplasmas have been associated with fewer diseases but some, such as corn stunt and citrus stubborn, can cause serious losses. Spiroplasmas occur widely in insects, with which they often form mutualistic relationships. On very rare occasions they have been recorded infecting humans.

Overall, mollicutes are similar to viruses in their epidemiology, being transmitted by insect vectors, plant propagation, and grafting. Unlike viruses, they are sensitive to antibiotics such as tetracycline, and treatment can lead to remission of symptoms.

Protozoa as Plant Pathogens

Parasitic protozoa are mostly thought of in the context of devastating human diseases such as malaria and sleeping sickness but in fact, a number of ailments of tropical perennial crops are associated with the presence of flagellate protozoa, often in phloem tissues. These include a wilt disease of coffee, heart-rot of coconuts, and sudden wilt of oil palm.

The flagellates, classified as *Phytomonas*, can be transmitted from plant to plant by insect vectors. A disease of cassava occurring in northern Brazil, typified by chlorosis and poor root development, may also be due to such pathogens, as large numbers of flagellates occur in the latex ducts of affected plants. No natural vector has been identified, but the disease can be transmitted by grafting.

Viruses as Plant Pathogens

Viruses are much simpler in structure than cellular microbial pathogens, consisting only of a nucleic acid core surrounded by a protein coat or capsid; within infected cells, they occur either as individual particles or as crystalline aggregates (Figure 3.1). At first sight, viruses might appear ill-equipped to act as pathogens, due to their extreme dependence on living cells. However, their diverse and effective methods of transmission between hosts, coupled with efficient replication once established within living cells, ensure that they are potentially devastating plant pathogens.

Viral parasitism is unique, in that the parasite is incorporated into the metabolism of the host cell. After gaining entry into a living cell, the nucleic acid component of the virus is released from its protein coat. The viral genome is then translated and replicated, and numerous new virus particles are assembled from the newly synthesized nucleic acid and protein. A virus can thus be visualized as a set of instructions for making more virus, packaged in a protective coat. In contrast to fungi and bacteria, viruses do not attack the structural integrity of their host tissues, but instead subvert the synthetic machinery of the host cell, acting as "molecular pirates."

While viruses are simple parasites, they exhibit great diversity in their structure, morphology, and mode of replication. Six major groups of viruses have been defined, based on the nature of their genome and mode of replication (Table 3.4). Viruses contain a DNA or RNA genome in either single- or double-stranded form; where the nucleic acid is single-stranded, the polarity of the strand may be the same as messenger RNA or complementary to it. The genome may be a single molecule (monopartite) or divided into two or more pieces (bipartite or multipartite). The precise sequence of events during replication within the host cell varies depending on the type of nucleic acid present; group III is noteworthy as replication involves copying RNA to DNA via the enzyme reverse transcriptase (RT). This group includes the retroviruses, such as the human immunodeficiency virus (HIV), where the single-stranded positive-sense RNA viral genome is copied to DNA and then integrates into the host genome. To date, no plant-infecting viruses of this type have been discovered. A single family of plant viruses, such as cauliflower mosaic virus (CaMV), are also grouped here; in this case, the double-stranded DNA genome encodes several proteins including the RT enzyme, and replication is by reverse transcription from a messenger RNA template.

Within each group, additional characteristics such as particle morphology, genome components and sequences, host range, mode of transmission, and serological properties (i.e., relatedness of virus protein) are used to further classify viruses into families, genera, and species. For instance, tobacco mosaic virus is placed in the family *Tombusviridae*, genus *Tobamovirus*, species *tobacco mosaic virus*.

Table 3.4 Virus groups based on genome type (International Committee on Taxonomy of Viruses [ICTV])

Group	Genome	Example	Comments
Group I	dsDNA	None in plants	Replicate without an RNA intermediate
Group II	ssDNA	Geminivirus, e.g., African cassava mosaic virus (ACMV)	Two small circular genome components
Group III	dsDNA-RT[a]	Caulimovirus, e.g., cauliflower mosaic virus (CaMV)	Single circular dsDNA genome
	ssRNA-RT[a]	None in plants	Includes retroviruses such as HIV
Group IV	dsRNA	Wound tumor virus (WTV)	
Group V	ssRNA−	Rhabdoviruses, e.g., lettuce necrotic yellows virus (LNYV)	Negative-sense single-stranded RNA genome
Group VI	ssRNA+	Tobacco mosaic virus (TMV)	Positive-sense single-stranded RNA
		Cauliflower mosaic virus (CaMV)	Majority of plant virus families (>10) are found in this group
Group VII	ssRNA subviral agents	Viroids	ssRNA genome that does not encode proteins

[a] Replication includes synthesis of RNA from DNA by reverse transcriptase.

The large majority of plant viruses so far described (Table 3.4) belong to group VI, and contain a single strand of RNA which, once freed from its coat protein, can act directly as messenger RNA in the synthesis of further virus particles. This messenger must code for at least two components: an RNA replicase enzyme and new virus-coat protein. The first plant virus discovered, tobacco mosaic virus (TMV), serves as an example; TMV is a helical rod-shaped particle containing a single strand of RNA approximately 6400 nucleotides long (Figure 3.5). Translation of the RNA strand yields four polypeptides differing in molecular weight. The two largest (P126 and P183) are initiated from the same start site, with the longer polypeptide arising by readthrough of a stop signal. The two other proteins (P30 and P17) are produced via synthesis of smaller copies of parts of the TMV genome, described as subgenomic RNAs. Thus, even with a relatively simple viral genome, the molecular events accompanying expression can be quite complex.

By introducing mutations at precise points in the RNA, one can obtain clues as to the function of the different protein products. Changes in the region coding for the large polypeptides interfere with virus replication. Mutations in the P30 coding region inhibit the movement of virus from cell to cell. The smallest product P17 is the coat protein, and is needed in large quantities as each virus particle contains more than 2000 capsid subunits. Production of a subgenomic RNA is a mechanism allowing rapid synthesis of multiple copies from a small template. This type of functional analysis is useful not only in

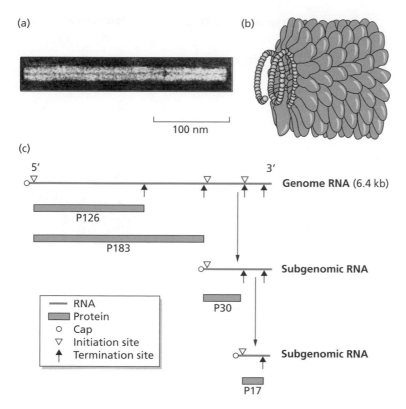

Figure 3.5   Structure of tobacco mosaic virus (TMV). (a) Electron micrograph of intact TMV particle. (b) Organization of particle with helical arrangement of coat protein subunits around the RNA. (c) RNA genome, subgenomic RNAs and the four proteins produced.

understanding the virus life cycle but also in devising new ways of engineering plants for resistance to viruses (see p.331).

In other plant viruses, the viral nucleic acid must first be transcribed to yield messenger RNA, from either an RNA or DNA template. The polymerase enzyme necessary to achieve this transcription may already be present in the intact virus particle.

Viroids as Plant Pathogens

Viroids are the smallest known infectious nucleic acids. They differ from viruses in the size of their RNA genomes and in their lack of a protein coat. These compact circular RNA molecules comprise only about 250–400 nucleotides, which is 10 times smaller than a typical RNA virus genome. This amount of genetic information is minimal, and apparently does not code for even a single protein. This raises interesting questions as to how these agents replicate or modify host cells to cause disease.

The first disease definitely attributed to a viroid was potato spindle tuber (PSTVd), and at least 30 other diseases are now known to be caused by these infectious agents. Viroids are currently divided into two families, with either a rod-like or branched secondary structure

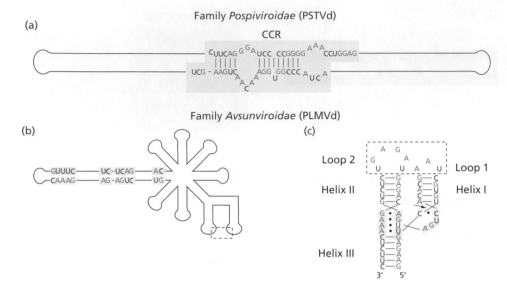

Figure 3.6 Viroid structures, showing (a) rod-like secondary structure with conserved central region (CCR) proposed for the family Pospiviroidae, example Potato Spindle Tuber Viroid (PSTVd), (b) Branched secondary structure proposed for family Avsunviroidae, example Peach Latent Mosaic Viroid (PLMVd), with (c) hammerhead ribozyme region enlarged. A ribozyme is a ribonucleic acid enzyme that is active in transfer of amino acids, or cleavage or joining of RNA molecules. *Source:* Flores et al. (2014).

(Figure 3.6). Despite their small genomes with minimal information content, viroids are responsible for some lethal plant diseases, including coconut cadang cadang (CCCVd) which has destroyed millions of trees in the Philippines. It is also possible that cryptic viroids may be present in many apparently healthy plants.

As viroids lack any protein-coding function, they must rely on host transcription and processing machinery to replicate. Viroid replication takes place either in the nucleus (family *Pospiviroidae*) or chloroplasts (family *Avsunviroidae*) via a rolling circle mechanism with RNA intermediates. The branched viroids possess secondary structures with catalytic activity, known as ribozymes, that play a part in the replication process. Viroids can move from cell to cell through plasmodesmata and systemically via the phloem, but the cellular factors involved in viroid trafficking are not yet known.

It is now known that small RNA molecules play important regulatory roles in cell biology, such as the small interfering (siRNAs) involved in gene silencing. Specific viroid sequence or structural elements may interact with host factors to alter plant gene expression and hence alter host developmental processes. Gene silencing by small RNAs derived from viroids might therefore explain at least some of their pathogenicity.

Where did viroids come from? One theory, based on similarities between these molecules and introns, the pieces spliced out of precursor RNA during processing, proposed that they represent rogue molecules that have somehow escaped from regulated cell functions. With the recent discovery of the diverse regulatory and catalytic activities of small RNA molecules, it is now suggested that viroids might be relics of an ancient RNA world that predated replication based on DNA and proteins.

Variation in Microbial Pathogen Populations

The ability of microorganisms to change and quickly adapt to new circumstances is a key factor in their survival and success as pathogens. Mutation and recombination are familiar ways of generating genetic variation, but one of the revelations of genomics is that there are other processes driving the evolution of microbial populations, allowing them to infect new host species, subvert host immunity, and adapt to the chemicals we use to control them.

The relative importance of the various mechanisms for the maintenance of variation, and recombination of genes, differs between the microbial groups (Table 3.5). Viruses would appear to have few options other than mutation, although recombination can occur between viral genomes where there is more than one piece of nucleic acid per particle, or where more than one strain of a virus simultaneously infects a host cell. Like viruses, mutation and recombination during replication have been identified as the major mechanisms driving evolution and adaptation of viroids to hosts and environmental factors. The sexual cycle is clearly important for many fungi and for some bacteria. Heterokaryosis (the long-term co-existence of two or more genetically different nuclei in a common cytoplasm) and the parasexual cycle are peculiar to the fungi. Fusion between the hyphae of two distinct individuals followed by nuclear migration can create a heterokaryon. In the parasexual cycle, this may be followed by nuclear fusion and genetic recombination to produce novel variants without the normal steps involved in sexual reproduction

Table 3.5 Mechanisms maintaining variation in pathogen populations. The number of pluses indicates the likely importance of each mechanism for each type

	Fungi	Oomycetes	Bacteria	Viruses
Mutation	+	+	+	+
Sexual recombination	+++	+++	+	−
Heterokaryosis	+++	+	−	−
Parasexual cycle	++	−	−	−
Hybridization	+	++	−	−
Cytoplasmic factors	+	+	+++	−
Horizontal gene transfer	+	+	++	−
Nucleic acid recombination[a]	−	−	−	++

[a] Between virus genomes.

taking place. Different isolates of the same pathogen may also differ in ploidy or in the number of chromosomes per nucleus. For instance, potato late blight, *Phytophthora infestans*, is usually diploid, but tetraploid and polyploid isolates occur. The rice blast fungus, *Magnaporthe grisea*, shows wide differences in chromosome number between strains, due mainly to the occurrence of very small mini-chromosomes, which partly explains the highly variable nature of this pathogen in the field. The first fully sequenced genome of the wheat leaf blotch pathogen *Zymoseptoria tritici* revealed 21 chromosomes, of which eight are dispensable, without showing any apparent effects on the fungus. Such chromosomes, while not essential, might provide a reservoir of genetic variation extending the adaptive options for the pathogen.

Cytoplasmic genetic elements are important in both bacteria and fungi. Most pathogenic bacteria contain extrachromosomal self-replicating circles of DNA known as **plasmids**. One or more copies of a plasmid may be present, and they can also move from cell to cell, thereby transmitting information and increasing the genetic flexibility of the population. With many animal pathogenic bacteria, plasmids encode important virulence functions, such as adhesion to host cells and toxin production. Less information is available for plant-pathogenic bacteria, with the notable exception of *Agrobacterium* species which harbor tumor-inducing (Ti) or root-inducing (Ri) plasmids (see p.205). Genes located on plasmids may also influence the host range, nutrition, and survival of plant-infecting bacteria. For instance, some genes affecting virulence, antibiotic production, and resistance to toxic chemicals, such as copper, can occur on plasmids. Genetic equivalents of plasmids, some of which may originate from viruses, have also been discovered in fungi. Variability is further enhanced by transposable elements (**transposons**), so-called "jumping genes" which can insert at different sites in the genome, or move from plasmid to chromosome or vice versa. Transposable elements (TEs) can in some cases make up a sizeable proportion of the genome; in the plant-pathogenic fungus *Leptosphaeria maculans*, almost one-third of the genome comprises TEs, and invasion of these elements has most likely led to chromosomal rearrangements and other dynamic processes affecting adaptation to the *Brassica* host plant.

It is now clear that within a particular pathogen species, a series of processes, including mutation, recombination, gene loss and duplication, and expansion of particular gene families, such as those encoding pathogenicity functions, can all contribute to genetic flexibility. But there is increasing evidence that genetic interactions between species can also profoundly affect pathogen behavior. **Hybridization** between species combines two genomes and can lead to the creation of completely new genotypes. For instance, a destructive new *Phytophthora* attacking alder (*Alnus*) trees in Europe emerged during the 1990s and was shown to comprise a hybrid, *Phytophthora alni*, most likely between *Phytophthora cambivora* and another species, subsequently segregating as a group or "swarm" with different phenotypes. Such events can promote rapid evolution of new pathotypes. Comparative analysis of pathogen genome sequences has also revealed the presence of genes, or gene clusters, that appear to have originated from outside the usual gene pool, transferred from closely or more distantly related species, or even different kingdoms, such as bacterial genes transferred to fungi or oomycetes. This process, known as **horizontal gene transfer** (HGT) or **lateral gene transfer** (LGT) can lead to new properties such as production of enzymes affecting substrate utilization and metabolism, or synthesis of toxins causing disease in new hosts.

There is evidence that some fungi infecting woody tree hosts may have acquired the battery of enzymes necessary to break down complex cell wall polymers by transfer from saprophytic species causing wood decay. A particularly interesting example of extension of the host range of a pathogen by HGT concerns tan spot disease of wheat, caused by the fungus *Pyrenophora tritici-repentis*. This disease first emerged during the 1940s and the virulence of the fungus was subsequently shown to be associated with production of a small secreted protein, Tox A, active against susceptible wheat varieties (see Chapter 8, Figure 8.9). Analysis of the genome sequence of *P. tritici-repentis* and another wheat pathogen, *Parastaganospora nodorum*, has revealed that both contain an almost identical gene encoding the toxin, and it is likely that emergence of tan spot as a damaging disease of wheat was due to *P. tritici-repentis* acquiring the toxin gene from *P. nodorum* by HGT in the relatively recent past.

Many pathogenic bacteria contain genomic regions with clusters of genes involved in disease development known as **pathogenicity islands**. These are often associated with mobile genetic elements such as transposons, and can transfer from plasmid to chromosome, or between bacterial cells, converting formerly benign strains into virulent pathogens. This is further described in Chapter 8.

All in all, these recent revelations have led to a revision of the idea that microbial species are fixed entities, with a stable DNA content. Instead, the genome architecture and gene complement of microbial cells are highly dynamic, providing the genetic flexibility to quickly adapt to a changing environment, including the living host. Indeed, the apparently endless capacity for variation among pathogens remains a major stumbling block in the search for sustainable methods of disease control.

Dispersal of Pathogens

The problem of dispersal is a fundamental feature of the life cycle of all living organisms. As host plants are unevenly distributed in space and time, most plant pathogens require an especially efficient means of solving this problem. Dispersal is usually thought of in terms of spatial spread (Figure 3.7), but the time scale over which this takes place is also important. Local dispersal may be completed in a few minutes or even seconds, while in other cases spread may take place over hours, weeks or even years.

Many pathogens spread through crops in a spectacular manner. Some can cover long distances at remarkable speeds. In general, the most dramatic examples involve pathogens with air-borne propagules. Pathogens which infect subterranean organs often exhibit a restricted pattern of dispersal through the soil. These generalizations apply particularly to pathogens which have developed their own independent processes for dispersal. Pathogens which rely on another, vector organism (see later in this chapter) for spread are at the mercy of these agents, in respect of the distances and speeds achieved.

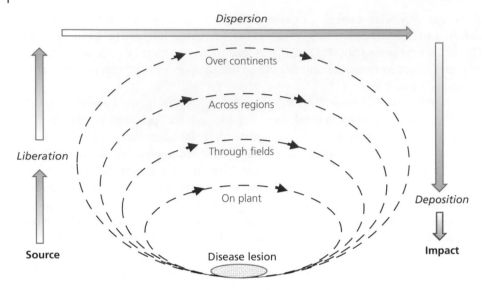

Figure 3.7 The processes involved in pathogen dispersal and the scale on which they operate.

There are many different dispersal mechanisms and routes, both active (depending on a biological response to environmental conditions) and passive, in which cells, spores or other propagules are carried by wind, water or other agents (Figure 3.8). Some mechanisms are both elegant and highly efficient in that they are closely adapted to the biology of the host. The synchronization of spore release in certain pathogens, such as *Venturia* and *Claviceps*, with bud-break or flowering is essential if the propagules are to find vulnerable host tissues. Other pathogens merely saturate the environment with propagules in an apparently haphazard and wasteful manner. For example, the fruit bodies of bracket fungi causing heart-rots of trees release literally billions of spores over several months or years. Only occasionally will a spore land on the exposed wood of a newly damaged tree, where it can germinate, colonize, and cause extensive tissue destruction.

Aerial Environment

Dispersal through the atmosphere involves three distinct processes: liberation of the propagules, dispersion, and finally deposition (Figure 3.7). In the first phase, the pathogen itself may participate actively but during the two latter processes, propagules are transported passively.

Liberation has two aspects: the release of propagules from the parent colony and their take-off from the host surface. Take-off is of special interest as it involves the boundary layer phenomenon. This is a zone of still air which surrounds all surfaces including leaves, stems, and petals, as well as the crop canopy, inasmuch as this forms a definite layer. The thickness of the boundary layer varies according to a number of factors, particularly wind speed and turbulence and the size and shape of the surface. Some pathogens, notably many fungi, have evolved sophisticated mechanisms which enable their spores to pass through the boundary layer into the turbulent air beyond. Release of spores may be passive, for instance from aerial reproductive structures that extend above the leaf surface, or they are actively discharged from fruit bodies in response to particular environmental conditions,

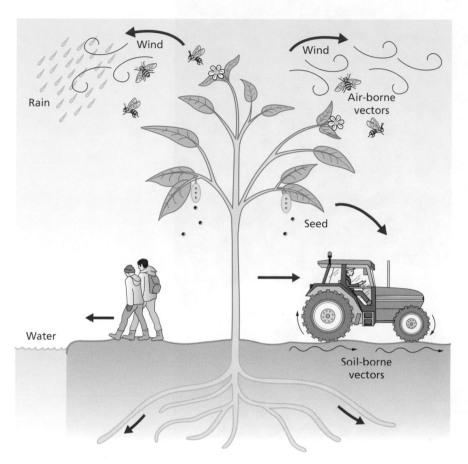

Figure 3.8 Some routes by which pathogens are dispersed.

such as changes in humidity (Figure 3.9). Hence the timing of spore release varies depending on the mechanism involved, and this results in different diurnal patterns of aerial spore populations.

Once released, spores may travel only a short distance or, in the presence of turbulent updrafts, can be carried to high altitudes. Other fungi, and many bacteria, take advantage of external dispersal agencies. Foremost among these are rain drops which splash or puff and tap spores from wet and dry surfaces respectively (Figure 3.10). The potential importance of rain drops as agents of dispersal can be simply illustrated. In regions with a rainfall of 100 cm per year, which is typical of north-west Europe, the north-east United States and many other regions, each square meter of ground will be hit each year by about 1000 million rain drops. Each drop can break up into as many as 5000 splash droplets which may carry spores through the boundary layer. These droplets, plus spores or cells, can bounce directly onto an adjacent leaf or plant or, if the water evaporates, the propagules can become air-borne. With some fungal pathogens, for instance *Z. tritici* on wheat, the risk of disease outbreaks can be directly related to rainfall events.

Wind-driven rain is especially important in the dispersal of plant pathogenic bacteria. Experiments simulating the combined effect of wind and rain on the dispersal dynamics of

Figure 3.9 Active discharge of ascospores of *Sclerotinia sclerotiorum* from sexual fruit bodies (apothecia) in response to changes in humidity. *Source:* Photo courtesy of Jon West/Stephanie Heard.

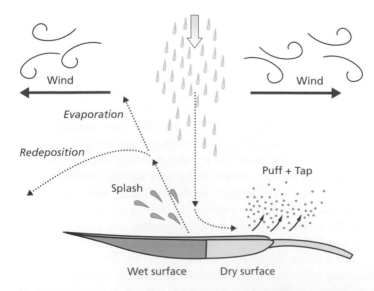

Wind

Evaporation

Wind

Redeposition

Puff + Tap

Splash

Wet surface   Dry surface

Figure 3.10 Dispersal of spores or bacterial cells by rain drops from wet and dry leaf surfaces.

the citrus canker pathogen *Xanthomonas axonopodis* pv. *citri* showed that large numbers of bacterial cells are initially blown downwind from infected trees, and while most splash droplets settle out within a few meters, some cells can travel much further, possibly in aerosols. Hence attempts to eradicate the disease set the distance for removal of potentially exposed trees at more than 500 m. However, extreme weather events such as hurricanes may disperse such pathogens over much longer distances, measured in kilometers, and can quickly spread infection across whole regions.

Liberation is followed by dispersion. This may be rapid and local, as when spores travel in splash drops, or alternatively may involve transport in the upper atmosphere for long distances over several weeks. Many fascinating questions concerning this stage in the propagule's journey remain to be answered. Can spores cross major geographical boundaries such as mountain ranges and oceans? Do sequential disease outbreaks across a region signify the movement of pathogen inoculum? What factors limit the survival of spores in the upper reaches of the Earth's atmosphere? Most microbial cells, for instance, are sensitive to exposure to ultraviolet (UV) light. Modeling the long-range transport of plant pathogens is a complex science intimately linked to meteorology. Nevertheless, there is good evidence that some pathogens can disperse through the atmosphere on a continental scale. Urediospores of rust fungi are thick-walled and relatively resistant to UV, desiccation and low temperatures, and have been trapped at high altitudes. The first occurrence of yellow stripe rust (*Puccinia striiformis*) in New Zealand in 1980 is considered most likely due to the wind transport of urediospores from eastern Australia.

Other unanswered questions concern deposition onto a plant surface. In calm conditions, diffusion brings propagules into contact with the boundary layer. On entering this layer of still air, they fall onto the surface under the influence of gravity. In windy conditions, propagules can be impacted on to leaf hairs and other projections. This process is not, however, very efficient and it is only of any importance for large spores which become impacted on small, narrow surfaces at high wind speeds. The number of spores deposited in this way is increased if the air currents are turbulent.

Rain is also important during the deposition phase as rain drops collect the air spora and deposit it on plant surfaces. Both large and small propagules are washed out from the atmosphere. Less is known about deposition due to electric charges or temperature gradients, but both may be important for some pathogens and/or crops.

Soil Environment

Patterns of spatial dispersal of soil-borne pathogens are generally restricted, at least in the short term. Some spread through crops only slowly, forming circular patches or **foci** within which most plants become diseased. Other pathogens which primarily attack subterranean tissues are not restricted to the soil and emerge briefly from below ground solely for the purpose of dispersal. The take-all fungus *Gaeumannomyces* spreads by mycelial growth along roots but it can also produce an annual crop of air-borne ascospores from reproductive structures embedded in stubble. Ascospores were almost certainly responsible for the outbreaks of take-all on newly reclaimed polders in Holland. Other soil-borne pathogens are passively if irregularly dispersed as a result of soil erosion and flash flooding. Agricultural practices such as plowing can also aid dispersal, and contamination of machinery or the wheels of vehicles that move between fields and farms can move soil-borne pathogens to new areas (Figure 3.8).

Whilst soil may not offer the same opportunities for rapid, long-distance dispersal as the aerial environment, it does in part compensate by providing a better medium for microbial growth. Among microbial pathogens, only the fungi possess the capabilities to really exploit this advantage. Two aspects are involved: the pathogen must have the capacity to obtain nutrients from soil substrates to continue growth, and also be able to withstand antagonism from the soil microflora and fauna. Pathogens vary widely in their potential for saprotrophic growth in soil. Some, such as *Plasmodiophora*, are unable to grow at all away from the living host, whilst others, for example *Pythium* and *Rhizoctonia*, can behave as saprotrophs utilizing plant residues for part of their life cycle. This enables them to spread extensively through soil by hyphal growth.

As mentioned above, fungi infecting forest trees, such as *Armillaria* and *Fomes*, produce rhizomorphs that extend along roots and spread from host to host via root contact, but these cord-like structures are also able to grow through soil for considerable distances. In this instance, nutrition is being supplied by the previously parasitized host rather than by utilizing substrates in the soil.

The capacity of plant-pathogenic bacteria to grow in soil is probably limited, although some can metabolize compounds typically found in plant debris, and others produce bacteriocins or antibiotics, chemicals which specifically inhibit the growth of rival bacteria.

Vectors

A great diversity of animals feed on green plants. During their feeding activities, these herbivores move from plant to plant, sometimes over long distances. Not surprisingly, pathogens exploit some of these organisms as agents of dispersal. Any organism which transmits or disperses a pathogen is termed a **vector**. Some examples are shown in Figure 3.11.

Two main types of pathogen–vector relationship may be identified. In the first, the pathogen is carried externally, on the body or mouthparts, or within the digestive tract of the vector. The pathogen does not cross membrane barriers to enter or exit the vector, and does not multiply within the vector. In effect, the vector is contaminated with the pathogen. However, a distinction may be made among these vectors according to the extent to which their lifestyle is matched with that of the pathogen. In some cases, the pathogen is spread sporadically by a variety of vectors. Alternatively, the interaction may be quite specific and require a close match between the partners for effective transmission to take place. In the second type of pathogen–vector relationship, the pathogen crosses membranes to enter the body of the vector and may multiply within it. The vector is infected by the pathogen and may be capable of transmitting this infection to plant hosts for the remainder of its life. Sometimes, the pathogen may actually be passed on to the progeny of the vector. These remarkable dovetailed patterns of co-evolution have resulted in dispersal systems which are highly efficient.

Vectors are frequently thought of as only being involved in the transmission of viruses. However, whilst it is true that vectors are of supreme importance in the infection cycle of numerous viruses, they are also essential for the dispersal of many bacteria and fungi. For example, bacterial wilt of cucurbits, *Erwinia tracheiphila*, is spread by beetles that feed on

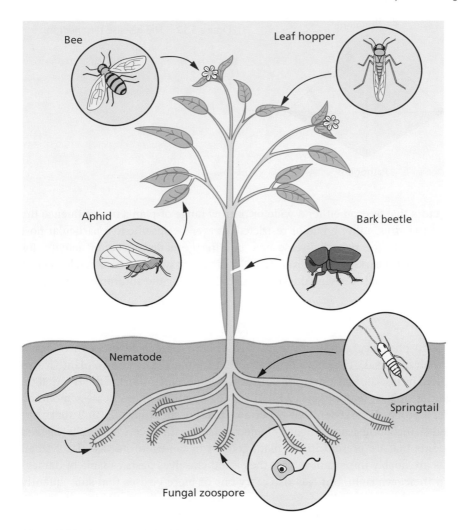

Figure 3.11 Vectors exploited by plant pathogens.

leaves and introduce the pathogen into xylem vessels, while the fungus responsible for Dutch elm disease, *Ophiostoma novo-ulmi*, is carried from tree to tree by bark beetles. The fireblight bacterium *E. amylovora* exploits pollinating insects to disperse inoculum between flowers. Many fungi produce spores which can survive the digestive processes in animal guts. Ingestion and subsequent egestion may thus result in spores being dispersed and deposited within a nutritious fecal pellet. Phytoplasmas are another group of pathogens disseminated in nature by insects, usually leafhoppers, in which they can also multiply; the relationship is thus analogous to that seen with certain viruses.

Where pathogen dispersal is largely dependent on a vector organism, the disease triangle (see p.3) becomes a tetrahedron (Figure 3.12).

To analyze disease in these cases requires an understanding of vector behavior, and the interactions between the vector, the host plant, the pathogen, and the environment. For

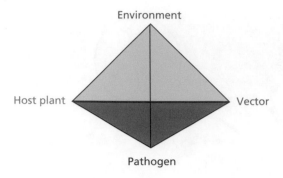

Figure 3.12   The disease tetrahedron.

instance, vectors that feed on either a wide or narrow range of plants may influence the host range of the virus, either broad or restricted, or even be specific to a particular host species. Additionally, the risk of disease may be directly related to vector activity; for instance, outbreaks of barley yellow dwarf (BYDV) disease in winter cereals can be related to the movement of aphids carrying the virus into crops during the autumn (see p.99).

Virus Vectors

Due to their morphological simplicity, many viruses have come to depend entirely upon vectors for their dispersal. It has been estimated that around 70% of known plant viruses are transmitted by insects, and among these the most important group are the Hemiptera, which includes aphids, leafhoppers, whiteflies, and psyllids. These possess modified mouthparts, a proboscis or stylet, that penetrates plant tissues and cells and sucks out their contents, including from phloem elements. Hence, they can both take up and transmit agents that are present inside plant cells or in sap from the vascular system. Such insects are doubly important in that they are themselves pests capable of damaging crops in their own right, but can also carry one or more viruses that subsequently infect the crop.

Three important stages in virus–vector interaction can be defined: **acquisition**, the time taken for the virus to be taken up by the vector; **latent period**, the time between acquisition and the vector becoming infective to other hosts; and **inoculation**, the time required for the virus to be introduced into a new host. The relationship between the vector and the virus can also be defined in terms of the period of time that the vector remains infective (**retention**), varying from transient to the whole life of the insect. In the latter case, the virus may replicate inside the vector, a process termed **propagative**, and even be transmitted through eggs to its progeny. Such cases are of particular biological interest as the host range of the virus includes two different kingdoms, both plant and animal.

The different relationships between viruses and their vectors, with particular reference to patterns of infectivity, are indicated in Table 3.6.

Further details of the biology of virus–insect interactions, based on the example of aphids, are shown in Figure 3.13. Nonpersistent viruses attach to the tip of the insect stylet, interacting with putative receptors in the cuticle. This may involve a domain present in the coat protein of the virus particle. Alternatively, additional virus-encoded proteins, known as helper components, may participate in binding of the virus to the receptor. Examples

Table 3.6 Virus–vector relationships, based on insects in the order Hemiptera

	Persistence of virus in vector		
	Nonpersistent	**Semipersistent**	**Persistent**
Location in vector	On or in mouthparts (stylet-borne)	On or in mouthparts (stylet-borne)	Internally, e.g., in hemolymph
Vector–virus specificity	Mostly low	Intermediate	Usually high
Latent period	Several seconds or minutes	30 minutes to several hours	12 hours to several days
Retention of infectivity	<10 hours	10–100 hours	>100 hours
Multiplication of virus in vector	No	No	Yes, in some cases

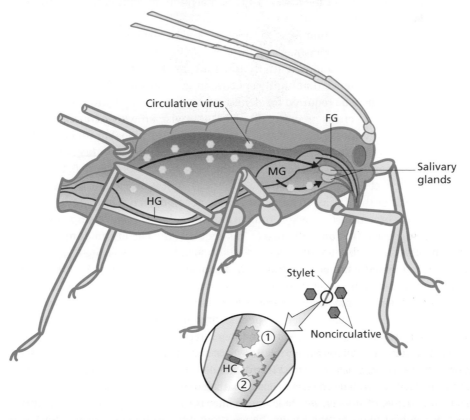

Figure 3.13 Different routes of plant viruses in their aphid vectors. The gut is shown in blue, and the salivary glands in grey. The red arrows show the pathway of circulative viruses (yellow particles) within the aphid, across the gut epithelium into the hemolymph and ultimately to the salivary glands. Non-circulative (non-persistent) viruses (red particles) are shown at their attachment sites at the tip of the stylet. The virus may bind directly to receptors (1) or via additional viral proteins (2) known as "helper components" (HC) that assist binding. FG = foregut, MG = midgut, HG = hindgut. Adapted from Blanc et al. (2014).

include the aphid transmitted potato virus Y (PVY) group, in which the helper protein is known as HCPro, and the P2 protein of CaMV. Persistent viruses pass into the gut of the insect, cross the gut epithelium by an as yet unknown mechanism, and enter the hemolymph. From here, they ultimately reach the salivary glands. When the aphid starts to probe a host plant with its stylet, virus particles are then injected into host cells along with the saliva, initiating infection. The pathway described above explains the longer time required by persistent viruses from acquisition to inoculation, compared with nonpersistent viruses. These contrasting types of virus transmission are described as **stylet-borne** versus **circulative**.

Whiteflies are another insect group of particular importance as vectors of virus diseases of vegetable crops. Several emerging diseases are transmitted by the whitefly *Bemisia tabaci*, that has been extending its range, possibly due to climate change. This insect feeds on a wide variety of host species, contributing to its importance as a virus vector. Transmission of different viruses by *B. tabaci* is either semipersistent or persistent, with the virus retained in the foregut of the insect or the hemolymph. In some cases, the virus can replicate in the whitefly.

Broadly similar vector relationships to those seen with viruses are also found with some other pathogens. Phytoplasmas are phloem limited and therefore only phloem-feeding insects can acquire and potentially transmit them. Leafhoppers are the main vectors, and there is evidence that highly specific interactions between phytoplasmas and insect receptors may be required for penetration of the gut and entry into salivary glands. The devastating citrus greening disease (otherwise known as huanglongbing, HLB), that is currently spreading worldwide and threatening global production of citrus crops, is caused by three related phloem-limited bacteria (*Candidus Liberibacter* species) predominantly transmitted by a psyllid vector *Diaphorina citri*. Originally from Asia, this insect has now invaded the Americas, and is one major factor driving the recent spread of the disease. The pathogen has been detected in the gut, salivary glands and other organs of the vector, indicating a circulative relationship, and some studies have suggested that once the insect has acquired the bacterium, it may remain infective for life.

Another small, obligate bacterial pathogen, *X. fastidiosa*, causal agent of Pierce's disease of grapevines and various other diseases of fruit trees, occurs in xylem tissues and is transmitted by xylem-feeding leafhoppers. In this case, the bacteria are quickly acquired and transmitted over an extended period (persistent), but infectivity is lost after molting, suggesting that the pathogen is limited to the foregut region.

These sophisticated and fascinating relationships and modes of transmission by insects indicate a degree of complexity derived from extended co-evolution of the pathogens and their vectors. The tripartite interactions (host × pathogen × vector) illustrated in Figure 3.12 have a further dimension, as virus infection can alter the nutritional properties and attractiveness of the host plant for the vector (hence promoting feeding and acquisition), and also infection with a plant virus has in rare cases been shown to alter vector behavior. A recent study found that a thrips vector persistently transmitting a tomato virus fed more intensively on host plants than virus-free insects, hence increasing the chances of virus inoculation and consequently survival and spread of the pathogen.

Other Virus Vectors

Some viruses are transmitted by mites rather than insects, for instance wheat streak mosaic virus (WSMV) that is vectored by the wheat curl mite. Transmission is semipersistent and may involve helper proteins similar to those found in aphid transmission. A few soil-borne viruses are spread by parasitic nematodes and fungi that colonize plant roots (Figure 3.11). Grapevine fan leaf virus (GFLV), causal agent of a degenerative disease of grapevines worldwide, is transmitted by the migratory nematode *Xiphinema index* that feeds on growing root tips. Molecular analyses suggest that the viral coat protein is the specific determinant of transmission by the nematode vector. Lettuce big vein (LBVV) and tobacco stunt (TSV) viruses are transmitted by the parasitic root-infecting fungus *Olpidium*. The virus particle appears to adsorb to the outer surface of the fungal zoospore and hence transmission is nonpersistent.

This is in contrast to another group of soil-borne viruses that are vectored by the root pathogen *Polymyxa*, originally believed to be a fungus but now classified as a plasmodiophorid, more closely related to protists. *Polymyxa* also produces zoospores but in this case, the viruses are carried inside the spore, and also persist in the resting spores that are produced inside infected roots. Hence, the viruses can survive for long periods in contaminated soil. Examples include beet necrotic yellow vein virus (BNYVV), the causal agent of the important rhizomania disease of sugar beet, transmitted by *Polymyxa betae*, and several viruses infecting cereals, such as barley yellow mosaic virus (BaYMV) and barley mild mosaic virus (BaMMV), spread by the related species *Polymyxa graminis*. Proteins produced by these viruses possess distinct transmembrane domains that may facilitate passage across the zoospore membrane and hence aid entry into the vector. Zoospores move most effectively in waterlogged soils, and some recent horticultural practices (hydroponics), in which plants are grown in a circulating liquid film of nutrients, provide favorable opportunities for the rapid dispersal of zoospore-borne viruses.

Two interesting routes of virus dispersal involve the parasitic angiosperm dodder and transmission via pollen. Dodder has been used experimentally to transmit viruses between different host species, but it is doubtful if it is of any practical significance under field conditions. Pollen offers a potentially ideal route for virus spread, because it is widely dispersed and the virus need never be exposed to the external environment. Contaminated pollen may transmit virus through the stigma to either the seed parent or the seed, or to both. From an agricultural viewpoint, it is fortunate that such an efficient transmission system is uncommon. Among cases which are known are a number of fruit crop viruses, such as prunus necrotic ringspot virus (PNRSV) and raspberry bushy dwarf virus (RBDV), which are both spread effectively as a result of cross-pollination from affected plants.

In those viruses which are seed-borne, it appears that contamination of the seed from the pollen is as important as contamination from the mother plant, with both processes occurring with equal frequency in most cases of seed-transmitted viruses.

Humans as Vectors

Human activities have had a profound influence on pathogen dispersal. People travel further and faster than any other potential vector, and also deliberately move plants and seed from place to place. Seeds may harbor a wide variety of pathogens which are either present internally or as surface contaminants. For example, more than 30 different pathogens have been recorded on or in soybean seeds, most of them carried on the seed coat. However, 13, including some fungi, bacteria, oomycetes, and seven viruses, can occur within deeper tissues and hence are more difficult to eradicate. Originally, seeds of many crop plants were saved on the farm or distributed only locally, but nowadays there is an international trade in seed between countries and even continents. In this way, pathogens are often imported and become established in regions where the conditions for disease development may be more favorable than in the country of origin.

Another activity which has, often unwittingly, increased the efficiency of pathogen dispersal is vegetative propagation. Potatoes, soft fruits, ornamentals, and many tropical plantation crops are regularly propagated by vegetative methods. Viruses and phytoplasmas have been especially favored by the use of grafts, buds, runners, cuttings, and tubers. Crops such as citrus have suffered particularly through viruses being spread by such propagation methods; for instance, citrus tristeza has spread to many countries due mainly to movement of infected bud wood for grafting. In Africa, the destructive mosaic viruses of cassava have been distributed by propagation via stem cuttings, creating disease foci from which further spread by the whitefly vector occurs.

People also act as vectors by carrying disease from crop to crop on themselves, machines, in foodstuffs or in soil or plant debris. In every respect, the threat due to such movement has increased significantly in recent years. Modern air travel means that most places in the world can be reached within a day of departure, and hence spores or other propagules are more likely to be viable than when sea voyages separated continents by weeks. Outbreaks of Jarrah dieback, a devastating disease of native Australian forests caused by the soil-borne pathogen *Phytophthora cinnamomi*, often originate along roadsides, suggesting that spores have been carried by traffic. Farm machinery has become larger and more costly, and is often shared between farms, so it too presents extensive opportunities for transport of propagules from field to field and farm to farm. The pattern of infection of sugar beet crops by rhizomania disease during the first confirmed outbreaks in the UK suggested that following introduction of the virus, contaminated soil was spread within fields by mechanical cultivation.

International trade can result in all sorts of plant material being moved from place to place (see p.129). The movement of container-grown plants poses a particular problem, as infected plants, or contaminated soil, or both, can be introduced. Sudden oak death, caused by *Phytophthora ramorum*, probably first arrived in Europe as a single introduction around 1990, but has subsequently spread rapidly from country to country in the nursery trade. A similar biosecurity challenge is faced in the USA, where *P. ramorum* has devastated natural forests in California and Oregon, and surveys have detected the pathogen at nurseries distributing potential hosts nationwide.

The import of timber can also pose a threat; the aggressive strain of the Dutch elm disease pathogen, *O. novo-ulmi*, responsible for the epidemic in western Europe from 1970 onwards, was almost certainly introduced in elm logs shipped from North America. Other potential sources include foodstuffs. Propagules in animal feed may pass through the gut unharmed and then reenter the soil in manure. Corn smut, *Ustilago maydis*, is spread in this way in the USA as a result of the use of maize cobs as food for pigs. Resting spores of

Polymyxa, the fungal vector of the virus causing rhizomania, can pass intact through sheep fed on infected crop debris, thus increasing the risk of pathogen dispersal.

Pathogen Survival

Whenever pathogens are not growing within a host, they face problems of survival in a potentially hostile environment. The extent of this problem for any particular pathogen depends on the duration of time between available hosts, and on the relative hostility of the environment. At one extreme, dispersal during epidemics may involve only brief periods when spores or other propagules are away from their hosts. Other pathogens survive between annual crops planted in successive growing seasons or in rotations when suitable hosts are available only every third, fourth or even fifth year. In the intervening periods, environmental extremes reduce the chance of pathogen survival. Drought, waterlogging, and extremes of temperature can all reduce the viability of dormant pathogens. The propagules constituting pathogen inoculum are also subject to antagonism by other microorganisms, which can debilitate or even destroy them.

A summary of the main survival strategies adopted by microorganisms is given in Figure 3.14. In some instances, for example when viruses become established in alternative hosts or vectors, or when fungi resort to saprotrophic growth in plant residues, it may be argued that the term "survival" is inappropriate. However, plant pathologists who are concerned with potential threats to crops use the term loosely to cover all the possibilities shown in Figure 3.14.

The fungi utilize by far the most extensive range of survival mechanisms, and many adopt several strategies which occur in parallel to increase their chances of enduring unfavorable periods. Fungi produce a variety of specialized structures which allow long-term perennation, ranging from single-celled **chlamydospores** to complex aggregations of hyphae termed **sclerotia**. Both are notable for their longevity and resistance to environmental extremes. Chlamydospores are individual cells with thick, pigmented walls and conspicuous lipid food

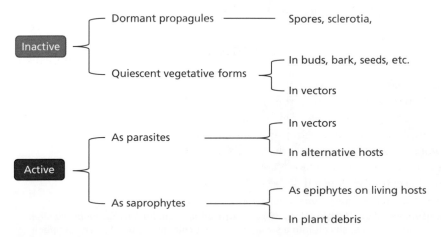

Figure 3.14 Survival strategies adopted by plant pathogens.

reserves. They develop directly from hyphal cells or from asexual conidia stranded in habitats where germination cannot immediately occur (Figure 3.15). Chlamydospores are frequently formed by fungi living in soil, most likely in response to changes in nutrient availability, but perhaps also due to the activities of antagonistic microorganisms. Sclerotia are multicellular structures with an outer rind of thick-walled, pigmented cells enclosing an inner medulla of thin-walled, hyaline cells. Sclerotial development appears to be more complex in that it is

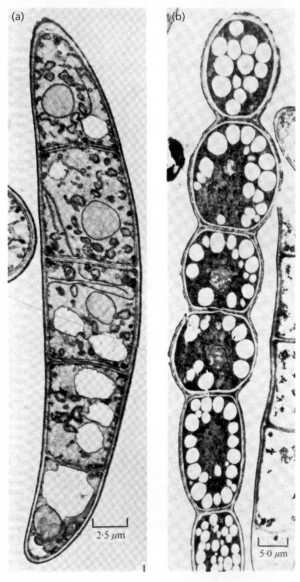

Figure 3.15 Electron micrograph sections of (a) a multicellular conidium of *Fusarium culmorum* and (b) the development of each cell into a swollen, resistant chlamydospore. *Source:* Campbell and Griffiths (1974).

induced in certain species by mechanical stress, such as damage to hyphae or barriers to extension growth, and in others by nutritional factors such as changes in the carbon/nitrogen ratio and the availability of mineral ions and vitamins.

Some pathogenic bacteria infecting humans, insects and nematodes, such as *Bacillus* and related groups, are capable of forming highly resistant endospores. Generally, these are not found in plant-pathogenic bacteria. Instead, these pathogens often persist as vegetative cells on or in the host, including in perennial tissues such as woody organs and buds. Many viruses also survive within the perennating organs of their main or alternative hosts, while others persist within vectors.

Dormancy

The switch from vegetative growth to the formation of chlamydospores or other survival structures may be due to changes in the host or in the environment. In both instances, the end-result is the development of structures exhibiting **dormancy**, which may be defined as a rest period interrupting development. Ideally, this period should be of sufficient duration so that reactivation of the dormant propagules coincides with renewed host activity. Precise systems of control are especially important where the pathogen has a limited ability to search for new hosts. Specific host signals may trigger the growth of dormant propagules. For example, the roots of onions and related *Allium* species exude flavor compounds known as alkyl-cysteine sulfoxides; these are in turn converted by soil microorganisms to a mixture of alkyltriols and sulfides, which specifically induce germination of dormant sclerotia of the onion white rot fungus, *Sclerotium cepivorum*. The presence of these chemicals in soil represents a biochemical "signature," whereby the pathogen is able to detect the presence of a suitable host plant. Key stages in pathogen development may also be linked to the life history of the host plant. Ascocarp maturation in the ergot fungus *C. purpurea* is controlled by environmental factors which also determine the time of flowering of its grass hosts. Hence, as flowers form and open the air spora contains numerous ascospores which mimic pollen and infect through the exposed host stigma.

Dormancy is maintained in some propagules by **constitutive** factors. These are internal controls which must be overcome before propagules can respond to favorable external conditions. Internal checks on germination include permeability barriers and inhibitory substances. Constitutive mechanisms can be overcome by a variety of external influences such as repeated freezing and thawing, exposure to high temperatures, or alternate wetting and drying. Dormancy is also controlled by external or **exogenous** factors. These include temperature, water, and nutrients. Thus, constitutive factors ensure that the propagule remains dormant for a minimum period which is appropriate to the life cycle of the pathogen, whereas exogenous factors are responsible for the synchronization of renewed activity with susceptible stages in host development.

A diagrammatic representation of these processes as they affect fungal sclerotia is given in Figure 3.16. The initial population gradually declines due to the effects of a range of factors affecting survival. At first, germination is prevented by constitutive controls. Once these are overcome, there is a requirement for a favorable combination of exogenous factors. When sclerotia germinate, they give rise to either vegetative mycelia or reproductive structures forming spores.

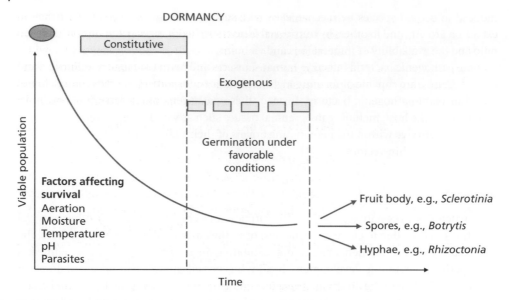

Figure 3.16 Survival and germination of fungal sclerotia under natural conditions.

Interest in germination processes has frequently centered on the possible role played by exogenous nutrients. Some propagules have a requirement for carbohydrates or similar nutrients, whereas others have sufficient endogenous reserves for prepenetration growth.

Fungistasis

While soil is a potentially favorable habitat for microbial growth, it is a highly competitive environment. The majority of propagules deposited in soil, such as fungal spores, do not germinate. This inhibition has been termed **fungistasis**. No fully satisfactory explanation exists as to exactly how fungistasis operates but it is linked with the availability of nutrients, as dormant spores usually germinate if glucose or other nutrients are added to the soil. Similar inhibitory effects probably operate among the microbial populations on the aerial surfaces of green plants.

The pathological significance of fungistasis lies in the fact that developing roots release soluble carbohydrates and other nutrients from their apices and from the sites of emerging lateral roots. These exudates have been shown experimentally to be sufficiently concentrated to overcome fungistasis. Pathogens then commence active growth and have only a short distance to travel to the root surface. Sugars have also been shown to leak out onto aerial plant surfaces. The response to these nutrients is essentially nonspecific as exudates from both host and nonhost plants can stimulate germination.

Survival in Hosts or Vectors

Pathogens also survive in a quiescent vegetative state among perennating host tissues. Survival on or within the host ensures that the pathogen is ideally placed to reinfect newly formed host tissues in the subsequent growing season. The powdery mildew pathogens of perennial hosts such as apple trees or roses overwinter in dormant buds which form in the autumn. Similarly, the leaf curl fungus *Taphrina* survives in crevices in the bark of peach and almond trees. Plant-pathogenic bacteria are particularly well suited by their size and shape to exploit sheltered microhabitats provided by buds, bark, and lenticels. The fireblight pathogen *E. amylovora* overwinters in cankers on twigs and trunks, and resumes pathogenic activity in the spring. It has also been observed that bacteria may be able to withstand environmental stress and chemical treatments by entering a so-called "viable but nonculturable state," a kind of physiological dormancy. Seeds offer opportunities in this context as they may be contaminated superficially with propagules or infected internally with pathogens which have become dormant. Viruses can persist between crops within vectors which may themselves be inactive. Two well-established examples of this strategy involve tobacco rattle virus (TRV), which is able to remain infective in dormant nematodes for at least a year, and BYMV, which can persist for at least 10 years in the resting spores of its vector *Polymyxa graminis*. A number of viruses can remain viable in an inactive form for prolonged periods. Viable TMV has been recovered from dried leaves which had been stored for 25 years, and, not surprisingly, can therefore be spread through crops by workers who smoke cigarettes!

Potential pathogens may be present in actively growing host tissues for an extended period without causing any visible symptoms. Factors which induce these dormant or **latent** infections to become active include a decrease in the concentration of a toxic substance or a change in the chemistry of host tissue. Most examples involve pathogens affecting fruits, where the changes associated with ripening have been shown to stimulate a quiescent pathogen which initially infected the immature tissues. During ripening, there are decreases in the level of tannins and phenols, increases in soluble sugars, and fruits soften due to the induction of endogenous pectolytic enzymes. Such ripe fruits are more susceptible to attack by soft rot bacteria and fungi, an observation confirmed by studies on nonripening mutants of tomatoes, which remain resistant to postharvest pathogens.

Alternative Hosts

Survival can also involve parasitism in alternative hosts or vectors; these bridge the gap between two successive crops of an economically important host. A clear distinction should be drawn between such hosts and the **alternate** host species infected during the life cycle of some rust fungi. The relationships between these fungi and their alternate hosts are highly specific and essential for the completion of the pathogen life cycle. For instance, black stem rust (*P. graminis*) and yellow rust (*P. striiformis*) of cereal alternates to barberry, *Berberis*, while white blister rust of pines *Cronartium ribicola* alternates to currants or gooseberries, *Ribes* species. The origin of such complex life cycles involving unrelated plant hosts is an intriguing biological mystery.

Most well-documented examples of alternative hosts involve either viruses or fungi. Some viruses in particular have extremely wide host ranges, including both cultivated and wild species. Cucumber mosaic virus (CMV), for example, has been recorded in more than 700 hosts from at least 80 plant families. This virus causes severe stunting and yellowing of lettuce, and was found in one area to be present in 12 out of 14 weed species growing in the vicinity of lettuce crops. The virus was not causing any recognizable symptoms in most of these weeds, but they would undoubtedly serve as a source of infection as the main aphid vector feeds indiscriminately on numerous plants.

Wild hosts are often of particular importance when crop plants are introduced into new regions. *Cacao* is indigenous to the tropical forests of South America, yet the main areas of commercial cocoa production are in West Africa. The devastating virus disease cocoa swollen shoot virus (CSSV) most likely originated in the African crop through spread, by mealy bug vectors, from unrelated rainforest trees and shrubs harboring the pathogen.

Survival as Saprotrophs

Some facultative parasites, including both bacteria and fungi, may be able to survive for long periods on the surfaces of their hosts without penetrating underlying tissues. These **epiphytes** are capable of continuing growth even on the limited nutritional resources present on the plant surface. Research has shown that bacteria such as *P. syringae* often survive on leaf surfaces; following heavy rain, these populations multiply (Figure 3.17) and may enter the leaf through stomata to cause disease. Most commonly, facultative parasites compete with populations of free-living decomposers. Success in this competition depends partly on the speed of germination and the subsequent growth rate. An ability to utilize a wide variety of nutrient substrates is also essential if the organism is going to be able to grow extensively in a natural habitat such as soil. During growth, the organism will have to tolerate the presence of other microorganisms in its immediate environment, although it may restrict the growth of competitors by producing toxic metabolites. All these attributes are embraced in the concept of **competitive saprophytic ability**. This has proved useful when comparing the behavior of pathogens which survive as saprotrophs.

Two contrasting fungal pathogens illustrate these points. *Rhizoctonia solani* attacks juvenile and senescent tissues of a wide range of hosts, and once these have been killed and decomposed, the fungus is able to grow freely through the soil. The take-all fungus *G. graminis* only attacks cereals, causing extensive root necrosis, and persists within lesions on host debris. The extent to which *Gaeumannomyces* can survive successfully in this way is dependent upon the prevailing environmental conditions. Warm, moist, well-aerated soils promote the decomposition of roots and straw, with a consequent reduction in *Gaeumannomyces* inoculum. The nitrogen regime in the soil is also very important as cereal tissues have a very high carbon to nitrogen ratio and this alone can limit the rate of their decomposition.

Longevity

Data concerning the longevity of particular pathogens under natural conditions are often unreliable. An outbreak of a soil-borne disease in an area which has not carried a particular crop for a known number of years has often been taken as proof of the ability of the pathogen to survive through the intervening period. Clearly, such information is of practical

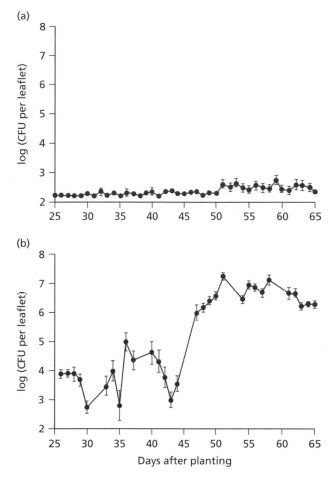

Figure 3.17 Daily changes in populations of *Pseudomonas syringae* on bean leaves during two growing seasons. (a) During season A, the population remained at a low level throughout the sampling period. (b) During season B, large increases followed rainfall on days 35 and 44 and sustained wet weather after day 44. *Source:* After Hirano et al. (1995).

significance, but it is usually impossible to relate it to the survival of particular propagules, such as spores or sclerotia. Data on the survival of such propagules rely on experimentation. Given that some propagules can persist for several decades, it requires little imagination to see the practical difficulties attending studies of this sort (Figure 3.18). In addition, such experiments involve declining populations (Figure 3.16) and increasingly sensitive sampling techniques may be required to detect smaller and smaller numbers of viable spores.

Table 3.7 shows the results of an experiment in which sclerotia of the onion white rot fungus *S. cepivorum* were buried for up to 20 years at different depths in field soil. Even after this extended period, the majority of sclerotia from the deeper layers of soil were still viable, a finding which virtually rules out crop rotation as a practical means of countering this disease.

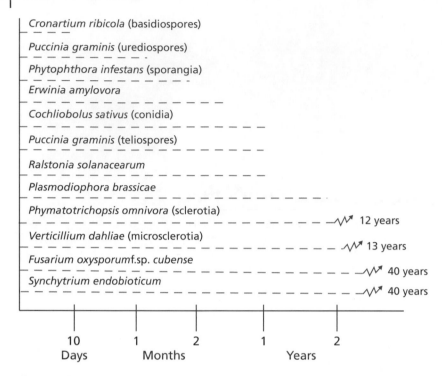

Figure 3.18   Survival periods recorded for some pathogens in the field.

Table 3.7   Survival of sclerotia of *Sclerotium cepivorum* buried at different depths in soil

Depth of burial (cm)	Recovery of viable sclerotia (%)	
	15 years	20 years
0	42	28
7.5	90	88
15.0	96	72
22.5	88	79

Source: Coley-Smith et al. (1990).

Further Reading

Books

Hull, R. (2013). *Plant Virology*, 5e. Washington, DC: Academic Press.
Jackson, R.W. (ed.) (2009). *Plant Pathogenic Bacteria: Genomics and Molecular Biology.*
   Norfolk, UK.: Caister Academic Press.

Watkinson, S., Boddy, L. Money, N. (eds.) (2015). The fungi, 3e. London. Academic Press.

Webster, J. and Weber, R. (2007). *Introduction to Fungi*, 3e. Cambridge, UK: Cambridge University Press.

Weintraub, P.G. and Jones, P. (eds.) (2010). *Phytoplasmas: Genomes, Plant Hosts and Vectors*. Oxford, UK: CABI.

Reviews and Papers

Antunes, L.C.M., Ferreira, R.B.R., Buckner, M.M.C. et al. (2010). Quorum sensing in bacterial virulence. *Microbiology* 156: 2271–2282.

Aylward, J., Steenkamp, E.T., Dreyer, L.L. et al. (2017). A plant pathology perspective of fungal genome sequencing. *IMA Fungus* 8 (1): 1–45. https://doi.org/10.5598/imafungus.2017.08.01.01.

Baltrus, D.A., McCann, H.C., and Guttman, D.S. (2017). Evolution, genomics and epidemiology of *Pseudomonas syringae*: challenges in bacterial molecular plant pathology. *Molecular Plant Pathology* 18 (1): 152–168.

Crous, P.W., Hawksworth, D.L., and Wingfield, M.J. (2015). Identifying and naming plant-pathogenic fungi: past, present, and future. *Annual Review of Phytopathology* 53: 247–267.

Dean, R., van Kan, J.A.L., Pretorius, Z.A. et al. (2012). The top 10 fungal pathogens in molecular plant pathology. *Molecular Plant Pathology* 13 (4): 414–430.

Flores, R., Gago-Zachert, S., Serra, P. et al. (2014). Viroids: survivors from the RNA world? *Annual Review of Microbiology* 68: 395–414.

Kamoun, S., Furzer, O., Jones, J.D.G. et al. (2015). The top 10 Oomycete pathogens in molecular plant pathology. *Molecular Plant Pathology* 16 (4): 413–434.

Lindow, S.E. and Brand, M.T. (2003). Microbiology of the phyllosphere. *Applied and Environmental Microbiology* 69 (4): 1875–1883.

Mansfield, J., Genin, S., Magori, S. et al. (2012). Top 10 plant pathogenic bacteria in molecular plant pathology. *Molecular Plant Pathology* 13 (6): 614–629.

Marcone, C. (2014). Molecular biology and pathogenicity of phytoplasmas. *Annals of Applied Biology* 165 (2): 199–221.

Scholthof, K.-B.G., Adkins, S., Czosnek, H. et al. (2011). The top 10 plant viruses in molecular plant pathology. *Molecular Plant Pathology* 12: 938–954.

Simmonds, P. and Aiewsakun, P. (2018). Virus classification – where do you draw the line? *Archives of Virology* 163 (8): 2037–2046.

Thines, M. (2014). Phylogeny and evolution of plant pathogenic oomycetes – a global overview. *European Journal of Plant Pathology* 138 (3): 431–447.

Vivian, A., Murillo, J., and Jackson, R.W. (2001). The roles of plasmids in phytopathogenic bacteria: mobile arsenals? *Microbiology* 147: 763–780.

Wingfield, M.J., de Beer, Z.W., Slippers, B. et al. (2012). One fungus, one name promotes progressive plant pathology. *Molecular Plant Pathology* 13 (6): 604–613.

4

Disease Assessment and Forecasting

Plant pathogens are shifty enemies.

(E.C. Stakman, 1940)

In crop husbandry, the fate of the individual plant is usually irrelevant. The farmer or forester deals with millions of individuals and only when sufficient numbers succumb to disease is action deemed necessary. In the field, therefore, disease is a phenomenon to be considered in terms of populations.

Analysis of a disease outbreak requires correct identification of the agent responsible and accurate methods for estimating the amount or level of disease present. Ideally, this analysis should also include estimates of the future progress of the disease, so that eventual effects on the yield or quality of a crop can be predicted. This is essential if economic options for disease control are to be properly evaluated. Disease assessment therefore includes a number of interrelated activities (Figure 4.1).

Monitoring the occurrence of a pathogen on a particular crop species over several years can provide clues as to the factors regulating its incidence and severity. This information can then be used to devise predictive systems which forecast the incidence of future outbreaks of the disease. The development of such forecasting systems is especially important for those diseases of economically important crop plants which regularly increase to epidemic proportions.

Disease Detection

An essential first step in disease assessment is correct identification of the causal agent. In many cases, macroscopic symptoms, often combined with microscopic examination of specimens, or isolation of an agent onto culture media may be sufficient to detect the presence of a known pathogen. If there is uncertainty, further, more rigorous tests, such as the application of Koch's rules (see Chapter 3), are necessary to identify the disease agent. For viruses and phytoplasmas, electron microscopy or inoculation of indicator plants may be used.

The problem with these more traditional approaches to disease diagnosis is that they are time consuming and labour intensive, requiring, for instance, a delay of days or even weeks

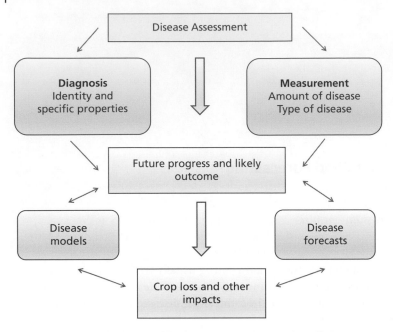

Figure 4.1 Activities involved in disease assessment and prediction.

while cultures grow or test plants develop symptoms. Also, the sensitivity of some tests may be inadequate to detect small amounts of a pathogen. This is especially important when determining the level of infection in propagation material, such as seed stocks, cuttings, or tubers. To detect loose smut *Ustilago nuda* infection in cereal seeds requires detailed microscopic examination of at least 1000 individual embryos, a laborious procedure even with refinements such as the use of sensitive fluorescent stains to visualize the pathogen. Relatively few contaminated seeds may be sufficient to initiate a severe disease outbreak.

There is much interest, therefore, in developing more rapid and sensitive methods for disease diagnosis and pathogen detection. Ideally, such methods should be applicable in the field, at the point of first contact with the problem, and without the need to send samples to a laboratory. Advances in imaging technology are one option; it is now possible to use portable digital microscopes to capture images on tablets or mobile phones and communicate the results to a specialist. But the more widely applicable techniques have originated from fundamental advances in molecular biology and genomics, which are now revolutionizing medical and agricultural diagnostics.

Molecular Diagnostics

An ideal diagnostic test should possess the properties listed in Table 4.1. Sensitivity and accuracy are vital but to find wide application, a test should also be cheap, easy to perform, and give swift results.

There are now a variety of molecular and biochemical methods available for pathogen detection and identification (Table 4.2). The most sensitive and specific methods for detecting pathogens are currently based either on antibodies, which recognize particular antigens, or

Table 4.1 Properties of an ideal diagnostic test

● Accurate	High specificity for target pathogen
● Sensitive	Able to detect small amounts of pathogen
● Simple	Does not require complex laboratory facilities
● Rapid	Gives early detection of pathogen
● Cheap	Affordable and disposable
● Safe	Nonhazardous for operator

Table 4.2 Molecular and biochemical diagnostic methods for pathogen detection and identification

Method	Target	Comments
Serology-based methods		
Immunofluorescence	Pathogen antigen(s)	Requires microscopy
Enzyme-linked immunosorbent assay (ELISA)	Pathogen antigen(s)	Devices now available for use in field
DNA-based methods		
DNA microarrays (based on gene-specific DNA fragments)	Pathogen-specific gene sequences	Can be multiplexed to detect several pathogens
Polymerase chain reaction (PCR)	Pathogen-specific gene sequences	Several formats, widely used, e.g., real-time PCR
Loop-mediated isothermal amplification (LAMP)	Pathogen-specific gene sequences	Does not require thermo-cycler
Next-generation sequencing (NGS)	Multiple DNA and RNA sequences	Requires rapid DNA sequencing platform
Biochemical methods		
Gas chromatography-mass spectrometry (GC-MS)	Volatile organic compounds (VOCs)	High specificity but mainly lab-based
MALDI-TOF mass spectrometry	Pathogen proteins/peptides	Rapid, sensitive, but high initial cost

nucleic acid probes, which target genomic sequences characteristic of the pathogen. An alternative approach is to look for a specific pathogen by-product, for instance a microbial toxin, or some other chemical "signature" that indicates the presence of infection.

Serological tests utilizing the specificity of antigen–antibody reactions are by no means new, having for instance been applied for many years in the diagnosis of plant viruses. Such methods were originally based on polyclonal antisera containing a mixture of antibodies, which often reduced the reliability of the test due to nonspecific cross-reactions. The development of monoclonal antibodies, which have narrow specificity against a single type of antigen, has greatly improved the accuracy of such tests. Monoclonal antibodies can, for instance, discriminate between different bacterial species, whereas polyclonal sera often react with several species within a genus.

Along with improvements in the specificity of serological tests, there have been advances in sensitivity, so that tiny amounts of an antigen can now be detected. One widely used method is the enzyme-linked immunosorbent assay (ELISA), in which samples suspected of containing a pathogen are reacted with antibody in wells on a test plate, and bound antigen is then detected by a further antibody linked to an enzyme; the latter gives a color reaction once substrate is added (Figure 4.2a). As well as being very sensitive, this test is also quantitative, as the intensity of color is proportional to the amount of pathogen antigen in the original sample. It can therefore be used to estimate the level of infection in plants, tissue samples, or seeds. Although mainly used for virus detection, ELISA tests are now available for many bacteria and phytoplasmas, and an increasing number of plant-pathogenic fungi and oomycetes.

Serological methods based on ELISA can also be adapted for use in hand-held diagnostic devices, based on lateral flow technology (Figure 4.2b), similar to the widely used home pregnancy test. A sample of diseased tissue mixed with buffer is introduced into the well and then flows by chromatography through the test strip, where it encounters an antibody specific for a pathogen antigen. A positive result is seen as a colored line at the zone of

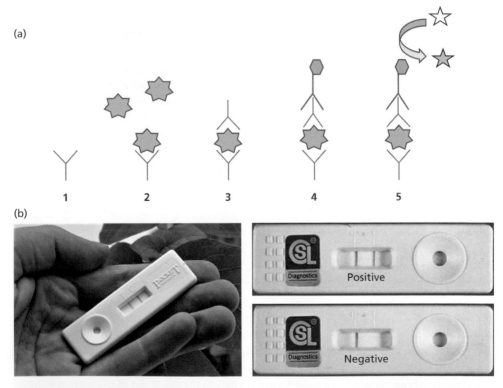

Figure 4.2 Serology-based pathogen detection. (a) A sandwich ELISA assay. 1. Plate is coated with capture antibody. 2 Sample is added and antigen (pink star) binds to capture antibody. 3. Detecting antibody is added and binds to antigen. 4. Enzyme-linked secondary antibody is added and binds to detecting antibody. 5. Substrate is added and is converted by enzyme to colored product (yellow star). (b) A lateral flow device. Sample extracted in buffer is added to the well (right) and flows through the assay cell reacting with pathogen-specific antibody (first colored line = positive) and plant-specific control antibody (second colored line). A negative test gives only one reaction line (with plant-specific antibody). A result is obtained in the field within two minutes. (b courtesy of Pocket Diagnostic® lateral flow immunoassay rapid test).

interaction between antigen and antibody. A range of lateral flow diagnostic tests are now commercially available for plant pathogens, giving swift results on the spot, and are of particular value where rapid detection of a disease is a key consideration.

Serological methods can also be adapted for use in microscopy, by labeling antibodies with dyes which fluoresce under ultraviolet (UV) light; bacterial cells, fungal hyphae, or virus particles can then be specifically visualized in tissue samples.

Diagnosis Based on Nucleic Acids

Nucleic acid probes which hybridize with DNA or RNA sequences characteristic of a particular pathogen are a more recent development arising from recombinant DNA technology. Not surprisingly, the first application of such probes was with viruses and viroids, which exist for part or all of their life cycle as nucleic acid strands. Hybridization with a complementary copy of part of the viral genome can therefore be used to directly detect the pathogen. It is also possible, however, to detect more complex disease agents such as bacteria and fungi using an appropriate DNA probe. One approach is to find DNA sequences that are abundant in the genome, for instance the repetitive DNA which codes for ribosomal RNA. Copies of such regions can then be used to specifically detect very small amounts of pathogen in a sample. The sensitivity of nucleic acid hybridization assays is comparable to that of serological tests such as ELISA, and, like ELISA, the results can be quantified. The limits for detection by nucleic acid-based methods can, however, be greatly enhanced by using the polymerase chain reaction (PCR) that synthesizes multiple copies of a particular nucleotide sequence. Once optimized, this amplification method may be able to detect only a few cells of the target pathogen.

There are now several nucleic acid-based techniques for disease diagnosis available in different test formats. With microarrays, fluorescent-labeled pathogen sequences are hybridized to specific capture probes arranged on a solid surface, such as a microscope slide. One advantage of microarrays is that one can simultaneously detect more than one pathogen, if different colored fluorescent labels are used.

In PCR, complementary oligonucleotide primers target a specific sequence in the pathogen nucleic acid, which is then amplified by the enzyme Taq polymerase in a thermocycling reaction. The amplified product may then be cut with a restriction enzyme to produce a specific molecular fingerprint when run on a gel (Figure 4.3c). Alternatively, if a fluorescent probe is included, the progress of the reaction can be measured to give a quantitative readout in real time (real-time polymerase chain reaction, RT-PCR). Hence, the amount of pathogen nucleic acid present in the original sample can be estimated. By using different probes with contrasting fluorochromes, the assay can be multiplexed and hence detect more than one pathogen. RT-PCR is currently the technique of choice for molecular diagnosis of many diseases, but some specialist kit is needed to do the assay, such as a thermo-cycler. Current developments are therefore focusing on nucleic acid amplification techniques that can be done at one temperature (i.e., isothermal methods). An example of this is LAMP (Loop-mediated isothermal amplification) that has already been successfully used to detect viruses, phytoplasmas, bacteria, and fungal pathogens.

A limitation of all the above techniques is that they rely on prior knowledge of the pathogen responsible for the disease, based on known antigens or nucleotide sequences. They cannot detect a new or unknown disease agent. A negative result might suggest that a novel

Figure 4.3 Coconut lethal yellowing disease (LYD) caused by a phytoplasma. (a) Symptoms of foliage yellowing progress from the older leaves upward through the crown, yellow leaves turn brown, desiccate, and eventually fall. (b) Coconut plantation killed by the disease. (c) PCR-RFLP profiles of phloem samples from coconut palms infected with the LYD phytoplasma in Jamaica (LYJ-1,2), Florida (LYF-1-3), and Honduras (LYH-1-3) showing 16S rRNA sequence heterogeneity among phytoplasma strain populations. *Source:* Harrison et al. (2002).

pathogen is responsible, but further painstaking analysis is usually necessary. An alternative approach, now possible due to the increasing speed and ever-decreasing cost of the technology, is simply to extract all the DNA or RNA from an infected plant and sequence it using a next-generation sequencing (NGS) platform. In this exercise, one ends up with sequences from the host plant, and also from any pathogens that may be present, including novel disease agents. The main challenge is then to analyze the large amount of sequence data produced, using a range of bioinformatic tools. These should filter out any plant

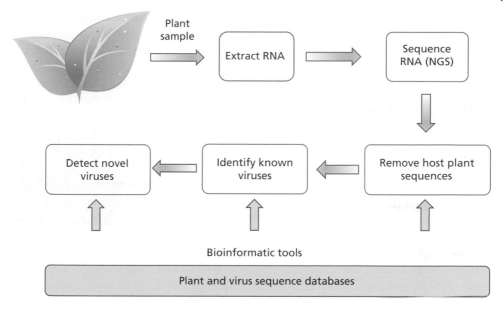

Figure 4.4 Next-generation sequencing (NGS) workflow based on RNA-seq for virus detection in plants. *Source:* Adapted from Jones et al. (2017).

sequences, and leave a mixed population of sequences (or **metagenome**) reflecting the presence of any microorganisms or pathogens.

Using the example of an RNA virus, or viruses, total RNA is prepared from the infected host and reverse transcribed to cDNA that is then sequenced (Figure 4.4). The cDNA can first be hybridized with cDNA prepared from an uninfected host to remove plant sequences and hence enrich for suspected virus. The remaining sequences are then aligned with known virus sequences in databases, revealing the presence of any previously described viruses, or alternatively variant forms or even completely new virus species. For instance, using this protocol with a tomato plant infected with Pepino mosaic virus (PepMV), around 20% of the cDNA was found to correspond to PepMV and the rest to the tomato host. The uninfected host cDNA hybridization enriched the sample to 60% virus sequences, from which the whole PepMV sequence was successfully assembled. Applying this technique to an indicator plant inoculated with an unidentified mottling disease agent, a large number of sequences related to, but distinct from, cucumber mosaic virus (CMV) were found, and the complete genome of a previously unknown *Cucumovirus* was assembled. The power of NGS as a diagnostic tool for plant viruses is now widely acknowledged, but the extent of variation in virus genomes revealed by this approach is posing challenges for current virus taxonomy.

It is also possible to use sophisticated analytical techniques to detect chemical products from pathogens, or infected plants (Table 4.2). Volatile organic compounds (VOCs) are often released by plants as a result of damage or infection, and these can be detected and characterized by gas chromatography combined with mass spectrometry. Such chemicals can serve as biomarkers of disease, and in some cases may be diagnostic for particular pathogens. Protein profiles from pathogens or diseased samples can also be analyzed using

MALDI-TOF mass spectrometry. This approach is already being explored for diagnostic use in clinical medicine, and may also have potential as a relatively rapid, high-throughput approach for detecting plant pathogens. However, the method requires access to suitable databases of reference spectra.

These new technologies are already revolutionizing disease diagnosis, but some obstacles remain. The costs associated with molecular techniques are falling, but they require trained staff, specialized equipment, and controlled conditions to give reproducible results. Hence, the application of new diagnostic tests is mainly laboratory based, for example in plant disease clinics. Ideally, diagnostics should provide rapid results at the point where decisions on disease control are made, rather than awaiting feedback from a clinic. This is especially the case in developing agriculture, where the infrastructure and facilities to support effective programs of disease surveillance are often not present. Nonetheless, provided these logistical and economic limitations can be solved, one can envisage increased use of rapid diagnostic kits in the field. Thus, the grower might directly test a crop by processing samples with a cheap disposable kit which within minutes can detect and identify a specific pathogen. A further advantage of such simple but sensitive tests is the possibility of detecting a pathogen even before symptoms become evident in a crop, thus providing an early warning system. To complete this idealized scenario, one must also envisage good networks of communication so that the response to a disease outbreak can be effectively coordinated.

Disease Assessment in the Crop

Disease diagnosis is focused on the question "What is it?" However, the identification of a pathogen in a particular crop may in itself be of little interest to the grower. The likelihood of the disease increasing, of it spreading to adjacent fields, and the possible losses it might cause are issues of greater concern. This aspect of disease assessment is concerned with the question "How much disease?" As infectious diseases are dynamic in nature, the question inevitably includes an element of prediction, whereby the future progress and possible outcome of a disease outbreak are considered (Figure 4.1). Such issues transcend the interests of individual farmers and demand the collection of data on a regional or national basis.

Measuring Disease

There are several scales by which disease can be estimated. **Disease prevalence** is a broad measure of the amount of disease in a particular crop in a geographic region. In a medical context, it is defined as the number of cases in a population at a given moment in time. Precise information of this kind is not usually available for crop plants, and measurement is based instead on regional disease surveys, often on a seasonal or annual basis. **Disease incidence** is a measure of the proportion of plants infected within a particular crop, often in a single field or plantation, while **disease severity** is a measure of the degree of infection of an individual plant. Figure 4.5a shows a risk map based on annual records of Septoria tritici leaf blotch on wheat crops in England and Wales over a 10-year period. Such maps can inform growers and crop advisors of the likelihood of a particular disease

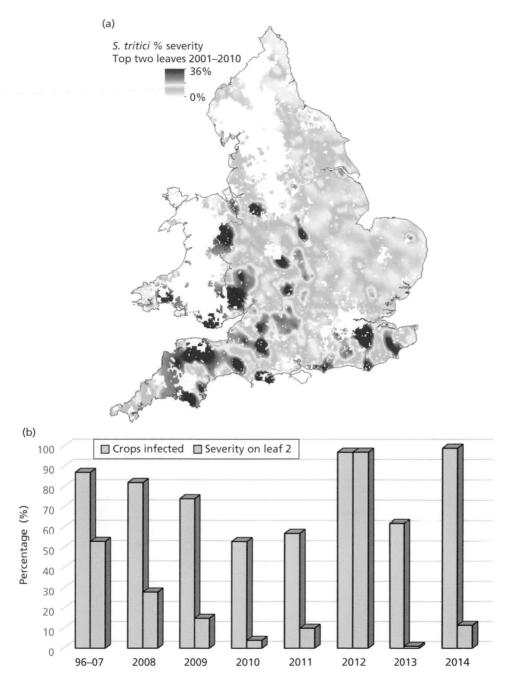

Figure 4.5   Septoria tritici leaf blotch in England and Wales. (a) Risk map based on occurrence of the disease 2001–2010. The disease is more common in the wetter western regions. (b) Prevalence and severity of the disease from 2008 to 2014, compared with the 1996–2007 average. Data show % crops infected and disease severity on leaf 2. The disease was widespread and severe during the exceptionally wet 2012 season. *Source:* Data provided by the Defra Winter Wheat Disease Survey.

occurring so they can plan accordingly. Figure 4.5b shows annual data for the same disease based on the prevalence of infection in surveyed crops and estimates of severity measured as amounts of disease on leaf 2. The extent of infection by this pathogen varies year on year, depending mainly on the weather, and especially frequent rainfall in spring and early summer, as occurred in the unusually wet season of 2012.

Monitoring the distribution of a plant disease is especially important when an invasive pathogen spreads to a new area. Asian soybean rust (*Phakopsora pachyrhizi*) was first recorded in the USA in 2004, and since then scouting of crops has provided growers with regularly updated information on where the disease has occurred (Figure 4.6). The survey also includes the use of a network of sentinel plots deployed across different states that are monitored to aid early detection of rust outbreaks and subsequent progress of the seasonal epidemic. Such plots are planted with susceptible soybean varieties, and may also include other legumes and wild hosts that can harbor the disease. Overall, this coordinated program has not only helped to manage an emerging threat to an important crop, but also increased knowledge of the epidemiology of the pathogen in North America.

The extent of infection on individual plants is usually based on visual assessment in the field, often aided by use of an appropriate assessment key (Figure 4.7). Such keys have been produced for many pathogens and their use for particular diseases has been standardized to ensure that assessments can be compared. Visual keys, while helpful in providing a standard disease scale, may still suffer inaccuracies due to variation between operators, so there has been increasing interest in new automated imaging technology to

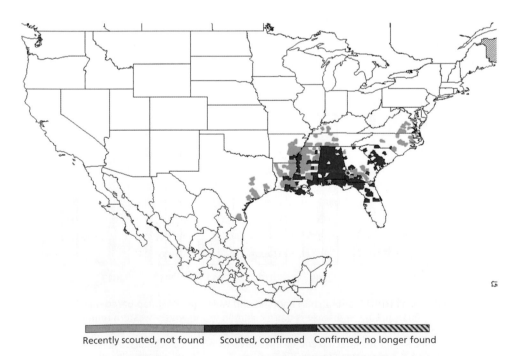

Figure 4.6 Distribution of Asian soybean rust (*Phakopsora pachyrhizi*) in the USA in September 2013. The maps are regularly updated on a website throughout the season. *Source:* Sikora et al. (2014).

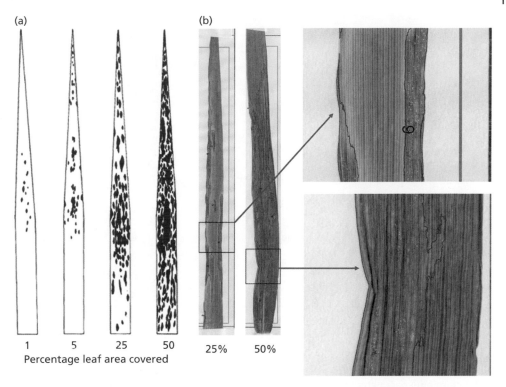

Figure 4.7 Assessment of Septoria tritici leaf blotch of wheat (*Zymoseptoria tritici*) using visual keys and image analysis. (a) Key to aid visual estimates of disease severity (from James, 1971). (b) Digital photos of infected leaves in which an automated image analysis system is used to estimate lesion size and coverage (*purple outlines*). Magnified portions of images highlight pycnidia (originally seen as black dots) to quantify the reproductive capacity of the pathogen on this host. *Source:* Image kindly provided by Bruce McDonald (based on Karisto et al. 2018).

measure disease severity. Figure 4.7b shows examples of digital images of a wheat leaf infected by *Zymoseptoria tritici* in which automated assessment of the percent leaf area covered by lesions is determined, along with the number and size of pycnidia, the asexual reproductive structures of the fungus. This approach can be used to measure not only the severity of infection but also other parameters influencing the rate of epidemic development, such as the reproductive capacity of the pathogen. Applications have now been developed for use on tablets or mobile phones where symptom images can be matched with digital standard area diagrams and the results immediately communicated. Another advantage of this approach is that the digital image database can be continually updated and refined.

Imaging Plant Disease

Plant disease can be detected over a range of scales, from the microscopic (cells and tissues) to plots, fields, and landscapes, using a variety of platforms and sensors (Figure 4.8). These can be applied for several purposes, from measuring disease severity on single leaves or other organs, to high-throughput **disease phenotyping** of populations of plants

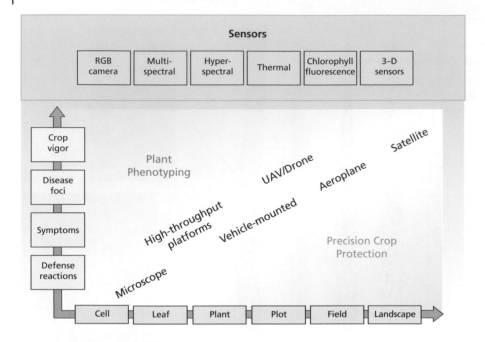

Figure 4.8 Sensor technologies used for automated detection and evaluation of host–pathogen interactions and crop health. These sensors can be used for precision agriculture applications and determining parameters of crop development and health status (plant phenotyping) on different scales from single cells to agroecosystems. *Source:* Adapted from Mahlein (2016).

to assess their relative resistance or susceptibility to a particular pathogen. As described earlier, visible light imaging with conventional cameras can substitute for assessments by eye, and give greater resolution and accuracy when combined with digital image analysis. Different regions of the visible spectrum can be used to obtain additional information on the state of health of plant tissues. For instance, plants undergoing biotic and abiotic stress often show changes in chlorophyll fluorescence emission. Hence, fluorescence imaging can be a useful indicator of disease, particularly at early stages of infection prior to the appearance of visible symptoms. The relatively new technology of hyperspectral imaging, which collects electromagnetic spectra at high resolution within an image, has the potential not only to detect disease but also to differentiate between symptoms caused by different pathogens. Infrared imaging can detect differences in leaf surface temperature caused by certain physiological changes, such as transpiration rate, and can therefore produce a thermal disease image.

While all these imaging techniques show promise for automating various disease parameters, they are not yet applicable to all disease states or situations. For asymptomatic or cryptic infections, it may still be necessary to measure alternative parameters, such as amounts of pathogen biomass, using microbiological or molecular methods. Furthermore, challenges remain in adapting some modern imaging techniques for use with field crops in varying environments.

Remote Sensing of Disease

Various types of sensors (Figure 4.8) also have potential for measuring disease in plant populations grown over large areas in fields or plantations. The idea of using optical sensors mounted on tractors, or remote platforms such as aircraft or satellites, is not new but is now attracting renewed interest with technological advances such as the advent of higher resolution cameras on unmanned aerial vehicles (UAVs) such as radio-controlled planes or drones (Figure 4.10). The ideal scenario is an automated imaging system able to discriminate between different diseases, pests, or weeds, that can be updated in real time and is linked to a global positioning system (GPS) able to direct precision application of an effective chemical exactly where and when it is needed, rather than over the entire field.

Aerial photography from manned aircraft has been in use for many years to monitor crops, and can detect certain types of disease, in particular those that occur in visible patches known as **disease foci** (Figure 4.9). Such surveys are also of value where crops are grown over large areas or inaccessible terrain, such as those used for forestry. Examples include monitoring the incidence of *Heterobasidion* rot (see Chapter 13, Figure 13.8) in pine plantations in the UK, and infestations of dwarf mistletoe in stands of spruce and other commercial conifers in the USA.

The use of aircraft or helicopters is expensive, hence the more recent development of drones that can be used on farm by a trained operator, including the growers themselves (Figure 4.10). A range of crop-specific information on plant development and health can be

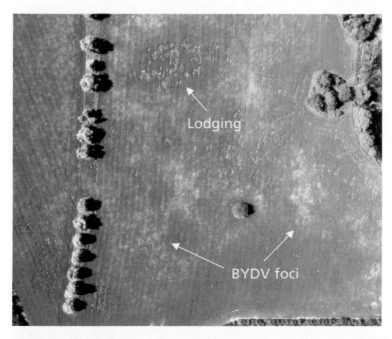

Figure 4.9   Aerial photo of a winter wheat crop in the UK in June, showing disease foci of barley yellow dwarf virus (BYDV) visible as lighter spots, and areas of crop collapse (lodging) caused by the eyespot fungi (*Oculimacula* spp.). *Source:* Greaves et al. (1983), courtesy of B.J. Walpole of the ADAS aerial photography unit.

(a)

(b)

Disease plots

NDVI

High

Low

Figure 4.10 Imaging plant health. (a) Octocopter drone carrying sensors. (b) Normalized Difference Vegetation Index (NDVI) image of cereal variety trial and diseased plots showing changes in reflectance due to crop senescence or leaf infection. *Source:* (a) provided by Rothamsted Research Visual Communications Unit and (b) courtesy of David Whattoff, SOYL Ltd, UK.

captured depending on the types of cameras and other sensors mounted on the drone. For instance, the ratio between near infrared and red reflectivity of plants can be measured to obtain a Normalized Difference Vegetation Index (NDVI) value that can distinguish between healthy, stressed, or dying plants (Figure 4.10). There are some barriers to wider uptake of this technology, including the large amounts of data requiring analysis, and regulatory constraints on the use of drones in national airspace.

At any one moment, a network of satellites equipped with sophisticated imaging equipment is circling the planet, acquiring meteorological data and images of the Earth's surface. How useful is this for surveying crop health? Until recently, the main limitation was the resolution of the images, with a pixel size of around 10 square meters at best. This is adequate for monitoring some indicators of crop stress, such as changes in canopy pigmentation due to drought, but less useful for detecting diseases, unless one is tracking the progress of destructive epidemics leading to large areas of senescing or dead plants. Recent improvements in resolution, estimated at about 1 square meter per pixel, suggest that satellite images could play a significant role in monitoring disease in both crops and natural vegetation. Beyond the technical limitations are issues of cost and availability of crop information in real time, given the other demands on commercial satellite use. Aerial detection systems are also subject to interference by cloud cover and other climatic factors. Nonetheless, there is a potentially important role for such technology in monitoring the incidence of invasive diseases that occur over very wide areas, such as soybean rust.

The objective of combining automated imaging of crop disease with precision application of control treatments has, to date, been most closely met by in-field, tractor-mounted systems. These have the advantage of higher resolution and are not affected by climatic conditions. Cameras or other sensors can be carried on the same vehicle as the spraying equipment controlled by GPS. Images captured by such systems can discriminate patches of weeds and disease foci, for instance early outbreaks of yellow rust, but there are limitations with more evenly distributed diseases. Research has shown that even where disease foci can be accurately mapped, it is vital to take the surrounding zone of infected but asymptomatic plants into account when applying a treatment.

Monitoring Virulence and Pesticide Resistance

While early detection of a pathogen in a crop is vital to ensure effective disease management, additional information on the properties of the pathogen may also be important (Figure 4.1). Two relevant features here are the virulence of the pathogen and the relative sensitivity of the pathogen to chemicals which might be used for control.

Where a pathogen occurs as a series of pathotypes (races) differing in virulence on particular host cultivars, identification of the pathotype involved in a disease outbreak is important to predict the likelihood of spread onto other crops in the area. For major pathogens, the race composition of epidemics is monitored continuously to enable a pattern to be built up over a season, as well as from year to year. This exercise is analogous to the screening programs conducted with human pathogens, such as the influenza virus, to alert doctors to sudden changes in the type of virus present. For instance, in Europe the yellow rust (*Puccinia striiformis*) and potato blight (*Phytophthora infestans*) populations are analyzed each year by collecting isolates from different regions and testing them on sets of differential cultivars. Alternatively, plots of differential cultivars can be grown at a range of sites and are regularly monitored for disease.

Other sampling methods include mobile trap nurseries, comprising trays of seedlings raised under disease-free conditions. Such trays can be exposed on the roof of a traveling vehicle, and are then incubated under conditions which favor disease development. These analyses provide valuable information on the prevalence of different pathogen genotypes, and hence which cultivars are at risk of infection; the same data can therefore be used to inform cultivar choice for the following season. Year on year, the assembled data plot trends in the genetic composition of the pathogen population and, more importantly, can reveal the emergence of novel genotypes with new virulence combinations (see Chapter 12, Figure 12.5). They also help to build up a picture of how a disease varies over time as conditions change and pathogens evolve. Until recently, yellow rust was not considered a significant risk in most areas of Germany, but it has now become a serious threat for growers, possibly linked to the evolution of more aggressive strains of rust able to grow at higher temperatures.

The response of a pathogen to any chemicals used to control disease also needs to be evaluated, especially where problems of resistance to such chemicals (see p.297) may have arisen. When, for instance, a fungus population occurs as a mixture of fungicide-sensitive and -resistant strains, the sensitivity of any strain responsible for a disease outbreak needs to be determined to enable choice of the most effective fungicide for control.

Until recently, virulence testing and evaluation of response to fungicides were both labor intensive, involving inoculation of a range of test plants and/or culture media containing fungicides. With the advances in diagnostic tests described earlier, it is now possible to directly detect virulence genes, or genes responsible for resistance to chemicals, utilizing DNA-based methods. This is described in more detail in Chapters 11 and 12.

Monitoring Pathogen Inoculum

Many plant-pathogenic fungi spread by air-borne spores, so rather than waiting until a disease outbreak occurs, is it possible to detect air-borne inoculum before infection takes place? In fact, several types of trap are available to sample air-borne particles including spores or pollen grains. The best known is the Hirst-type spore trap (Figure 4.11), named after its inventor, Jim Hirst. This continuously samples air by

Figure 4.11   Some examples of equipment used to monitor the risk of disease. (a) Hirst-type spore trap deployed in a flax crop. (b) Biosensor and automatic weather station in a field of oilseed rape. The biosensor (*arrow*) sucks in air and collects spores in media where they either germinate and release specific metabolites that can be detected or, in an updated format, are characterized by a DNA-based LAMP assay. *Source:* Courtesy of Jon West, Rothamsted Research. (c) Aphid suction trap sampling flying insects 12 m above ground. They are collected in liquid and can then be identified to species by microscopy or DNA-based assays.

sucking it through a narrow orifice and over a sticky tape mounted on a revolving drum; any particles present in the air stick on the transparent tape. The speed of revolution of the drum can be adjusted to alter the sampling period, for instance 24 hours or seven days. The daily or weekly spore catch can then be analyzed. Originally, this was done by examining each part of the tape under a microscope to find and identify fungal spores, a laborious process requiring skilled operators able to discriminate specific spore types amidst a background of other particles. Nowadays, DNA-based diagnostics can be used to detect and quantify inoculum of many pathogens. Molecular assays of trap samples can also discriminate specific genotypes of interest, such as different pathotypes or those with resistance to fungicides.

Alternative trapping devices include the rotorod, in which spores impact on rapidly revolving sticky rods, and cyclone traps that capture spores in liquid. For culturable micro-organisms, it is also possible to use techniques in which air is sampled directly onto or into culture media. The cultures can then be grown on and the microbes identified using the usual criteria.

As with the other disease monitoring techniques described in this chapter, automation and speed are important objectives. Research is now under way to design automated disease sensors that warn of the presence of a pathogen before disease occurs. These are further discussed under disease forecasting (see p.104).

Monitoring Vectors

Similar considerations apply to monitoring populations of vectors which may carry disease agents. In Europe, for example, aphid populations are monitored by a network of suction traps (Figure 4.11) which sample the atmosphere above crops and detect whether winged aphids are present (Figure 4.12). Collating data from different regions then allows a map to be compiled showing the distribution and movement of known virus vector species.

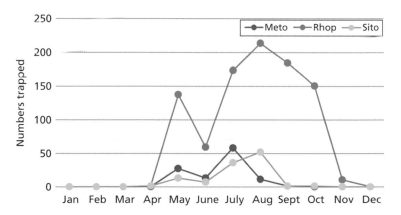

Figure 4.12 Numbers of three aphid species (*Metopolophium dirhodum, Rhopalosiphum padi*, and *Sitobion avenae*) trapped at Rothamsted UK during the 2016 season. The migration of *R. padi*, a principal vector of barley yellow dwarf virus, extends into the autumn, overlapping with the planting of winter cereal crops. *Source:* Data from the Rothamsted Insect Survey, funded by the Biotechnology and Biological Sciences Research Council of the UK.

The presence of a vector can indicate the potential for a disease outbreak, but additional information is usually required to accurately assess risk. Overall estimates of vector populations are useful, but a more valuable measure is the proportion carrying a pathogen, for instance a virus. This can be determined by feeding trapped insects on susceptible indicator plants which develop symptoms of disease if a virus is present. However, this is a relatively slow and laborious process. Once again, molecular methods may be used to enhance the speed and accuracy of such tests. For example, the different viruses causing yellows diseases of sugar beet can be detected in single aphids using an ELISA test based on monoclonal antibodies, while potato leafroll virus (PLRV) has been detected by a PCR assay sensitive enough to detect a single virus-carrying aphid among a population of 30 aphids.

Like plant pathogens, vectors can occur as a series of strains, or biotypes, and these may differ in their host range or ability to transmit disease. The whitefly *Bemisia tabaci* is an important vector of geminiviruses, responsible for diseases such as tomato yellow leaf curl and African cassava mosaic. The emergence of a new biotype of *B. tabaci*, which feeds on a wider range of host plants, has resulted in geminiviruses infecting previously unaffected crops in Europe. This new strain may also be an important factor in the severe epidemic of mosaic disease affecting cassava crops in West Africa. Hence information on the diversity of vector populations may be vital in assessing disease risk.

Disease Severity and Crop Loss Relationships

For disease assessments to be useful, the relationship between the amount of disease present and effects on crop yield and quality needs to be understood. This is one of the most problematic aspects of disease assessment, as in many cases the relationship is far from simple. It is important to understand the growth and physiology of the healthy plant, and how the pathogen affects these processes. The impact of disease will depend on several factors, including the growth stage of the crop, as well as overall disease severity. For instance, a wheat plant is likely to be killed outright by the take-all pathogen *Gaeumannomyces graminis* if infection occurs at the seedling stage, whereas later infections result in reductions in a range of different yield components (Table 4.3). With foliar diseases of cereals, the key factor is effects on green leaf area (GLA) and the duration of green tissues during

Table 4.3 Effect of take-all on yield components of winter wheat

	Infected plants	Noninfected plants	Reduction (%)
Ears per plant	1.6	1.8	11.1
Grains per ear	27.1	33.8	20
1000 grain weight (g)	29.5	46.3	36.3
Fertile spikelets per ear	14.1	15.9	11.3
Grains per fertile spikelet	1.9	2.1	9.9
Grain weight per plant (g)	1.3	2.8	54.9

Source: Green and Ivins (1984).

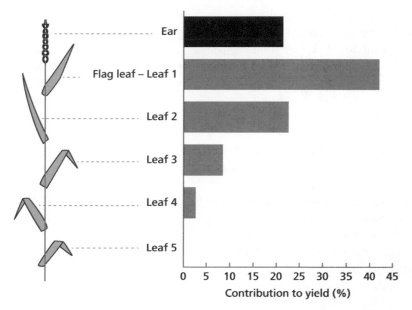

Figure 4.13 Contribution of the upper leaves and ear to final yield of wheat. Around 65% of grain yield originates from the flag leaf and ear. *Source:* AHDB Wheat Disease Management Guide 2016.

grain filling. Figure 4.13 shows that a large percentage of final yield of wheat is contributed by the top three leaves and the ear, so any pathogens that reduce the photosynthetic efficiency of the upper leaves, or attack the ear, will impact on final yield and grain quality. An example is shown in Figure 4.14 where yield loss due to brown rust in wheat can be related to the amount of disease present on the flag leaf, or on the side shoots, known as tillers. The amount of rust in this case was calculated by estimating the area under the disease progress curve (AUDPC); further explanation of this parameter is given in Chapter 5.

One of the best understood diseases in terms of effects on yield is late blight (*P. infestans*) of potato. Initially, a key was created which allowed visual assessment of disease severity in the field. These assessments were then related to actual yield losses by comparing diseased plants with others kept healthy by fungicide treatment. The tuber weight losses resulting from severe blight epidemics clearly depend upon the time in the season when infection occurs (Figure 4.15). This is predictable inasmuch as, unlike the cereal examples described earlier, bulking of tubers is a long-term process and destruction of foliage late in the season will have little appreciable effect on yield.

Crop loss is not only about quantitative impacts but also effects on quality. Cosmetic aspects are important with consumers and disfiguring diseases such as apple scab (see Chapter 2, Figure 2.3) and potato common scab *Streptomyces scabies* can, if severe, significantly reduce the market value of the crop. A further consideration is the effect of some pathogens on the storage properties of the produce. In the case of Sigatoka disease of bananas (*Mycosphaerella fijiensis*), one outcome of infection is to cause premature ripening of the fruit. The consequences of this for an export crop which is transported to distant markets should be obvious. Even more subtle effects assume significance in products in which quality is of paramount importance. Wine made from grapes affected by grapevine

(a)

(b)

Figure 4.14 Effects of brown rust (*Puccinia triticina*) of cereals. (a) Wheat crop infected by rust on the upper leaves. (b) Relationship between area under the disease progress curve (AUDPC) on the flag leaf or tillers of wheat and percentage yield loss. *Source:* Seck et al. (1988) with kind permission from Elsevier Science Ltd.

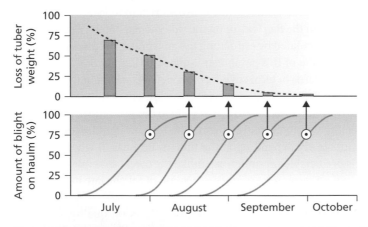

Figure 4.15 Relationship between late blight progress and yield loss of the potato cultivar Majestic. *Source:* Data from Large (1952).

leafroll virus is of inferior quality and color due to the low sugar and high tartaric acid levels present in diseased grapes at harvest. However, infection of grapes by the gray mold fungus *Botrytis cinerea* can actually improve their value; this so-called "noble rot" imparts a distinctive characteristic to Sauterne wines. It should be noted, however, that in most circumstances *B. cinerea* is considered a damaging pathogen of grapevines.

Other fungal pathogens are mainly important because they produce secondary metabolites which pose a health risk to consumers. The most notorious example is ergot of cereals, *Claviceps purpurea*, which replaces infected grain with black sclerotia which contain a cocktail of alkaloids (Figure 4.16a). Harvested grain containing sclerotia can contaminate flour during milling and cause ergotism in people consuming affected bread. There is zero tolerance for ergot contamination of grain destined for human use. Fortunately, this problem is now very rare, although ergotism is sometimes still recorded in livestock grazing on infected grasses. Other examples of **mycotoxins** which can get into the human food chain include carcinogenic aflatoxins produced by *Aspergillus* species growing in maize cobs, and a complex of *Fusarium* species causing head blight of cereals (Figure 4.16b). Under suitable conditions, these produce potent mycotoxins such as the trichothecenes and fumonisins, which have a range of toxic effects on humans and livestock. There are strict legal limits on mycotoxin levels in grain entering the food chain, and samples are routinely tested to assess potential risks to the consumer. If tolerance levels for human or animal consumption are exceeded, the grain may be downgraded or rejected altogether.

(a)   (b)

Figure 4.16   Fungal pathogens that can contaminate cereal grain. (a) Ergot on wheat ear, caused by *Claviceps purpurea*. The pathogen infects individual florets via the stigma and colonizes the ovary, eventually replacing the grain with a dark sclerotium that contains a mixture of alkaloids. Problems arise when the sclerotia contaminate harvested grain. *Source:* Photo courtesy of Jon West. (b) Wheat ear infected by *Fusarium* spp. causing head blight. The pathogen produces several trichothecene toxins that pose a risk to health. *Source:* Photo by John Lucas.

Disease Forecasting

As disease assessment includes a predictive element, it is closely linked with disease forecasting (Figure 4.1). Such forecasts aim to predict whether or not disease will actually occur, as well as estimating the likely extent of its progress through a crop. In its simplest form, forecasting merely relies on knowledge of whether a disease has occurred before in the area concerned. The prevalence of particular pathogens is strongly influenced by regional differences in environmental conditions, and risk maps can often be drawn up to show the likely distribution of particular diseases (Figure 4.5). Many farmers know through experience that crops grown in certain fields will be particularly prone to disease. This knowledge will inform decisions on which crops or crop varieties to grow. Patterns of occurrence of disease in the recent past can provide an indication as to the amount of inoculum in the environment. The presence of air-borne spores in the vicinity of crops can be monitored directly using spore traps to sample the atmosphere during critical periods (Figure 4.11). The presence of other host plants, such as the alternate hosts of rust fungi, or alternative hosts for viruses, may also be a risk factor.

Diseases which are introduced in seed or on other propagation material may be forecast by measuring the extent of any contamination in laboratory or greenhouse tests carried out prior to planting the crop. In many countries, such testing is routinely done to support certification schemes guaranteeing fitness for use. One problem with such schemes is that it may be difficult to relate levels of seed contamination to the subsequent development of epidemics, and hence to determine maximum limits above which seed must be either treated or rejected. In some cases, clear relationships have been established, for example with halo blight of beans (*Pseudomonas syringae* pv. *phaseolicola*). Seed batches were tested for the presence of the bacterium utilizing an immunofluorescence detection method. Subsequent field studies showed that *Phaseolus* seed samples should have less than 0.25% infection if serious epidemics are to be avoided. This level of seed contamination will give rise to about 0.0025% primary seedling infections, and if the subsequent apparent infection rate (r; see p.113) is 0.15 per infected plant per day, this will lead to about 4% infection in the mature crop, which is the maximum which can be tolerated before significant losses are experienced.

Other information which may have a bearing on disease forecasts includes soil type and local environmental factors, for example, microclimates created by undulating terrain or the proximity of woods and hedgerows. The intuitive approach of many farmers to disease avoidance may embrace considerations such as these.

Meteorological Forecasting Systems

It has been known for many years that outbreaks of certain diseases show a marked correlation with particular kinds of weather. An example is shown in Figure 4.17 in which the incidence of black mold disease of cashew trees shows a close relationship with seasonal rainfall. For many pathogens, the dominant factors regulating disease development are related to climate. In some instances, it is possible to establish direct links between climatic conditions and specific processes in the life of the pathogen, for example spore germination, reproduction on the host, or dispersal to new hosts. Furthermore, with some

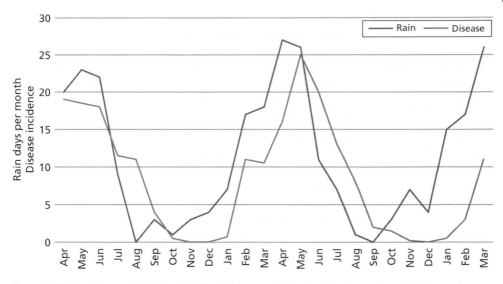

Figure 4.17 Incidence of black mold (*Pilgeriella anacardii*) in dwarf cashew plantations in north-east Brazil with monthly rainfall records 1994–1996. *Source:* Adapted from Cardoso et al. (2000).

Table 4.4 Factors important in the development of meteorological forecasting systems

1. The incidence of disease varies with time due, at least in part, to the influence of climatic factors

2. Factors affecting development of the pathogen and the disease have been experimentally defined

3. The relationship between disease severity and crop loss has been defined

4. Cost-effective control measures are available

5. A system is in place to issue warnings to growers

pathogens, the effects of climate can be narrowed down to one or a few specific factors, such as temperature, relative humidity or rainfall. Where the progress of an epidemic can be attributed directly to such factors, monitoring the weather may provide the basis for an accurate disease forecast. The components aiding development of such schemes are listed in Table 4.4.

One of the most intensively studied diseases in this respect is potato late blight, caused by *P. infestans*. In the UK, potato late blight warnings were initially based on the recognition of a "Beaumont period," defined as a period of 48 hours over which the temperature remains above 10 °C and the relative humidity does not fall below 75%. Subsequently, a simplified system was employed in which blight warnings were broadcast if the minimum temperature remains at or above 10 °C for 48 hours, during which the humidity is at 90% or more for at least 11 hours. These "Smith periods" allowed outbreaks of late blight to be forecast in maincrop potatoes in the UK with an acceptable degree of accuracy using meteorological station data. The combination of temperature and humidity specified allows the pathogen to reproduce, spread, and reinfect adjacent plants and hence to initiate an epidemic.

The significance of a Smith period will vary depending on the growth stage of the potato crop at which it occurs and the extent to which the pathogen is already established. This led to the recognition of zero dates, before which disease outbreaks are either rare or of little significance. However, in some seasons, blight outbreaks occurred before a Smith period was recorded, and the rules were therefore recently modified to include shorter periods of six hours humidity of 90% or more on two consecutive days. These risk factors are now known as Hutton Criteria after the research institute at which large datasets were analyzed to correlate unexpected outbreaks with particular weather conditions and refine the Smith system. Growers can now access such disease warning systems on websites updated throughout the growing season (see Further reading section).

Disease forecasts for late blight based on similar criteria were also developed in other European countries and the USA. The American system, known as BLITECAST, included the participation of individual farmers who used on-farm collection of weather data supplied to a central computer to calculate a weekly severity value summed from daily risk scores based on combinations of rainfall, relative humidity, and temperature. The output was an estimate of blight risk combined with a fungicide spray recommendation. In a German scheme, Phytprog, emphasis was placed on the provision of negative forecasts, so that no action needed to be taken to control the disease for a specified period. This can usefully limit the costs of chemical control measures as well as reducing the possible environmental impact.

Forecasting systems have also been developed for several other diseases caused by oomycete pathogens including tobacco blue mold (*Peronospora tabacina*), downy mildew of cucurbits (*Pseudoperonospora cubensis*), and grapevine downy mildew (*Plasmopara viticola*). Seasonal epidemics of the first two diseases start in southern states of the USA and spread northwards. Early season outbreaks are reported by a network of growers and extension scientists, while meteorological data are used to forecast the risk of further disease progress. This includes analysis of local and national weather and predicts the likely trajectory of spore clouds in the atmosphere. Both the reporting and forecasting of disease outbreaks are web based, with daily updates.

Grapevine downy mildew poses a serious threat to growers in Europe and the USA, although the degree of risk varies due to differences in the disease susceptibility of the main cultivars grown in each region. One US disease warning system is based on initial controlled environment experiments that determined the most favorable conditions for infection and sporulation of the pathogen, and continuous monitoring of ambient temperature, relative humidity, wetness, and rainfall to identify high-risk periods for spore production and infection. The system was evaluated over a seven-year period in vineyards in Ohio where fungicide treatments were applied either using a calendar spray schedule (every 14 days) or according to the risk model. The disease was effectively controlled in both scenarios, but use of the warning system resulted in an average reduction of three sprays per season. An alternative model, based on estimates of the risk of primary infections arising from oospores of the pathogen which survive between seasons, has been field-tested in Italy, again resulting in significant reductions in fungicide use without compromising disease control.

Such forecasting systems are constantly updated and refined to improve their accuracy and usefulness for growers. Initially, forecasts were based on regional weather patterns.

This gave a macroscale estimate of disease risk but did not take account of local variations due to altitude, topography, rivers, and other factors. The advent of portable microcomputers coupled to relatively simple environmental sensors for continuous data logging now permits much higher resolution of risk, including on-farm forecasts of disease. Such automation is now being extended to development of biosensors able to detect pathogen spores in the air in proximity to the crop (Figure 4.11), further refining local estimates of disease risk.

Long-Range Forecasts

All the forecasting systems described above involve models based on actual weather observations. However, the weather favoring a disease outbreak has already happened when forecasts are made, and control measures must be applied quickly following a disease warning. Delay will allow the pathogen to penetrate new hosts and so be less amenable to control by many chemicals. The discovery of fungicides which have a measure of curative activity has to some extent reduced the need for prior warning of pathogen spread, but there is still a relatively brief window in which a spray must be applied. The logistical difficulty of scheduling disease forecast-based treatments is one reason why adoption of these systems by growers has often been limited.

More sophisticated and commercially useful forecasts are likely to be obtained in the future by using synoptic weather charts, which indicate the likely sequence of meteorological events in succeeding days, weeks, or even months. Such information is already used in some forecasting systems, such as that employed for potato late blight in Ireland. Long-term forecasts are potentially more valuable than short-term systems, as control measures may then be integrated with other agricultural practices. Advances in digital technologies will also improve the accessibility and spatial resolution of disease alerts. Increasingly, disease forecasts are now internet based and accessed by mobile phone or tablet, hence accelerating disease reporting and communication of advice for growers. They are often incorporated into a **decision support system** (DSS) which may include additional aspects of crop management. This is further discussed in Chapter 14 on integrated disease control.

Further Reading

Books

Lacomme, C. (ed.) (2015). *Plant Pathology: Techniques and Protocols*, 2e. New York: Humana Press.

Reviews and Papers

Adams, I.P., Glover, R.H., Monger, W.A. et al. (2009). Next-generation sequencing and metagenomic analysis: a universal diagnostic tool in plant virology. *Molecular Plant Pathology* 10 (4): 537–545.

Caffi, T., Rossi, V., and Bugiani, R. (2010). Evaluation of a warning system for controlling primary infections of grapevine downy mildew. *Plant Disease* 94 (6): 709–716.

Fang, Y. and Ramasamy, R.P. (2015). Current and prospective methods for plant disease detection. *Biosensors* 4: 537–561. https://doi.org/10.3390/bios5030537.

Gent, D.H., Mahaffee, W.F., McRoberts, N. et al. (2013). The use and role of predictive systems in disease management. *Annual Review of Phytopathology* 51: 267–289.

Green, C.F. and Ivins, J.D. (1984). Late infestations of take-all *(Gaeumannomyces graminis* var *tritici)* on winter wheat (*Triticum aestivum* cv. Virtue): yield, yield components and photosynthetic potential. *Field Crops Research* 8: 199–206.

Ho, T. and Tzanetakis, I.E. (2014). Development of a virus detection and discovery pipeline using next generation sequencing. *Virology* 471: 54–60.

Mahlein, A.K. (2016). Plant disease detection by imaging sensors – parallels and specific demands for precision agriculture and plant phenotyping. *Plant Disease* 100 (2): 241–251.

Mumford, R.A., Macarthur, R., and Boonham, N. (2016). The role and challenges of new diagnostic technology in plant biosecurity. *Food Security* 8 (1): 103–109.

Mutka, A.M. and Bart, R.S. (2015). Image-based phenotyping of plant disease symptoms. *Frontiers in Plant Science* 5: 734. https://doi.org/10.3389/fpls.2014.00734.

Shakoor, N., Lee, S., and Mockler, T.C. (2017). High throughput phenotyping to accelerate crop breeding and monitoring of diseases in the field. *Current Opinion in Plant Biology* 38: 184–192.

Sikora, E.J., Allen, T.W., Wise, K.A. et al. (2014). A coordinated effort to manage soybean rust in North America: a success story in soybean disease monitoring. *Plant Disease* 98 (7): 864–875.

Small, I.M., Joseph, L., and Fry, W.E. (2015). Development and implementation of the BlightPro decision support system for potato and tomato late blight management. *Computers and Electronics in Agriculture* 115: 57–65.

Stewart, E.L. and McDonald, B.A. (2014). Measuring quantitative virulence in the wheat pathogen *Zymoseptoria tritici* using high-throughput automated image analysis. *Phytopathology* 104 (9): 985–992.

West, J.S. and Kimber, R.B.E. (2015). Innovations in air sampling to detect plant pathogens. *Annals of Applied Biology* 166 (1): 4–17.

Websites for Disease Forecasts

Potato late blight
www.syngenta.co.uk/blightcast

Cucurbit downy mildew
http://cdm.ipmpipe.org

Fusarium head blight
www.wheatscab.psu.edu

5

Plant Disease Epidemics

Ours is a military campaign against agents that destroy our plants. We cannot wage this campaign successfully without knowing the measure of the enemy's ability to destroy.

(K. Starr Chester, 1959)

Disease outbreaks due to infectious agents are dynamic events often characterized by rapid changes in the occurrence and distribution of pathogens within a host population. The science of **epidemiology** seeks to understand the processes underlying such temporal and spatial changes in disease incidence, both to identify causes and to provide a rational basis for disease control.

Early studies on the population dynamics of plant pathogens were mainly descriptive and concerned with identifying the environmental factors affecting disease. This era of epidemiology has now been superseded to a large extent by the quantitative analysis of epidemics using a mathematical approach. Simulation modeling of disease is a powerful tool, not only to predict the future progress of an outbreak, but also to evaluate the likely benefits of any control strategy. Hence, the apparently abstract world of mathematical modeling is, in this instance, closely linked with the practical business of disease management.

Epidemic Development

An **epidemic** can be simply defined as an increase in disease with time, or more completely:

change in disease intensity in a host population over time and space.

It should be added that this definition assumes a progressive increase in disease incidence within a particular population of host plants over a timescale which is relevant to the maturation of the crop. An epidemic occurring on a continental or global scale is described as a **pandemic**.

There are two components in disease increase: multiplication of the pathogen (i.e., increase in number of individuals) and spread of the pathogen to new hosts. Epidemiology

Plant Pathology and Plant Pathogens, Fourth Edition. John A. Lucas.
© 2020 John Wiley & Sons Ltd. Published 2020 by John Wiley & Sons Ltd.
Companion website: www.wiley.com/go/Lucas_PlantPathology4

is therefore concerned with the population dynamics of the pathogen and the factors affecting the spatial spread of disease.

A disease epidemic can only occur if three basic requirements are satisfied.

1) A large number of host plants are available at a suitable stage of development.
2) There is a source of virulent inoculum.
3) Environmental conditions are favorable for the growth and spread of the pathogen.

In natural plant communities destructive disease epidemics are rare (see p.20). Where host and pathogen have co-existed for thousands of years, it is reasonable to suggest that some form of equilibrium will have evolved between the two. The domestication of plants for human use, along with the gradual intensification of agriculture, has altered this balance. Cultivation of selected varieties of a single crop species has loaded the odds in favor of the pathogen by providing large stands of genetically uniform hosts. These monocultures provide an ideal situation for the rapid increase of any pathogen which is virulent on the crop genotype concerned. A spectacular example was the major epidemic of southern corn leaf blight, *Bipolaris maydis*, which affected around 85% of the total USA maize crop during the 1970 season. The progressive development of this disease is charted in Figure 5.1. Fortunately, the situation did not repeat itself in 1971 and an effective level of resistance has subsequently been reintroduced into the crop.

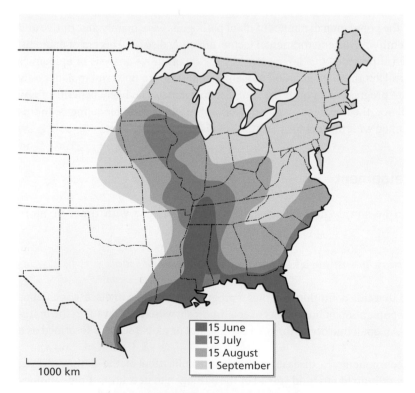

15 June
15 July
15 August
1 September

1000 km

Figure 5.1 Records of occurrence of southern corn leaf blight (*Bipolaris maydis*) in North America during 1970. *Source:* Zadoks and Schein (1979).

Inoculum, the portion of a pathogen that is transmitted to or grows into contact with a new host, may be present either as air-borne or soil-borne propagules, or may already occur in contaminated seed or on other plant propagation material. The infection process, and the subsequent development of the pathogen to the stage at which new inoculum is produced, is dependent upon the prevailing environmental conditions.

The two latter requirements, virulent inoculum and favorable environment, are sometimes included together in the concept of **inoculum potential**. This is essentially a measure of the amount of biological energy available for the colonization of the host. A distinction can be drawn between the intensity factor, that is, the number of infective propagules present, which is sometimes termed **inoculum density**, and the capacity factor or **infection potential**. The latter is a measure of the ability of individual propagules to cause disease under the prevailing conditions. Environmental factors influencing the progress of an epidemic include temperature, relative humidity, dew formation, rainfall, photoperiod, wind speed and direction, sunshine duration, and soil pH. In other words, anything which affects either the development of the pathogen or the performance of the host can influence a disease epidemic.

The Disease Growth Curve

By plotting the amount of disease in a crop against time, one can obtain graphs which provide useful information about the dynamics of an epidemic. Two contrasting types of growth curve can be distinguished (Figure 5.2). The first is characteristic of epidemics where all the infections occurring in a season are derived from inoculum present at the start of the season. This is typically the case with many soil-borne pathogens where an increase in diseased plants as the season advances is due to progressive contact between host roots and pathogen propagules present in the soil, rather than to reproduction of the pathogen. In the specific example shown, the fungus *Sclerotium rolfsii* attacking carrots, initial infection is from sclerotia surviving in the soil, but spread from plant to plant can occur by mycelial growth. The density of host plants therefore affects the shape of the curve. Such epidemics are termed **monocyclic**. In contrast, where the pathogen passes through several generations in one growing season, and each generation produces further inoculum, a **polycyclic** epidemic occurs. Examples of this type of epidemic are air-borne, foliar pathogens, such as the rusts, mildews, and potato late blight, where each spore is capable of initiating a lesion in which further sporulation occurs. On susceptible hosts under favorable environmental conditions the multiplication of the pathogen is very rapid and consequently the rate of disease increase is exponential. In the example shown, anthracnose disease of yams, caused by the foliar pathogen *Colletotrichum gloeosporioides*, differences in the slope of the curve are due to environmental differences between sites, and especially the amount of rainfall which influences the rate of spread and the infection process of this splash-dispersed fungus. This pattern of accelerating epidemic development is not commonly seen with monocyclic pathogens, although in perennial crops or where continuous cultivation of one crop is practiced, disease may increase dramatically over a number of years.

The epidemiologist James Vanderplank compared these two patterns of disease increase to the different types of interest earned on money invested in savings accounts. Monocyclic

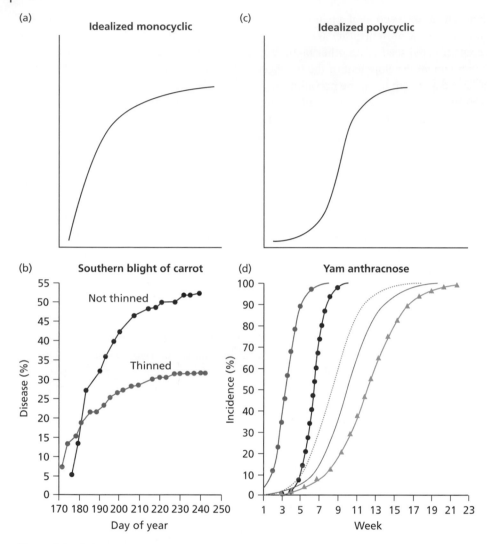

Figure 5.2 Epidemic growth curves for monocyclic (a,b) and polycyclic (c,d) pathogens. Idealized curves (*top*) are compared with examples (*below*) from real epidemics. (b). Disease progress of southern blight of carrots, caused by the soil-borne fungus *Sclerotium rolfsii* in plots where the crop was either thinned or left at high density *Source:* Smith et al. (1988). (d) Disease progress curves for foliar anthracnose of yams, caused by *Colletotrichum gloeosporioides*. Differences in the rate of disease development between five different sites in Barbados are mainly due to climatic factors, especially amounts of rainfall. *Source:* Sweetmore et al. (1994).

pathogens increase in a manner analogous to simple interest, while polycyclic pathogens can be described more accurately by equations for compound interest. The epidemics can also be contrasted in terms of the relative birth rates and death rates of the pathogens involved. The main features of each type of epidemic are summarized in Table 5.1.

These concepts are useful when comparing the dynamics of different epidemics but, as usual, there are exceptions to such generalizations. For instance, soil-borne pathogens like

Table 5.1 Some features of monocyclic and polycyclic epidemics

	Monocyclic	Polycyclic
Rate of increase	Simple interest model	Compound interest model
Reproduction	Long cycle, low birth rate	Short cycle, high birth rate
Survival	Long lived, low death rate	Short-lived, high death rate
Examples	Soil-borne diseases	Air-borne diseases

Pythium and many species of *Phytophthora* may reproduce rapidly on plant roots to release new inoculum in the form of zoospores capable of infecting further hosts.

Mathematical Description of Epidemics

With monocyclic pathogens, the disease growth curve (Figure 5.2) shows a steep early increase in incidence, followed by a gradual decline. With polycyclic pathogens, the growth curve is sigmoid, similar to the classic growth curve describing the increase in biomass of an individual organism or the multiplication of a bacterial culture. One can distinguish three phases: an initial lag, a phase of exponential increase, and a final decline.

What factors determine these shapes? With monocyclic diseases, the principal determinants of disease are the amount of inoculum present at the start of an epidemic and the number of host plants available for infection. This can be described by an equation, derived by assuming that if the number of infected individuals is q, then the rate of increase of disease dq/dt, at time t, is proportional to the number of uninfected individuals in the population. Thus, if the total population size is q_A, then the rate of increase is proportional to $(q_A - q)$, so that:

$$\frac{dq}{dt} = r\left(q_A - q\right)$$

(5.1)

where r is the **infection rate**.

This can be solved to give the following equation:

$$q = q_A - \left(q_A - q_0\right)e^{-rt}$$

(5.2)

where q_o is the initial number of infected individuals, and e is the exponential constant (the base for natural logarithms).

With polycyclic pathogens, the early lag phase is primarily due to the small amount of inoculum present at the start of an epidemic. In the case of potato late blight, the fungus overwinters in seed tubers or discarded tubers left in piles known as cull heaps. Epidemics are initiated by sporangia formed on shoots developing from such tubers. At first, there may be very few infected host plants in the population; a high proportion of these disease foci will be passing through the **latent period** between infection of the host and production of new infective propagules by the pathogen. Once the amount of inoculum has increased, however, the epidemic moves into the exponential phase (Figure 5.2).

The availability of inoculum is no longer limiting and there are still numerous disease-free hosts within the crop. Eventually, the rate of disease increase begins to fall and the epidemic declines. There may be several reasons for this, including the finite number of hosts available for infection and the onset of unfavorable environmental conditions. In the case of potato late blight, the pathogen may literally exhaust the supply of new host plants.

In Figure 5.2, sigmoid epidemic curves were obtained by plotting numbers of diseased plants against time. For this purpose, the quantity of disease may be measured in any convenient way, including number of plants affected, area of leaf or other tissue colonized, or biomass loss as compared with control plants. The model used to describe the disease curve for polycyclic pathogens assumes that the rate of increase of disease is proportional to the number of infected individuals (q), and also to the proportion of uninfected individuals ($1-q/q_A$) so that:

$$\frac{dq}{dt} = rq\left(1 - \frac{q}{q_A}\right)$$ (5.3)

During the initial phase of infection, the proportion of the population infected is small, so that the rate of increase of disease is proportional to q, and hence:

$$\frac{dq}{dt} = rq$$ (5.4)

and growth is exponential, with the number of disease individuals being given by:

$$q = q_0 e^{rt}$$ (5.5)

During this phase of growth:

$$\ln q = \ln q_0 + rt$$ (5.6)

where r is the infection rate, q_0 is the quantity of disease present in the crop at the start of the epidemic, and the relationship between $\ln q$ and time is linear (Figure 5.3).

The vital factor in these formulae, in terms of its significance in different host–pathogen–environment combinations, is the infection rate r. One can obtain a value for r in any field situation by measuring the amounts of disease present (q_1, q_2) at times t_1 and t_2 during the logarithmic phase of disease increase (Figure 5.3). Equation (5.6) may then be rewritten:

$$\ln q_1 = \ln q_0 + rt_1$$ (5.7)

and:

$$\ln q_2 = \ln q_0 + rt_2$$ (5.8)

If Equation (5.8) is subtracted from Eq. (5.7) then:

$$\ln q_2 - \ln q_1 = r\left(t_2 - t_1\right)$$ (5.9)

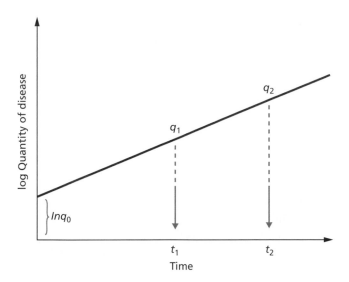

Figure 5.3 Disease increase with time plotted on a logarithmic scale.

or:

$$r = \frac{\left(\ln q_2 - \ln q_1\right)}{t_2 - t_1} \tag{5.10}$$

An important qualification must be made in applying these equations, as they only hold if there is no restriction of the development of the pathogen. In practice, of course, the population of healthy plants is declining as q increases. In this case, the full solution of the growth equation can be used which is the classic logistic curve:

$$q = \frac{q_A}{1 + \left(\dfrac{q_A - q_0}{q_0}\right)e^{-rt}} \tag{5.11}$$

This can be written in a form similar to Equation (5.6) in which $\ln q$ is replaced in the equation by $\ln(q/q_A - q)$ to give:

$$\ln\left(\frac{q}{q_A - q}\right) = \ln\left(\frac{q_o}{q_A - q_0}\right) + rt \tag{5.12}$$

Thus, instead of using the amount of disease, q, to estimate the infection rate, we use $q/(q_A - q)$, the ratio of the quantity of diseased material to the quantity of healthy material (the previous equation assumes that the quantity of healthy material is almost the same as the total amount of material, i.e., the amount of diseased host is very small). This ratio becomes $q/(100 - q)$ if disease is expressed in percentage terms.

We can then estimate r using:

$$r = \left(\ln\left(\frac{q_2}{q_A - q_2} \right) - \ln\left(\frac{q_1}{q_A - q_1} \right) \right) \cdot \left(\frac{1}{t_2 - t_1} \right) \tag{5.13}$$

or:

$$r = \ln\left(\frac{q_2\left(1 - q_1\right)}{q_1\left(1 - q_2\right)} \right) \cdot \left(\frac{1}{t_2 - t_1} \right) \tag{5.14}$$

Although the disease growth curve can provide useful generalizations about the progress of an epidemic, it is inevitably an oversimplification. We have assumed a uniform environment for multiplication and spread of the pathogen. In reality, environmental conditions fluctuate and the progress of an epidemic will be intermittent rather than continuous. Similarly, the spatial distribution of disease will be discontinuous and the rate of increase will differ depending upon the density of uninfected hosts and other factors. This temporal and spatial heterogeneity is reflected in more recent, nonlinear models of epidemic development.

The influence of the environment on an epidemic is complex, with both direct effects on the latent period and indirect effects on the relative resistance of host plants. Temperature has a profound effect on the incubation period between infection and production of further inoculum. For instance, with black stem rust, *Puccinia graminis*, the disease cycle (from spore to spore) takes 15 days at 10 °C, but only 5–6 days at 23 °C. Obviously, in this case the latent period is reduced at higher temperatures and the epidemic will progress more rapidly. Most fungal and bacterial pathogens require free water for infection, and the correlation between rainfall, surface wetness or relative humidity, and disease increase can be striking, and is used in many of the forecasting schemes already discussed in Chapter 4.

The progress of an epidemic may also depend on the date at which it begins. Practical experience shows that if a disease occurs before a certain crop growth stage, it will have a far greater impact than later outbreaks. This may simply be because of the time available for the pathogen to multiply and spread during the remainder of the growing season.

Further Disease Models

The mathematical analyses of epidemics described above laid the foundation for more concerted efforts to model disease enabled by the growing power of computers. There is now a wide variety of models designed for different purposes. These can be broadly classified as empirical or mechanistic (Figure 5.4), depending on their derivation.

Empirical models, such as those discussed in the previous section, are based on statistical methods such as regression analysis to estimate the response of a single variable, for instance disease intensity versus time. **Mechanistic** models aim to develop a more explicit, biologically realistic simulation of disease. One example is the HLIR disease model shown in Figure 5.5, which represents the different states of individuals in the host population as a series of boxes with estimates of the rate at which they change state, to determine the

Type of model

Empirical	Mechanistic
Description and derivation	Description and derivation
• Describes a relationship as accurately as possible	• Describes underlying processes
• Based on mathematical / statistical functions	• Based on biological assumptions
• Describes one variable (e.g. severity) of a system	• Describes multiple variables
• Specific to the data set the model is derived from	• Generalizable to other data sets
Example	Example
• Regression models	• HLIR models

Uses:
- Summarization of epidemic data
- Parameter estimation
- Testing hypotheses (control strategies, optimal monitoring and surveillance designs)
- Forecasting (predicting spread of epidemics, decision support systems, consequences of climate change)

Figure 5.4 Disease models and their uses.

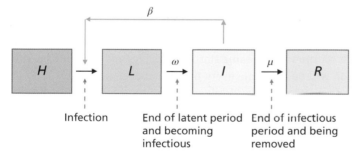

Figure 5.5 Diagrammatic representation of the HLIR disease model. Boxes show the state variables Healthy (H), infected but in the Latent period (i.e., not yet infectious) (L), Infectious (I) and Removed (R). ω is the probability per time unit that a latently infected individual transfers into the infectious category, and μ is the probability that an infectious individual is removed. β is the transmission rate establishing new infections.

overall progress of a disease. Similar models are used in medical epidemiology, but in this case the state variables are described as susceptible (S), infected (I) or recovered (R). Such models are widely applicable, as the contents of the boxes and the links between them can be varied to reflect the specific biological processes leading to disease development. Figure 5.6 presents a typical output from an HLIR model showing dynamic changes in the proportions of the host population in different states over time.

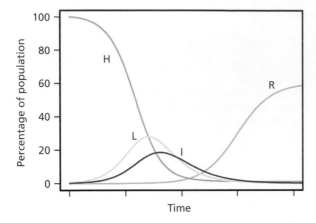

Figure 5.6 Example output from an HLIR model showing proportions of a host population in Healthy, Latent, Infectious, and Removed categories over time.

One important result with this model is that disease will increase, and an epidemic will occur, only if:

$$\beta H_0 / \mu > 1$$

The expression on the left of the equation is known as the basic reproduction number R_0. Determining this is of practical value in allowing comparisons to be made between epidemics and, in particular, how different control interventions might result in $R_0 < 1$ (see later in this chapter).

The HLIR model in Figure 5.5 estimates the expected (or average) disease progression in a large population, in which every individual is equally likely to become infected. Furthermore, if the model is set up with the same starting variables and the same parameters, the disease progression will be the same every time the model is run; this kind of model is therefore described as deterministic. While this is a useful tool to see whether disease is likely to increase or decrease, and to test what methods can be used to slow down a disease outbreak, it will not accurately reflect an actual epidemic. To do this, the model needs to account for both spatial factors (individuals are more likely to become infected if they are close to each other) and chance variation (even if two individuals are close to each other, it won't always lead to infection). Models incorporating such variation are known as stochastic.

Spatial Spread of Disease

As the pathogen population increases in size during an epidemic, it also expands in space. Mathematical models have been developed to describe this process. The simplest models of disease spread assume that the amount of disease at a given distance x from the inoculum source can be described by an empirical function, for example:

$$\ln y = \ln a - b \ln x \qquad (5.15)$$

Figure 5.7 Disease gradient for bean rust, *Uromyces phaseoli*, dispersing from a single infected source plant. *Source:* After Mundt and Leonard (1985).

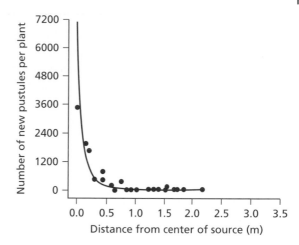

which is known as the Inverse Power Law model, or the alternative Negative Exponential model:

$$\ln y = \ln c - dx \tag{5.16}$$

Figure 5.7 shows a typical disease gradient obtained for a rust fungus dispersing from a single infected source plant in a field. There is a steep decline in disease with distance from the source, reflecting the fact that most of the air-borne spores do not travel far. However, depending on factors such as wind speed or air turbulence, inoculum such as spores or water droplets carrying bacterial cells may be carried, on rare occasions, over much longer distances, thereby establishing new disease foci far from the source.

These empirical models do not take account of the biological mechanisms involved in disease spread. More complex mechanistic models have been proposed, based on diffusion–reaction equations, where, for large populations, inoculum can be considered to behave as a gas. Alternatively, stochastic models which incorporate probabilities of infection, dependent on the distance from an inoculum source, can be applied. The equations used in such models are complex and beyond the scope of this book.

Model Validation

An important step in any model development is validation by comparison with data from actual epidemics, so that the accuracy and predictive power of the model outputs can be evaluated. Often, models are developed based on a specific set of conditions, and may not be valid when applied to a different disease situation.

Downy mildew of grapevines, caused by the oomycete *Plasmopara viticola*, is an important disease in all grape production areas with frequent rains. The pathogen survives between seasons as dormant oospores in leaf litter and soil, that germinate once conditions are favorable to initiate epidemics. A detailed mechanistic model of these primary infections was developed by simulating key stages in the pathogen life cycle, from oospore maturation and germination, through sporangium production, survival, and zoospore release, to splash dispersal of these swimming spores by rain, leaf infection, and subsequent lesion formation. The model provides predictions of disease outbreaks in vineyards, so the timing of fungicide treatments can be optimized. The model was initially evaluated in vineyards throughout Italy, and subsequently under different conditions in Canada. A high degree of accuracy (around 90%) was found, but there was a small proportion of false positives where disease was predicted but did not occur, illustrating the kind of stochastic variation that may affect biological systems.

A second example concerns the model LATEBLIGHT which simulates the effect of weather, host growth, cultivar resistance, and fungicides on the development of the potato blight pathogen *Phytophthora infestans* on foliage. The model was validated using data from late blight epidemics occurring in the Peruvian Andes, where it accurately predicted disease progress. When epidemics in Ecuador, Mexico, Israel, and the USA were compared with model predictions, in the majority of cases a close fit with the data was found, but there were discrepancies for some host cultivars and one location. It was concluded that the model can be applied to a wider range of conditions than those for which it was designed, but some limitations have still to be resolved. Discrepancies between model predictions and real disease outbreaks may themselves be of value in identifying key gaps in our knowledge of host–pathogen–environment interactions.

Disease models therefore have a number of potential uses. Where clear relationships can be defined between disease outbreaks and variables such as climatic factors or the presence of pathogen inoculum, more accurate forecasts of the risk of disease can be made, as already discussed in Chapter 4. Alternatively, one can simulate the progress of disease in a host population so that the effects of weather or control measures such as sanitation, fungicide treatment or planting more resistant crop varieties can be evaluated. Other model outputs include estimates of the likely spatial and temporal spread of an epidemic, so that monitoring and surveillance methods can be designed to increase the likelihood of early detection. It should also be noted that models have proved of educational value in demonstrating the key principles of epidemiology and disease management to farmers and students alike.

Population Structure of Epidemics

The mathematical methods and models discussed so far in this chapter are concerned primarily with the population dynamics of plant pathogens, and developing quantitative estimates of disease progress. There are also, of course, important qualitative questions about epidemics, concerning, for instance, the original source(s) of infection and the specific properties of the pathogen involved in the disease outbreak. Does the multiplication and spread of disease within a field or across a region represent an increase in a single pathotype or are several different pathotypes involved? In other words, what is the composition of the pathogen population during epidemic development? These questions focus on the genetic structure of the pathogen population, the uniformity or diversity of pathogen genotypes present, and the extent of gene flow between them.

Until recently, information on pathogen population structure was based on certain characteristics, or phenotypes, which could be easily determined. Two examples, already mentioned in Chapter 4, are the virulence of different pathogen isolates on a range of host cultivars differing in resistance, and sensitivity to chemicals such as fungicides. Sampling the pathogen population during a disease outbreak, at different sites and different stages of epidemic development, provides insights into the extent of variation present. These surveys help to determine the influence of particular selection pressures, such as host resistance genes or fungicide treatments, on the pathogen population (see Chapters 11 and 12).

The advent of techniques for detecting variation at the genomic level, based on differences in DNA sequences, means that it is now possible to analyze pathogen populations with a much higher degree of resolution. DNA fingerprinting can, for instance, detect different variants within a population, and sometimes even distinguish between different individuals. A whole range of different molecular markers are now available for analyzing genetic diversity in populations, including isoenzymes, amplification of random or defined DNA sequences via the polymerase chain reaction, and DNA sequencing.

DNA markers have proved a valuable tool for estimating the degree of gene flow between different populations, and for determining the extent to which sexual recombination is occurring. With *Zymoseptoria tritici* (*Mycosphaerella graminicola*) on wheat, for example, comparison of pathogen populations between seasons has revealed a high level of genetic diversity, confirming that epidemics originate from sexual ascospore inoculum in which recombination has taken place. This contrasts with epidemics of several other foliar pathogens of cereals, such as many rusts and powdery mildews, where asexual spores are often the primary inoculum and disease is due to multiplication and spread of a single, clonal population of the pathogen.

Global Pandemics

The capacity of microbial pathogens to multiply and disperse, sometimes over long distances, has led to the worldwide spread of some plant diseases, often aided by human activities such as international travel and trade (see p.129). The origins and migration routes of these invasive species were often a source of debate, but the more recent use of genetic and genomic criteria to assess the relatedness of populations occurring in different regions or continents has provided new insights into how these pandemics developed.

Global Epidemiology of Potato Late Blight

The genetic center of origin of the late blight pathogen, *P. infestans*, is believed to be in Central or South America. In Mexico, there is a high degree of genetic diversity in the pathogen population, due to the occurrence of both mating types of the pathogen (A1 and A2), which allows sexual reproduction to occur. The first global migration of the pathogen during the 1840s involved only the A1 mating type. Additional evidence based on isozyme and DNA markers suggests that this pandemic, and until recently most subsequent late blight epidemics, have been caused by one dominant genotype of the pathogen, spreading outside Mexico as a single clonal lineage.

Since the 1970s, however, the A2 mating type has been detected in many different countries, and the composition of the late blight population has undergone a dramatic change.

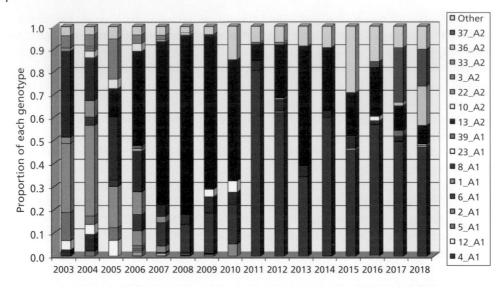

Figure 5.8 Annual incidence of *Phytophthora infestans* genotypes in UK populations showing overall diversity, emergence, and increase of the 13-A2 type (blue) from 2005 onwards, and more recent increase in 6-A1 (pink). A new clone 37_A2 (dark green) resistant to the fungicide fluzinam was first recorded in 2016 and has since become more common. *Source:* Data courtesy of AHDB and David Cooke, James Hutton Institute.

The original dominant genotype has been displaced by new genotypes with different isozyme patterns and DNA fingerprints, and the genetic diversity of the late blight population has increased. This indicates that a second, global migration of the pathogen has taken place.

The emergence of new strains that are more aggressive, or differ in virulence or sensitivity to fungicides, has posed a continuing threat to potato production in many countries. Figure 5.8 shows results from recent surveys done in the UK. The *P. infestans* population is a mixture of genotypes but the proportion of each changes over time, with one strain dominating for a period, and then being displaced by another. The 13-A2 strain (Blue 13) that emerged in 2004 quickly became dominant in many European countries and has since been recorded in the Middle East, China, and India, where it caused a severe epidemic in 2014 with serious consequences for producers. The most likely explanation for the appearance of new variants of the blight pathogen on different continents is shipping of seed tubers infected by the pathogen.

Recent History of Dutch Elm Disease

Two pandemics of Dutch elm disease have occurred in the northern hemisphere over the past century. The first spread widely in Europe and North America but caused only limited mortality in the elm population. A second, more serious outbreak started some time during the 1940s and has been spreading ever since. Detailed analysis of populations of the pathogen from different countries, using biological characters as well as DNA markers, has clarified the nature of these different outbreaks, and provided fascinating insights into the evolution of a plant disease epidemic.

First, it has been demonstrated that the two outbreaks were caused by two related but distinct fungal species. The initial epidemic involved *Ophiostoma ulmi*, and the second a more aggressive species, *O. novo-ulmi*, which has been responsible for the high mortality rate among elms. Furthermore, the current pandemic is due to two distinct strains or races of *O. novo-ulmi*. One, the Eurasian (EAN) race, appears to have arisen in eastern Europe, and subsequently spread east and west. The other, known as the North American (NAN) race, is believed to have evolved in the USA on susceptible elms and was introduced into western Europe in the 1960s on imported timber. Since then, it has spread eastwards so that in places the two races occur together. These highly aggressive strains, now classified as two subspecies, have rapidly replaced the original, nonaggressive species.

The use of molecular markers has, in this case, helped to clarify the possible origins and history of a disease epidemic. It has also provided information on the genetic structure of the pathogen population at different stages of epidemic development. Where the disease is spreading into a new area, the so-called epidemic front, there is low genetic diversity, suggesting that the population is clonal. In areas where the disease has been established for longer, the population is more heterogeneous, linked to the development of variant types of the fungus which differ in their ability to undergo genetic exchange. It is now known that occasional hybridization can occur between the two aggressive subspecies, with formation of highly pathogenic hybrids that can invade established populations.

The high rates of mortality among elms in many countries not only altered landscapes and local ecology, but has also led to changes in the dynamics of the epidemic itself. A proportion of diseased elms are subsequently able to regenerate from surviving roots. These saplings remain free from disease for several years, but succumb again once they reach an age at which they become attractive to the bark beetle vectors that transmit the fungus. Hence, in areas now devoid of mature elms, outbreaks continue to occur on a cyclical basis (Figure 5.9).

Yellow Rust of Wheat

The rust fungi, with their rapid infection cycles and pigmented air-borne spores, which can remain viable while dispersing in the atmosphere over long distances, have always had the potential to be highly invasive pathogens. As international travel and trade have increased, new opportunities for dissemination of these pathogens on a worldwide scale have arisen. In recent years, several pandemics caused by rust species infecting important crops such as wheat, soybeans, and coffee have occurred (see Chapter 1, Table 1.5), so that many regions and countries that were previously free from these diseases are now affected by them. Yellow (stripe) rust, *Puccinia striiformis* f.sp. *tritici* (Pst), of wheat is one such example.

Yellow rust is typically a disease of temperate regions favored by cool, wet conditions. The expansion in range of the disease has occurred as a result of migration of the pathogen, but also the emergence of strains that are more aggressive and able to infect and propagate at higher temperatures. The disease now has a global distribution (Figure 5.10). Genotyping of isolates from Pst populations in different countries using microsatellite markers has revealed their degree of relatedness and defined six groups with a different distribution and lineage. It is now possible to reconstruct the likely origin and migration routes of the pathogen. The center of diversity of Pst is in the Himalayan region of Asia, where the sexual

Figure 5.9 Hedgerow elm infected by Dutch elm disease showing chlorosis (yellowing) of foliage (background) followed by necrosis and dieback of branches (foreground). *Source:* Photo by John Lucas.

stage is present and recombination takes place. The pathogen population in most other areas such as north-west Europe, the Americas, and Australia has a clonal structure. The European population appears to have spread to the Americas in the early 1900s. In 1979, Pst was introduced into Australia from north-west Europe, possibly by spores carried on the clothing of visitors traveling by air. Subsequently, the pathogen spread to New Zealand, believed to be via air-borne spores blown across the Tasman Sea. The rust incursion into South Africa in 1996 closely resembles Pst populations in the Mediterranean region. The Pst groups in Pakistan and China most likely originated and diverged from the ancestral Himalayan population at a much earlier time.

The co-evolution of Pst with its wheat host is a dynamic and ongoing process. Since 2006, some new races have spread in Europe that have virulence to previously resistant wheat varieties, and also a different crop, triticale. Two, termed Warrior and Kranich due to their virulence on wheat varieties of these names, quickly became the predominant races in many European countries. Genotyping these two new variants showed that they most likely arose from a sexually recombining population in the Himalayan region, while the triticale aggressive race was more similar to populations in the Middle East and Central

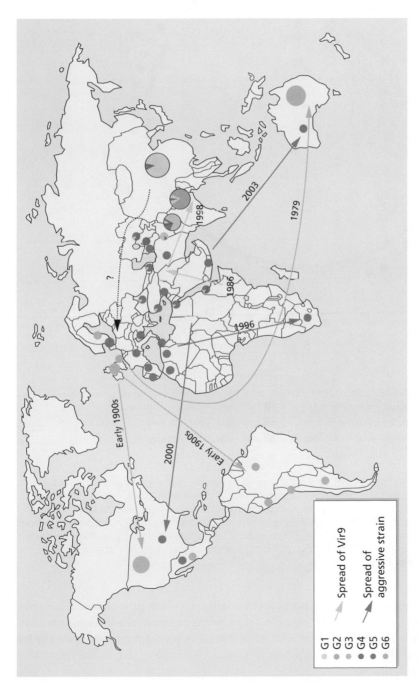

Figure 5.10 Global spread of yellow rust of wheat, *Puccinia striifomis* f.sp. *tritici*, showing likely routes by which the six main lineages were dispersed. *Source*: Ali et al. (2014).

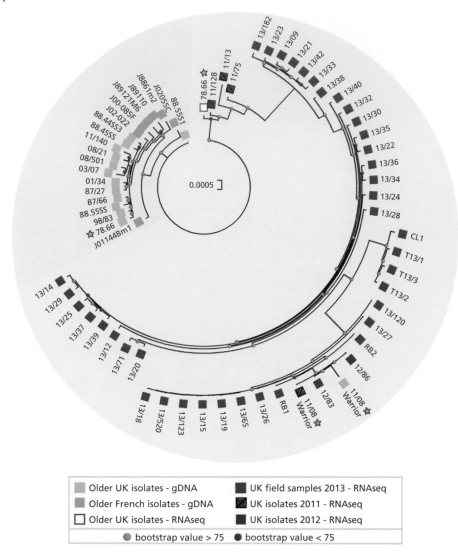

Figure 5.11 Genomic analysis of field isolates of wheat yellow rust, *Puccinia graminis* f.sp. *tritici*, showing genetic relatedness of strains and replacement of original UK population by new and more diverse strains sampled in 2013. Historical strains clustered on light blue background, new lineages on yellow. Strains related to Warrior pathotype located around light blue square at bottom right. Analysis of some older isolates based on genomic DNA, more recent samples RNA sequencing of infected wheat leaves. Radiating lines connecting clusters indicate genetic distance. *Source:* Based on Hubbard et al. (2015).

Asia. Hence, novel pathogen genotypes of exotic origin are replacing the previous European races, and in the process posing a new challenge for plant breeders and crop protection scientists. There is also evidence that new, more aggressive and heat-tolerant strains of Pst have been responsible for severe epidemics in the USA, and a second incursion into Australia (Figure 5.10).

Genomic approaches such as next-generation sequencing (NGS) are now being applied to the analysis of field populations of pathogens, including yellow rust. An example is shown in Figure 5.11. Leaves of wheat and triticale infected by Pst were collected from various field sites in the UK in 2013 and RNA sequencing was done to obtain the transcriptome. Fungal versus host plant sequences could be discriminated by comparison with the reference genomes of Pst and wheat. A set of older field isolates from the UK and France were also sequenced. This analysis showed that the historical isolates were closely related and mainly clustered together, while the 2013 UK population was only distantly related to the older population and included several diverse lineages, confirming the recent incursion of exotic strains. A subset of the 2013 field isolates were similar to a sequenced Warrior type, indicating that the Warrior pathotype was common in the UK at that time.

Relevance of Epidemiology to Control

Rational disease control measures are aimed at altering one or more of the three prerequisites for an epidemic to occur, namely susceptible hosts, virulent inoculum, and a favorable environment. For instance, one can plant resistant crop varieties, inoculum can be reduced by crop sanitation or seed sterilization, and the environment can be altered in such a way that infection no longer occurs. The use of fungicides or other chemicals may be regarded as examples of ways in which such unfavorable environments can be established. These applied aspects of epidemiology will be discussed further in Chapters 11–14.

The distinction between monocyclic and polycyclic diseases also has relevance to methods of control. With monocyclic disease, the most effective approach is often to reduce the initial inoculum as there is a direct relationship between inoculum density concentration and the amount of disease produced (see Chapter 13, Figure 13.3). Fungicidal seed treatments are effective in controlling several cereal smut diseases because they kill most of the seed-borne inoculum. However, when dealing with polycyclic diseases, merely reducing the initial inoculum is usually ineffective. This is because in these diseases, even a tiny amount of inoculum can quickly multiply to damaging proportions. Control measures for polycyclic diseases are usually designed to prevent the pathogen from generating fresh inoculum, thereby reducing the rate of infection. This is usually achieved by using fungicides or by growing less susceptible or resistant hosts in which the disease cycle is slowed down or prevented. Figure 5.12 shows the effects of various fungicide sprays on the progress of a potato blight epidemic. Although none of the treatments actually prevented the epidemic from occurring, three of them (Bordeaux mixture, zineb, maneb) extended the lag phase sufficiently to prevent significant reductions in crop yield. When the phase of exponential increase in disease is delayed in this way, bulking up of the tubers can take place before foliar damage reaches a critical level (see Chapter 4, Figure 4.15).

Hence, the real value of quantitative estimates of disease progress is that the effect of different control strategies on epidemic development can be evaluated. Such estimates are also useful in correlating amounts of disease at a particular stage in the season with differences in cultivar resistance, and with eventual effects on crop yield. Parameters such as the area under the disease progress curve (AUDPC) can, for instance, be used to compare

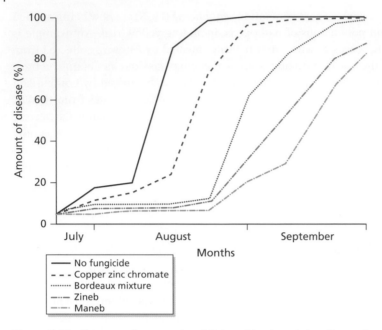

Figure 5.12 Progress of a potato late blight epidemic and the effects of two copper-based (Bordeaux mixture and copper zinc chromate) and two dithiocarbamate fungicides applied weekly from mid-July. *Source:* Data from Hooker (1956).

epidemic development on different genotypes of the same crop. This is a simple descriptor of an epidemic which takes account of the time of disease onset and the shape of the curve, by calculating the area beneath the curve over a specified period. Epidemics which start earlier, and increase more rapidly, will obviously give higher AUDPC values than those which are delayed and develop more slowly. Comparisons with yield data can then be used to define critical thresholds for economic loss, and the need for control actions such as fungicide treatment.

Plant Biosecurity

An essential feature of the various strategies described above is that, while the pathogen is not completely eradicated, the progress of an epidemic is delayed. An alternative approach to disease control is to try to prevent the pathogen from coming into contact with the crop in the first place. The main aim in this case is to avoid introducing pathogens into new geographical areas where they may become **endemic**, that is, permanently established. A regular feature of the domestication and exploitation of plants has been the introduction of crops into new regions. In some instances, such crops have remained free of their most damaging pathogens and ideally this situation should continue. The cultivation of rubber in Malaysia provides a good example. This tree is a native of the Amazon basin and was introduced into Malaysia, via the Royal Botanic Gardens at Kew, in the 1870s. Fortuitously, the destructive leaf blight disease caused by *Microcyclus* (*Dothidella*) *ulei* was not present in the introduced stock and strict measures have ensured that the area has remained free from the disease ever since. In South America, leaf blight has limited the commercial exploitation of the rubber tree.

With the increasing globalization of trade in plants and plant products, the potential risk of spreading pathogens to new countries or continents has increased. This poses a threat not only to crops but also the native flora and fauna. Where there has been no previous history of co-evolution between hosts and pathogens, disease outbreaks may be particularly severe, due to the absence of any natural resistance in the host population. Hence, it is vital to adopt measures to avoid introducing new pests and pathogens that might be present in imported plants, seeds, or fruits, as well as products such as timber or wood packaging. The expansion of worldwide travel has also increased the opportunities for disease agents to move across natural barriers, such as oceans and mountain ranges, and to breach national boundaries. Recent experience in the UK, with a series of invasions by exotic forest pathogens and insect pests, highlights the problem (Figure 5.13). In many cases, these disease agents were introduced on living plants or timber, and are now at large in the environment threatening a range of native trees. Preventing such biosecurity breaches has now become a global economic as well as political priority.

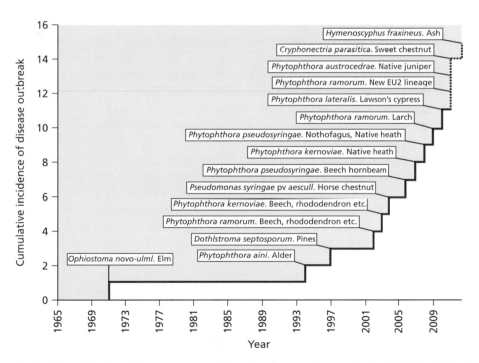

Figure 5.13 Timeline of first occurrence of invasive forest pathogens in the UK 1970–2012. After introduction of the new more aggressive strain of the Dutch elm fungus in 1970, there was a period of more than 20 years in which no new outbreaks were recorded, but since 1993 there has been an exponential increase in the occurrence of new species on a range of native tree hosts. Many of these are known to have been or suspected of being introduced on nursery stock or plant products. The graph shows only fungal, and oomycete pathogens. A similar increase has been recorded with bacteria and insect pests. *Source:* Courtesy of Joan Webber/Clive Brasier, Forest Research.

Quarantine and Other Biosecurity Measures

An effective system for reducing the risk of introducing pests and pathogens requires a range of measures that can be described as offshore, border, and onshore. To begin with (**offshore**), it is important to identify the potential biotic threats and their likely impact should an introduction occur. This can be done by undertaking a pest risk analysis (PRA) using all available information on the biology and epidemiology of the agent and its current distribution. In the UK, a Plant Health Risk Register has been compiled listing potential invaders that now numbers close to 1000 agents. This is too large a number to deal with effectively at one point in time, so it is important to prioritize those that pose the greatest immediate threats to agriculture, horticulture, and forestry. Appropriate surveillance and detection methods can then be put in place. At this stage, intelligence collected in other countries is vital, so ideally a network of plant health organizations should be involved. Information on the movement and proximity of invasive species that are expanding their range can help to preempt potential problems. International cooperation in plant protection is an increasingly important aspect of biosecurity.

Measures regulating the movement of plant materials such as germplasm also begin offshore where the exporting country must provide evidence that the material to be imported is free from pests and pathogens. This may include documentation such as "plant passports" produced by the exporter certifying that the appropriate plant health criteria have been met. The specific requirements depend on the crop. In Australia, for instance, only strawberry plantlets from approved overseas suppliers that have produced them via tissue culture and kept them in insect-proof environments are allowed entry.

The next set of measures concern **border security**. Trained plant health inspectors use a range of methods to detect contamination in imported goods at ports and airports. These include visual inspection as well as diagnostic assays of plant samples using high-throughput microbiological or molecular tests. It may also be possible to detect volatile chemicals associated with decay or pest damage with dogs or electronic sensors. The challenge at this stage is to sample in sufficient quantity to stand a good chance of identifying a biological threat, given the scale and complexity of the current plant trade. There may also be loopholes in the inspection of some nonregulated commodities such as wood packaging, which should be heat-treated and debarked prior to import.

Postentry (**onshore**) procedures often involve holding plants in a quarantine facility for a period of time to establish their pest-free status prior to release. This is particularly important for viruses and phytoplasmas that may be at low concentrations and asymptomatic until later stages of disease development. New surveillance tools such as NGS can improve the speed and accuracy of pathogen detection but also raise issues in terms of the amounts of data generated and false positives due to the presence of nonpathogenic close relatives. However, NGS has the potential to detect previously unknown or novel variant strains of plant pathogens.

In summary, maintaining adequate levels of biosecurity integrates a number of processes and procedures that have been described as the five Ps – predict, prevent, protect, prepare, and partner (Figure 5.14).

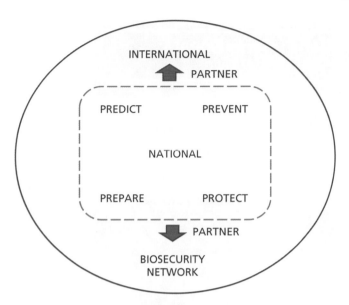

Figure 5.14 The five Ps of biosecurity, including national activities and international partnerships. *Source:* Courtesy of Nicola Spence, Defra.

Legislation for Plant Health

For quarantine to be effective, there must be strict rules governing the movement of plants and their products between countries. There also needs to be a level of awareness of the threat posed by imported pests and diseases among not only those who deal with plant materials but the general public as well. Figure 5.15 shows brochures used in the UK and New Zealand to alert travelers to the rules concerning plant and other biosecurity risk imports.

Legislative measures exist at both a national and international level. Agencies such as the Food and Agriculture Organization (FAO) of the UN aim to set international standards and to develop guidelines for quarantine in member countries by means of agreements such as the International Plant Protection Convention (IPPC). Individual countries also have their own legal framework to restrict the spread of disease. As an example, potato wart disease (Figure 5.16) has been controlled in the UK by a series of legislative measures which date from 1908. Initially, this pathogen was made a scheduled pest, which meant that farmers had to notify outbreaks to the Ministry of Agriculture. Subsequent government orders prohibited the import of wart-affected potatoes and compelled farmers to grow only resistant cultivars on contaminated land. There is also a zero tolerance policy for seed potatoes, and if an outbreak occurs, the affected land is scheduled and cultivation of potatoes is prohibited. At present, wart disease is rarely found in commercial crops in the UK and outbreaks are mainly confined to plants grown in allotments and gardens.

Continued vigilance is required to prevent other potentially serious diseases from becoming established in new areas. When outbreaks do occur, a stringent eradication policy is often adopted. A good example of this is citrus canker, caused by the bacterium

Figure 5.15   Border biosecurity. (*Top*) Sign at Guarulhos International Airport, São Paulo, Brazil. (*Bottom*) UK poster and New Zealand information leaflet alerting travelers to the rules on importing biological materials. *Sources:* Poster courtesy of UK Animal and Plant Health Agency (APHA) and European Plant Protection Organisation (EPPO). Brochure courtesy of Biosecurity New Zealand).

Xanthomonas axonopidis pv. *citri*. Citrus canker can be managed by a combination of chemical sprays, host resistance, and cultural measures, but as the pathogen is spread on fruit, it is important for producers to be free from the disease to avoid restrictions in export markets. In the early years of the twentieth century, a coordinated campaign, with regulatory powers imposed by government and the state, eradicated canker from citrus groves in Florida. The state then remained free of the disease for more than 50 years, largely due to port-of-entry inspections which intercepted infected material. In the 1980s, a new form of the pathogen was discovered on nursery stock in Florida, raising fears that the disease would become widely distributed. A strict quarantine was imposed on affected nurseries, with all citrus stock being burned. Where trees potentially exposed to the pathogen had been sent to new sites, all other trees within 40 m of such stock were also destroyed. By 1987, almost 20 million citrus trees had been removed, and the initial eradication policy had been revised to take account of the

Figure 5.16 Wart disease of potatoes, caused by *Synchytrium endobioticum.* Infection leads to abnormal cell division and enlargement, and wart-like outgrowths on tubers. *Source:* UK Crown Copyright. Courtesy of Fera.

perceived risk of further outbreaks. The disease subsequently reemerged during the 1990s. There is continued monitoring of the situation, and to reduce risk, citrus imports were restricted to states on the Pacific coast such as Washington and Oregon, where the crop is not grown. Similar eradication campaigns for citrus canker have been carried out in Australasia, South Africa, and parts of Brazil.

The costs and disruption caused by such programs have proved controversial, and there has been much debate about their effectiveness. Success in eradicating an introduced pathogen generally depends not only on an effective strategy but also a high level of awareness and compliance among both the industry concerned and the general public. With the ever-expanding reach of the internet and social media, it is now possible to directly involve large numbers of people in citizen science activities, including projects recording and mapping the incidence of pests and diseases. For instance, information on the recent arrival and spread of the ash dieback disease (*Hymenoscyphus fraxineus*) in the UK was in many cases gathered by members of the public using website or mobile phone applications (see www. observatree.org.uk). Such projects can greatly increase the scope of surveillance and provide an early warning system for new disease outbreaks.

Trade Versus Biosecurity?

There is an obvious tension between international trade promoting the free movement of goods and people, and the need to maintain high levels of biosecurity. At a time when many pathogens and disease vectors are already extending their ranges due to climate change, it will be difficult to resolve these conflicting goals. The World Trade Organization (WTO) Agreement on the Application of Sanitary and Phytosanitary Measures requires that member states conform to the quarantine rules and standards developed by the IPCC, but also states that such rules should not be used as an unreasonable barrier to trade. Instead, a balance should be struck between risk, based on scientific criteria, and the economic benefits of the trade involved.

It should also be recognized that in less developed countries, there is often an informal trade in seeds and propagation materials between growers. In Africa, for instance, virus diseases such as brown streak of cassava can be spread locally by insect vectors, but over longer distances by unregulated exchange of planting material. The solution to such problems requires both a system for production and distribution of disease-free material, but also greater awareness among growers of the factors underlying disease spread. As is often the case in practical crop protection, a combination of robust science and appropriate mechanisms for communicating the key messages is required.

Further Reading

Books

Madden, M.P., Hughes, G., and van den Bosch, F. (2007). *The Study of Plant Disease Epidemics*. St Paul, Minnesota: APS Press.

Zadoks, J.C. and Shein, R.D. (1979). *Epidemiology and Plant Disease Management*. Oxford: Oxford University Press.

Reviews and Papers

Ali, S., Gladieux, P., Leconte, M. et al. (2014). Origin, migration routes and worldwide population genetic structure of the wheat yellow rust *Puccinia striiformis* f. sp *tritici*. *PLoS Pathogens* 10 (1): e1003903. https://doi.org/10.1371/journal.ppat.1003903.

Alonso Chavez, V., Parnell, S., and van den Bosch, F. (2016). Monitoring invasive pathogens in plant nurseries for early-detection and to minimise the probability of escape. *Journal of Theoretical Biology* 407: 290–302.

Aylor, D.E. (2003). Spread of plant disease on a continental scale: role of aerial dispersal of pathogens. *Ecology* 84 (8): 1989–1997.

Brasier, C.M. (2008). The biosecurity threat to the UK and global environment from international trade in plants. *Plant Pathology* 57 (5): 792–808.

Brown, J.K.M. and Hovmoller, M.S. (2002). Epidemiology – aerial dispersal of pathogens on the global and continental scales and its impact on plant disease. *Science* 297 (5581): 537–541.

Cooke, D.E.L., Cano, L.M., Raffaele, S. et al. (2012). Genome analyses of an aggressive and invasive lineage of the Irish potato famine pathogen. *PLoS Pathogens* 8 (10): 1–14. https://doi.org/10.1371/journal.ppat.1002940.

Fry, W.E., McGrath, M.T., Seaman, A. et al. (2013). The 2009 late blight pandemic in the eastern United States – causes and results. *Plant Disease* 97 (3): 296–306.

Gilligan, C.A. and van den Bosch, F. (2008). Epidemiological models for invasion and persistence of pathogens. *Annual Review of Phytopathology* 46: 385–418.

Graham, J.H., Gottwald, T.R., Cubero, J. et al. (2004). *Xanthomonas axonopodis* pv. *citri*: factors affecting successful eradication of citrus canker. *Molecular Plant Pathology* 5: 1–15. https://doi.org/10.1046/j.1364-3703.2004.00197.x.

Hubbard, A., Lewis, C.M., Yoshida, K. et al. (2015). Field pathogenomics reveals the emergence of a diverse wheat yellow rust population. *Genome Biology* 16: 23. https://doi.org/10.1186/s13059-015-0590-8.

Hulme, P.E. (2009). Trade, transport and trouble: managing invasive species pathways in an era of globalization. *Journal of Applied Ecology* 46 (1): 10–18.

Madden, L.V. (2006). Botanical epidemiology: some key advances and its continuing role in disease management. *European Journal of Plant Pathology* 115 (1): 3–23.

MacLeod, A., Pautasso, M., Jeger, M.J. et al. (2010). Evolution of the international regulation of plant pests and challenges for future plant health. *Food Security* 2 (1): 49–70.

Martelli, G.P., Boscia, D., Porcelli, F. et al. (2016). The olive quick decline syndrome in South-East Italy: a threatening phytosanitary emergency. *European Journal of Plant Pathology* 144 (2): 235–243.

Parnell, S., Gottwald, T.R., Riley, T. et al. (2014). A generic risk-based surveying method for invading plant pathogens. *Ecological Applications* 24 (4): 779–790.

Schmale, D.G. and Ross, S.D. (2015). Highways in the sky: scales of atmospheric transport of plant pathogens. *Annual Review of Phytopathology* 53: 591–611.

Part II

Host–Pathogen Interactions

Far from being an insurmountable obstacle to the analysis of an organic system, a pathological disorder is often the key to understanding it.

(Konrad Lorenz, 1903–1989)

Green plants have co-evolved with microorganisms for many millions of years, during which time a diverse range of interactions have developed, from stable, mutually beneficial partnerships to harmful relationships leading to disease. According to the fossil record, the first land plants already possessed mutualistic associations with fungi in the form of mycorrhizas, and exploitation by parasitic microbes is also likely to have occurred. To survive constant exposure to potential pathogens and pests, plants developed efficient mechanisms to recognize and respond to such threats. Hence, they possess an innate immune system based on protein receptors that can detect "signature" molecules produced by pathogens and trigger active defenses to prevent invasion. In turn, pathogens have adapted and evolved strategies to counter or evade such defenses. A current focus in the study of host–pathogen interactions is to understand the molecular "cross-talk" taking place between microbial pathogens and their plant hosts, so that more effective and durable forms of resistance can be developed and deployed in crops.

Stages in Host–Pathogen Interaction

The invasion of a plant by a pathogenic microorganism is, in reality, a continuous process, but certain key steps can be identified (Table 1). First, the pathogen must locate the host; in many cases, this is a random process depending on chance contact between propagules and a susceptible plant. Many pathogenic fungi, for instance, produce air-borne spores which are carried by wind currents (see p.61). Similarly, bacterial cells may be splashed or blown between plants by wind-driven rain. In these cases, successful contact with a new host relies upon the huge numbers of propagules produced, which increases the probability that at least some will land on a suitable plant. Other pathogens, however, possess more sophisticated adaptations for locating the host, usually by responding to a particular signal. Such recognition cues may

Plant Pathology and Plant Pathogens, Fourth Edition. John A. Lucas.
© 2020 John Wiley & Sons Ltd. Published 2020 by John Wiley & Sons Ltd.
Companion website: www.wiley.com/go/Lucas_PlantPathology4

Table 1 Stages in host–pathogen interaction

- Locating the host
- Attachment/adhesion to host surface
- Penetration and entry to the host
- Colonization of host tissues
- Suppression or evasion of host defense
- Reproduction of the pathogen
- Dispersal from the host

trigger germination of a dormant propagule, or attract hyphae or motile cells toward the host. Zoospores of root-infecting oomycetes such as *Pythium* or *Phytophthora* swim along concentration gradients of sugars or amino acids exuded by plant roots. These chemotactic responses are often nonspecific, as both host and nonhost roots are attractive, but some recognition cues may be more finely tuned. Dormant sclerotia of *Sclerotium cepivorum* germinate in response to sulfur compounds released only by onion and a few related species (see p.75), while the tiny seeds of the parasitic angiosperm *Striga* respond to a germination stimulant, strigol, present in root exudates from host plants such as cotton. This compound is active at remarkably low concentrations. Pathogens dispersed by insects or other vectors take advantage of the host recognition mechanisms of the vector itself.

Once contact has been made, a complex sequence of interactions between host and pathogen is initiated (Table 1). The pathogen must be able to stick to the host surface and then breach or bypass the outer physical defenses such as the cuticle and underlying epidermal layer to gain access to host nutrients. The spores of biotrophic fungi, for instance, contain food resources sufficient only for germination and a limited amount of hyphal growth; unless the fungus is able to penetrate living host tissues during this brief phase of independence, it will die. This dependence on living cells reaches an extreme in the case of viruses. Multiplication can only occur if the virus gains access to the biosynthetic machinery of the host cell. By contrast, some necrotrophic parasites are able to survive for an extended period without entering the host.

Once entry has been achieved, the pathogen needs to grow and colonize tissues sufficiently to reproduce and generate propagules that can disperse and infect new hosts. The extent of colonization varies between pathogens with different lifestyles, but all face the challenge of suppressing or evading active host defense. Pathogens have an armory of offensive weapons to deploy during this phase of colonization, and the balance between these and the plant's immune response ultimately determines whether or not disease develops.

Pathogenicity and Plant Immunity

While there is a wide diversity of pathogenic agents able to attack plants, including fungi, oomycetes, bacteria, and invertebrates such as nematodes, some common themes have emerged concerning the kinds of molecular interactions taking place during infection. One is the molecular basis of microbial pathogenicity. What properties distinguish pathogens from nonpathogens? Another is the central role played by the plant immune system.

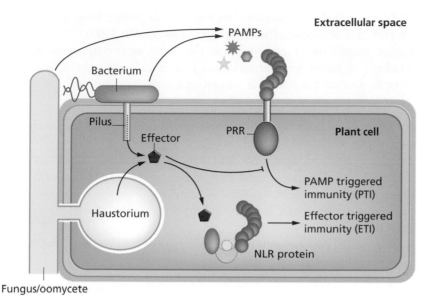

Figure 1 Simplified model of host–pathogen interactions involving fungi, oomycetes, and bacteria. Filamentous hyphae of fungi and oomycetes grow through intercellular spaces but may penetrate host cells by forming haustoria. Bacterial pathogens occupy intercellular spaces but can introduce molecules into host cells via a specialized secretion system with a thin pilus acting like a hypodermic syringe. Pathogen products (pathogen-associated molecular patterns, PAMPs) produced in the intercellular spaces can be detected by host cell surface-located pattern recognition receptors (PRRs) activating PAMP-triggered immunity (PTI). Pathogen effector molecules secreted into the host cell and targeting host defense can be detected by intracellular NLR receptor proteins, activating effector-triggered immunity (ETI). Both PTI and ETI involve a cascade of defense responses which is described in more detail in Chapter 9. *Source:* Adapted from Dodds and Rathjen (2010).

Beyond the physical and chemical barriers to invasion, there are two layers of surveillance able to detect the presence of a pathogen (Figure 1). The first consists of cell surface-located receptors tuned to recognize conserved structural elements of microorganisms, such as fungal chitin, or the protein flagellin from bacterial flagella. These are known as **pattern recognition receptors** (PRRs) and the molecules detected as **pathogen-associated molecular patterns** (PAMPs). For example, *Arabidopsis* possesses a PRR known as Flagellin Sensing 2 (FLS2) that binds the protein found in bacterial flagella and then activates a defense signaling pathway. This response is known as **PAMP-triggered immunity** (PTI).

The second layer of surveillance involves virulence factors produced by the invading pathogen, described as **effectors**. These are typically small secreted proteins delivered into host cells which target host defense pathways to suppress PTI. In turn, effectors may themselves be detected by a class of plant receptor proteins that contain nucleotide-binding and leucine-rich repeat domains (NB-LRR), now known as **NLR proteins**. These intracellular receptors confer resistance to a wide variety of microbial pathogens, viruses, and some invertebrate pests, and typically are the products of the R genes used for many years by plant breeders to introduce disease resistance into crop plants. The interaction of an effector with an NLR protein activates a strong defense response usually characterized by rapid host cell death (the so-called hypersensitive response). This second layer of the plant immune system is often described as **effector-triggered immunity** (ETI).

Like all biological models, the scheme shown in Figure 1 is an oversimplification of a more complex interaction involving other molecular players and partners, but serves as a useful framework for further discussion. It also raises a series of questions that will be addressed in the chapters that follow.

- How do pathogens get into plant hosts and access the resources required for growth and reproduction (Chapter 6)?
- How does infection by an invading microbe lead to the diseased state (Chapter 7)?
- What weapons are deployed by pathogens during invasion of the host. How does the successful pathogen evade or counter host defense (Chapter 8)?
- How do plants recognize microbial invaders and ward them off? What is the basis of plant immunity (Chapter 9)?
- Why do particular pathogens only attack a limited range of host plants? What factors determine host–pathogen specificity (Chapter 10)?

The approach taken in Part II is in part chronological, starting with the earlier ideas and discoveries relevant to the topic, and moving on to the more recent revelations arising from genetics, molecular biology, and genomics. As the science has advanced, some of the terminology has changed to reflect these advances. It has also spawned a series of abbreviations used as shorthand to cover many of the concepts. Many of these are defined when first used in the text, but to aid understanding a list of them is provided at the front of the book.

Further Reading

Books

Dickinson, M. (2003). *Molecular Plant Pathology*. New York: Garland Science (Taylor and Francis).

Reviews and Papers

Cook, D.E., Mesarich, C.H., and Thomma, B. (2015). Understanding plant immunity as a surveillance system to detect invasion. *Annual Review of Phytopathology* 53: 541–563.
Dodds, P.N. and Rathjen, J.P. (2010). Plant immunity: towards an integrated view of plant–pathogen interactions. *Nature Reviews Genetics* 11 (8): 539–548.
Guttman, D.S., McHardy, A.C., and Schulze-Lefert, P. (2014). Microbial genome-enabled insights into plant-microorganism interactions. *Nature Reviews Genetics* 15 (12): 797–813.
Hein, I., Gilroy, E.M., Armstrong, M.R. et al. (2009). The zig-zag-zig in Oomycete–plant interactions. *Molecular Plant Pathology* 10 (4): 547–562.
Thrall, P.H., Barrett, L.G., Dodds, P.N. et al. (2016). Epidemiological and evolutionary outcomes in gene-for-gene and matching allele models. *Frontiers in Plant Science* 6: 1084. https://doi.org/10.3389/fpls.2015.01084.

6

Entry and Colonization of the Host

Whenever the little seeds of rust come to rest upon the same stalk, finding some open mouths of the exhaling vessels, there they enchase their minute radical fibers, and there they infiltrate in such a manner, that they graft into the tender and delicate arteries peculiar to the plant.

(G. Targioni-Tozzetti, 1712–1783)

To gain access to host nutrients and establish a parasitic relationship, the microorganism must first pass through the external protective layers of the host. Plant pathogens enter their hosts in a variety of ways. Some penetrate directly through the intact surface covering of the plant. Others enter via natural openings or through regions where the external defenses are especially thin. The most important such route is through the stomata, but other zones of weakness include glands, hydathodes, lenticels, nectaries, and root tips. Many other pathogens exploit wounds resulting from physical or chemical damage or from the activities of animal pests. Wounds can also be self-inflicted, for instance by the abscission of leaves or during the emergence of lateral roots.

A distinction may therefore be drawn between pathogens which enter directly through the protective barriers of the host and those which bypass these defenses (Figure 6.1). Several major groups of pathogens, including the viruses, phytoplasmas, and many fungi causing postharvest diseases of fruit, are almost entirely dependent upon wounds to gain entry to the host.

The entry route is important in determining the nature of the initial host–pathogen interface formed. For instance, bacterial cells washed through stomata by rain can multiply initially in intercellular spaces, while pathogens penetrating directly through the host epidermis must cross cell walls and often grow within host cells. These differences affect the types of nutrients available to the pathogen, and also the molecular events involved in recognition of the pathogen by the host, or suppression or evasion of host defense by the pathogen.

The Infection Court

Let us assume that a pathogen has been successfully dispersed or has grown into contact with a potential host plant. Subsequent development on the surface of the host, penetration into the host, and the very early stages of establishment within host tissues comprise the

Plant Pathology and Plant Pathogens, Fourth Edition. John A. Lucas.
© 2020 John Wiley & Sons Ltd. Published 2020 by John Wiley & Sons Ltd.
Companion website: www.wiley.com/go/Lucas_PlantPathology4

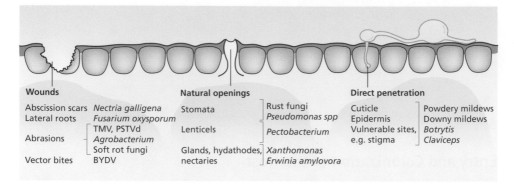

Wounds		Natural openings		Direct penetration	
Abscission scars	*Nectria galligena*	Stomata	Rust fungi	Cuticle	Powdery mildews
Lateral roots	*Fusarium oxysporum*		*Pseudomonas spp*	Epidermis	Downy mildews
Abrasions	TMV, PSTVd	Lenticels	*Pectobacterium*	Vulnerable sites,	*Botrytis*
	Agrobacterium			e.g. stigma	*Claviceps*
	Soft rot fungi	Glands, hydathodes,	*Xanthomonas*		
Vector bites	BYDV	nectaries	*Erwinia amylovora*		

Figure 6.1 Some entry routes for plant pathogens. PSTVd potato spindle tuber viroid; TMV, tobacco mosaic virus.

process of **infection**. This stage in the life cycle of the pathogen ends when the organism becomes dependent on the host for nutrients, at which point it begins to **colonize** tissues around the infection site.

The initial point of contact between the pathogen and the surface of the host is described as the **infection court**. In any discussion of host infection, it is useful to distinguish at the outset between the aerial and subterranean surfaces of the plant. In one respect, the problems confronting air-borne and soil-borne pathogens are similar, in that both must breach the outer defensive layers of the host, but there are major differences between the two environments. Soil exerts a buffering effect against extremes of temperature, water availability, and other environmental fluctuations. A propagule landing on an aerial plant surface is exposed to wide daily fluctuations in temperature and hazards such as desiccation.

The leaf and root surfaces of plants are termed the **phylloplane** and the **rhizoplane** respectively. The allied terms **phyllosphere** and **rhizosphere** describe the habitats adjacent to these surfaces. A further term, the **spermosphere**, is used for the zone surrounding seeds in soil. In recent years, a great deal has been learned regarding the influence of physical, chemical, and biological factors on pathogen behavior in these contrasting infection courts. Factors influencing the germination of fungal spores are of special significance and include humidity, duration of leaf surface wetness, temperature, light, pH, nutrient availability, and the quantity and quality of host exudates. Exudates from leaves and roots contain numerous chemicals such as sugars, amino acids, mineral salts, phenols, and alkaloids; any of these may stimulate or inhibit germination and/or growth of pathogens. Root exudates are particularly significant in determining the behavior of soil fungi and oomycetes which produce motile zoospores. These are chemotactically attracted to the elongating zone of host roots where they encyst prior to entry. The initial phases of development on the host surface represent a vulnerable stage in the life cycle of fungal pathogens, as witnessed by the efficacy of protectant fungicides in the control of many diseases.

Leaf and root surfaces are colonized by distinct microbial communities comprising mainly saprophytic species able to subsist on nutrient exudates or other substrates present in these habitats. The rhizosphere, in particular, is a zone of increased microbial activity that can extend several millimeters into the surrounding soil. Any potential pathogen may

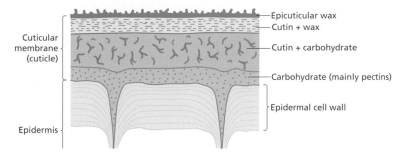

Epicuticular wax

Cutin + wax

Cuticular membrane (cuticle)

Cutin + carbohydrate

Carbohydrate (mainly pectins)

Epidermal cell wall

Epidermis

Figure 6.2 The external layers bounding herbaceous plant organs. *Source:* After Jeffree et al. (1976).

be subject to antagonism by other organisms in the rhizosphere, although in some cases their activity might actually provide opportunities for infection, such as wounds caused by root-feeding nematodes.

The principal components of the aerial surfaces of herbaceous plants are summarized in Figure 6.2.

In practice, there are considerable physical and chemical differences between the outer layers of various plant species, and even between different parts of the same plant. Hence, the cuticle may vary in chemical composition and thickness on leaves, flowers, and fruits. Variations are also found in different regions of the same organ, for example between the upper (adaxial) and lower (abaxial) surfaces of leaves. Other factors influencing the structure and composition of external layers include the conditions under which the plant has grown and the developmental stage that the plant has reached. Seedling tissues are particularly prone to infection by opportunist pathogens, such as the "damping off" oomycete *Pythium*, whereas mature plants are seldom attacked. A critical factor here is the relative ease with which the pathogen can penetrate the cuticle and epidermal layers in the young plant.

The outer cell walls of primary roots are usually impregnated with lipid materials, including suberin and cutin. These form a membrane comparable with the leaf cuticle, but is breached in places as lateral roots emerge. There is also some doubt as to whether such protective layers are present in the physiologically active apical region of the root. The root hair zone is especially vulnerable to pathogens, as it is in intimate contact with a large volume of soil. The necessity for efficient water and nutrient uptake by the root hairs means that substantial mechanical barriers, which would perhaps deter pathogens, are absent. Root cap cells secrete a mucilaginous gel which encloses the growing root and is a distinctive feature of the rhizosphere.

There are even greater differences between the surfaces of herbaceous tissues and the stems and roots of woody perennials. Periderm, commonly termed bark, is formed following secondary thickening and the accompanying increase in girth of the organs. It comprises three layers, with the outermost being composed of dead cork cells which have suberized walls. Suberin is a complex material containing mixtures of hydroxy acids and is very resistant to microbial attack. This substance, together with lignin and cellulose in cork cell walls and resins in their lumina, ensures that bark is virtually impregnable to invasion by microorganisms. Similar protective layers also form over abscission wounds and damaged tissues which are exposed when large branches are broken off in storms.

Adhesion

For air-borne wind- or splash-dispersed spores, the first problem is effective adhesion to the host surface. Epicuticular wax is hydrophobic and repels water droplets and any microbial propagules they contain; firm attachment is essential to prevent the pathogen losing contact prior to infection. Molecules aiding microbial attachment are often described as **adhesins**. Spores of many pathogenic fungi produce an extracellular matrix which surrounds the spore and binds tightly to plant cuticles, as well as to inert hydrophobic surfaces such as Teflon®, the coating of nonstick saucepans. This matrix may also protect spores from desiccation and provide a scaffold for immobilization of secreted enzymes. In the rice blast pathogen, *Magnaporthe oryzae*, mucilage is released from a terminal compartment of conidia following hydration (Figure 6.3a,b), serving as a form of biological "glue." Other pathogens which produce dry spores and which adhere to dry leaf surfaces, such as the rust and powdery mildew fungi, probably employ electrostatic mechanisms as well as adhesive materials to ensure attachment. Studies on the kinetics of adhesion show that it is often a two-step process: initial passive attachment due to interactions between the spore coat and the plant cuticle, followed by active synthesis and secretion of further adhesive material or structures to secure the spore more firmly. Following contact with the host surface, conidia of the powdery mildew fungi produce a small, short hypha (Figure 6.4a). This primary germ tube helps to anchor the spore to the host cuticle, but may also be important in taking up water to aid pathogen growth.

Bacterial pathogens also synthesize extracellular molecules which promote adhesion. With animal-pathogenic species, small hair-like processes known as fimbriae are important in attaching cells to epithelial surfaces. Similar proteins are known to bind to plant cell walls, but their significance in plant infection is not clear. Instead, secreted polysaccharides often play a role in adhesion. The crown gall pathogen, *Agrobacterium tumefaciens*, elaborates cellulose microfibrils which help to anchor the bacterium to host cells, as well as binding further bacteria. Some *Pseudomonas* spp. form aggregates of cells and more organized biofilms on the plant surface that aid survival and may provide protection from plant defense responses.

Direct Penetration

As shown in Figure 6.2, direct penetration of herbaceous tissues requires entry through layers of wax, cutin, pectin, and a network of cellulose fibrils impregnated with other cell wall polymers, before the pathogen makes contact with host protoplasm. This would seem to be a formidable obstacle. Nevertheless, many fungal and oomycete pathogens are able to enter their hosts in this way.

Direct penetration of the host is frequently associated with the development of hyphal modifications known collectively as infection structures. Some examples are shown in Figures 6.3 and 6.4. Once the spore has germinated, there follows a period of growth in which the germ tube extends over the leaf surface. This growth may appear to be random but some evidence points to the possibility that the germ tube is "searching" for a favorable site for penetration (Figure 6.4b). The length of germ tube developed varies but eventually extension growth ceases and the tip of the hypha swells to form an **appressorium**. This

(a)   (b)

(c)   (d)

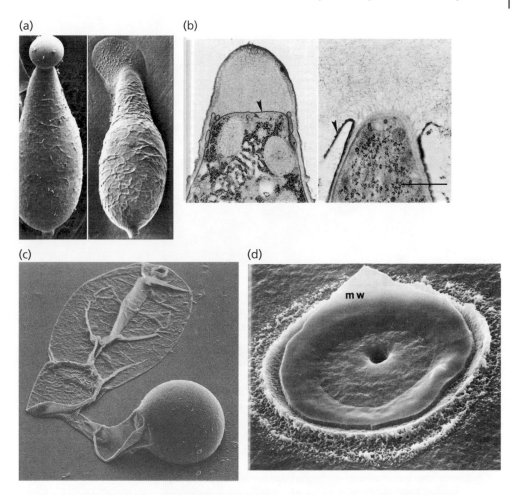

Figure 6.3 Early infection stages of the rice blast fungus *Magnaporthe oryzae*. (a) Conidium with apical droplet of mucilage (*left*) and conidium attached to substrate by adhesion of spore tip mucilage. ×2900. (b) Sections through spore tip with intact compartment containing mucilage (*left*) and fractured compartment (*right*) releasing the mucilage. Scale bar = 1 μm. (c) Empty collapsed conidium and germ tube attached to mature globose appressorium. A septum separates the appressorium from the germ tube. ×3000. (d) Remnants of appressorium attached to a polyethylene surface. The upper part of the cell has been removed by sonication. What remains is the appressorial pore, with the dent made by mechanical force of the infection peg clearly visible, and part of the smooth surrounding wall (mw) composed of melanin. Note the halo of extracellular matrix material around the attachment site. ×5500. *Source:* Braun and Howard (1994); Hamer et al. (1988).

spherical or ovoid structure increases the area of contact and attachment between the fungus and the host surface (Figure 6.3c,d). Penetration then takes place by the downward growth of a narrow hyphal thread or infection peg formed from the lower surface of the appressorium.

There has been much debate as to the actual mechanism of penetration. Early workers showed that many fungi will successfully penetrate artificial materials such as gold leaf, suggesting that the process is entirely mechanical. However, scanning and transmission electron microscope studies suggested that some pathogens degrade the cuticle and cell

(a)

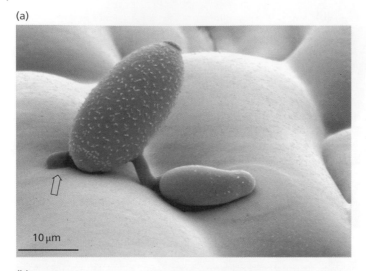

(b)

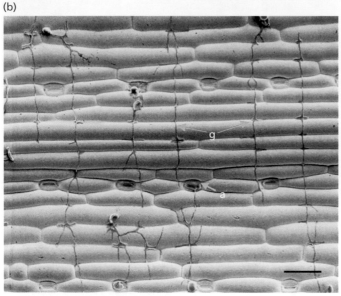

Figure 6.4 Early development of fungal pathogens on host leaves viewed by scanning electron microscopy. (a) Germinating spore of pea powdery mildew *Erysiphe pisi* showing small primary germ tube (*arrow*) and larger appressorial germ tube with appressorium developing on the host surface. Circular zone around appressorium suggests some effect on the cuticle. Scale bar = 10 μm. *Source:* Courtesy of Alison Daniels. (b) Germ tubes (g) of brown rust *Puccinia hordei* on barley leaves showing growth perpendicular to the orientation of epidermal cells, short branches formed at cell junctions, and an appressorium (a) formed over a stoma, which occur in rows. Scale bar = 100 μm. *Source:* Read et al. (1992).

wall polymers during penetration. Differential staining techniques have revealed localized dissolution of the cuticle and cell walls around infection pegs, implicating the action of hydrolytic enzymes in penetration by fungi.

The mechanisms employed by pathogenic fungi to penetrate plant surfaces have now been analyzed in more detail using molecular and genetic techniques. Several fungi are known to produce cutinase, an enzyme able to degrade cutin. Cutinase from the pea pathogen *Fusarium solani* f.sp. *pisi* was purified and used to raise antibodies specific for the enzyme. Labeled antibodies were utilized in electron microscopy to detect the presence of the enzyme at the penetration site. Inhibition of enzyme activity by the specific antibodies, or chemical inhibitors, prevented infection. Mutant strains of the fungus lacking cutinase activity were shown to be much less pathogenic than the wild-type strain, and pathogenicity could be restored by adding exogenous enzyme. Taken together, this evidence suggested that cutinase is a vital factor in breaching the host surface. The cutinase gene from *F. solani* was subsequently cloned and sequenced, and its regulation studied in detail. In germinating spores, cutinase synthesis is induced by breakdown products of cutin (Figure 6.5). Low levels of constitutive activity release cutin fragments which induce rapid expression of the cutinase gene. This system of regulation ensures that the enzyme is only synthesized in any quantity when pathogen spores contact a plant surface. Further evidence that cutinase is required for direct penetration was also obtained by introducing the *Fusarium* cutinase gene by transformation into another fungus, *Mycosphaerella*, which normally requires a wound to infect the host. Possession of the cutinase gene enabled the transformed fungus to penetrate intact host surfaces.

Subsequent experiments using gene disruption, a procedure in which the functional gene for cutinase is replaced by a defective copy unable to produce the enzyme, produced conflicting results. Some fungal transformants containing the disrupted gene, and in which cutinase synthesis was abolished, were still pathogenic to pea seedlings, and apparently able to penetrate the intact host surface. Other transformants showed reduced pathogenicity. Such differences might be accounted for by variation between fungal strains or the presence of more than one copy of a functional cutinase gene.

The demonstration that fungal enzymes such as cutinase are in some cases required for direct host entry does not preclude a major role for mechanical forces in penetration. Confirmation of the importance of such forces came, perhaps unexpectedly, from studies on the mode of action of certain fungicides. The compound tricyclazole effectively controls

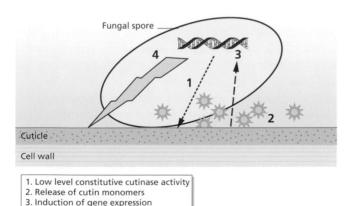

1. Low level constitutive cutinase activity
2. Release of cutin monomers
3. Induction of gene expression
4. High level enzyme activity

Figure 6.5 Model for induction of cutinase synthesis in spores of *Fusarium solani* f.sp. *pisi* on the host surface. Low levels of constitutive enzyme secreted by spores release breakdown products from the cuticle which induce high-level gene expression and cutinase production, aiding penetration of the plant. *Source:* Based on Woloshuk and Kolatukkudy (1986).

several fungi, including the rice blast pathogen *Magnaporthe oryzae*. This fungus penetrates rice plants directly from dark, pigmented appressoria; in the presence of the fungicide, the appressoria formed are nonpigmented, and no penetration takes place from them. The biochemical target of tricyclazole turns out to be melanin synthesis, so that production of the pigment is inhibited in treated appressoria. Normally, melanin is deposited in the appressorial wall, making it rigid and impermeable to solutes. As the appressorium matures, turgor pressure builds up inside the cell due to accumulation of osmotically active solutes such as glycerol until sufficient force is generated to drive the infection peg down through the cuticle. Functional appressoria formed on inert plastic surfaces actually leave a microscopic dent at the point where the peg projects (Figure 6.3d). In nonmelanized appressoria, the wall remains relatively thin, flexible and permeable, and the infection peg appears unable to breach the surface. Albino mutants of *M. grisea* which are unable to synthesize melanin are similarly incapable of achieving penetration from nonpigmented appressoria. Thus in this, and other similar pathogens with pigmented appressoria, mechanical force would appear to be the primary means for penetrating the host.

The early events in host infection by the rice blast fungus have recently been studied in great detail, from initial adhesion of the spore, through germling development on the host surface, penetration and subsequent invasion of plant cells. After the spore sticks to the wax layer, a germ tube emerges from one of the three cells of the spore and extends for 10–15 μm before polarized growth ceases and the hyphal tip differentiates into a swollen unicellular appressorium (Figure 6.3c). During this phase, the fungus is able to sense the hydrophobic surface and also constituents of the wax, so that both physical and chemical signals can induce appressorium formation. These trigger protein kinase and cyclic adenosine monophosphate (AMP) signaling pathways required for subsequent differentiation of appressoria. Development of a functional appressorium is also closely linked to the fungal cell cycle. The nucleus of the germinating spore cell migrates into the germ tube where mitosis occurs. One of the daughter nuclei enters the developing appressorium, while the other returns to the spore. Following this, cell division takes place and the three nuclei in the spore, along with the cell contents, then break down (a process known as autophagy) and are recycled to fuel further development of the infecting fungus. Within the appressorium, a reorganization of the cytoskeleton takes place to define a basal pore through which the considerable turgor pressure is focused, and the infection peg emerges to puncture the plant cuticle. Overall, this is a tightly regulated developmental program optimizing the chances of successful entry to the host.

While *M. oryzae* has the capability to generate the physical force to directly penetrate the host plant, analysis of the complete genome sequence of the fungus revealed up to eight genes encoding putative cutinase enzymes. Deletion of one of these had no effect on virulence, but mutating a second gene, *CUT2*, was shown to affect germling and appressorial development and to reduce virulence. Cutinase therefore plays a role in infection in *Magnaporthe*, most likely by influencing sensing and signaling on the host surface.

An overall conclusion from these elegant experimental studies is that both mechanical and enzymatic mechanisms are involved in direct penetration, but the relative contribution of each varies from pathogen to pathogen. In fungi that produce melanized appressoria, the predominant means is physical force, while in those forming nonmelanized appressoria, such as the powdery mildews and *Botrytis cinerea*, enzymatic digestion of the host cuticle and underlying cell wall may play a more important role.

Penetration of Plant Roots and Other Tissues

Root-infecting fungi can also form infection structures which in some cases are more complex than those produced by fungi attacking aerial tissues. The take-all pathogen *Gaeumannomyces graminis* invades the roots of cereals such as wheat. The fungus grows initially on the surface of the root as extending runner hyphae that produce short side branches that terminate in lobed structures known as hyphopodia (Figure 6.6d). These are larger than typical appressoria but structurally similar, with a melanized cell wall and basal ring through which narrow infection hyphae penetrate the epidermis and enter the root cortex. Melanized hyphopodia are also capable of generating high turgor pressure. *Rhizoctonia solani*, a versatile pathogen attacking a wide variety of host plants, forms both lobed appressoria and more complex aggregations of repeatedly branched hyphae, called infection cushions. The former tend to be produced on aerial tissues and the latter on roots and other subterranean organs. Multiple infection hyphae are produced from the lower surface of infection cushions, and enter the host.

The eyespot fungi, *Oculimacula* spp., infect the stem base of cereals by colonizing the seedling leaf (coleoptile) and then penetrating through successive leaf sheaths. Multicellular plates of mycelium, termed infection plaques, are produced on the surface of each leaf sheaf (Figure 6.6), and these act as compound appressoria, enabling the fungus to penetrate epidermal cells at numerous sites. Infection plaques can be induced in laboratory culture away from the host by growing the fungus beneath a hydrophobic membrane that mimics the narrow space between leaf sheaths. It seems likely, therefore, that this pathogen possesses a contact mechanism able to sense the particular features of the stem base infection court.

Fungi attacking perennial hosts, such as trees, in which the surface is protected by a layer of bark or periderm, often infect from compound structures. When rhizomorphs of *Armillaria mellea* encounter a suitable host, the concerted action of the numerous hyphae comprising these strands (see p.48) is often sufficient to penetrate the intact surface layers.

It was noted earlier that the root hair zone is particularly vulnerable to invasion by pathogens. The clubroot pathogen *Plasmodiophora brassicae* enters root hairs at an early stage in its life cycle (see p.12). The mode of entry of *Plasmodiophora* into host cells appears to be unique. Zoospores of the pathogen first encyst on the root hair wall. A pointed, bullet-shaped structure differentiates within the encysted spore This is then violently projected from within the cyst through the root hair wall, with the contents of the spore following into the host cell through the resulting puncture. This is a particularly dramatic example of mechanical penetration, as the injection process takes only about one second.

The bacterium *Rhizobium* also initiates infection through root hairs, one of the very few examples of direct penetration by a plant-infecting bacterium, although in this case the relationship is ultimately mutualistic, with the formation of nitrogen-fixing root nodules.

A further interesting example of a vulnerable site exploited by pathogens is the surface of the female organ, the stigma, which is adapted to trap pollen and permit penetration by pollen tubes to ensure fertilization. Several specialized pathogens, notably species of the ergot fungus *Claviceps*, produce air-borne spores which germinate on the host stigma to form penetration hyphae which mimic pollen tubes and extend downwards to invade the ovary. The period of susceptibility to infection is brief, due to the short time during which

Figure 6.6 Infection structures of stem- and root-infecting fungi. (a) Scanning electron microscope (SEM) image of infection plaques of the cereal eyespot pathogen *Oculimacula* spp. on the stem base of wheat. Bar = 20 μm. (b) SEM image of host surface beneath plaque showing multiple penetration sites. Bar =10 μm (c) Fluorescent confocal microscopy image of wheat coleoptile invaded by an isolate of *Oculimacula* expressing green fluorescent protein. Host cell walls fluoresce red. The underside of an infection plaque is visible at top right. Filamentous hyphae (*foreground*) are invading cortical cells and a vascular strand (*bottom right*). Bar =20 μm (d) Dark melanized hyphopodia of the take-all fungus *Gaeumannomyces graminis* formed on a hydrophobic polyester film. Bar = 20 μm *Source:* (a,b) Daniels et al. (1991); (c) Bowyer et al. (2000); (d) Money et al. (1998).

the stigma is receptive to pollination. As was discussed in Chapter 3, some viruses, for example bean mosaic virus, may be transmitted to the ovules of healthy plants through infected pollen. This is a particularly interesting case as the virus takes advantage of a normal event in the life cycle of the plant to circumvent the structural defenses of the host.

Penetration Through Natural Openings

Entry Through Stomata

The surface layers of plants possess a number of natural openings through which microbes can enter. The most important of these are stomata, via which many pathogens gain entry to their hosts. The detailed morphology of these structures might in some cases determine whether or not infection occurs. An early study compared stomatal structure in two species of citrus differing in susceptibility to the bacterial canker pathogens *Xanthomonas axonopodis* pv. *citri* and found that the conformation of the cuticle around the stoma either prevented or allowed the passage of water droplets containing bacterial cells. However, subsequent work was unable to show a clear correlation between stomatal morphology or density and resistance to citrus canker disease, suggesting instead that the rate of multiplication of the pathogen in leaf mesophyll tissues is the key determinant of disease severity. Stomata also provide the main site of entry for several important fungal and oomycete pathogens.

When a rust spore germinates on a cereal leaf, the germ tube grows at right angles to the long axis of the leaf (Figure 6.4b). This orientation of growth is a contact response to the surface topography of cells, known as **thigmotropism**; similar tropisms occur on inert plastic replicas of leaves. Experiments with artificial surfaces etched or scratched with precise patterns suggest that the fungus recognizes a repetitive series of ridges spaced at intervals similar to the width of epidermal cells. Hyphal growth across the long axis of the cereal leaf ensures that the germ tube will sooner or later encounter a stoma, as these occur in longitudinal rows (Figure 6.4b). Once a stoma is contacted, the rust germ tube differentiates an appressorium and an infection hypha enters the substomatal cavity. With many rust fungi, the signal for appressorium formation appears to be the shape of the stomatal guard cell, and in particular the stomatal lip. Bean rust, *Uromyces appendiculatus*, produces appressoria in response to small ridges about 0.5 μm in height, which corresponds closely to the dimensions of the stomatal lip of the host plant *Phaseolus*. On nonhost leaves, extension growth of the germ tube continues indefinitely until the endogenous nutrient reserves are exhausted and the germling dies. These precise morphogenetic responses to host surface features indicate that rust fungi possess a sophisticated contact-sensing system which aids location of natural openings for entry.

Several oomycete pathogens, including *Pseudoperonospora* on hops, *Plasmopara* on vines, and *Phytophthora* on potato, produce sporangia which germinate on host leaves by releasing motile zoospores. These zoospores are attracted to stomata where they encyst in a suitable position for their germ tubes to grow immediately between the guard cells. *Pseudoperonospora* zoospores are attracted to open stomata but not to closed stomata. This attraction is based in part on recognition of the morphology of the open apertures and partly on a chemical stimulus connected with gaseous photosynthetic metabolites.

Originally, it was thought that stomata are passive ports of entry for pathogens but recent studies have changed this view. When bacteria were inoculated onto *Arabidopsis* leaves, the stomata closed within two hours, suggesting that the plant somehow senses the presence of the bacteria and responds by shutting a potential entry route. This response was seen with both plant and human pathogenic bacteria (Figure 6.7). Interestingly, with the plant pathogen *Pseudomonas syringae*, the stomata reopened after a few hours, while with the human pathogen *Escherichia coli* they remained closed. This indicated that the plant-adapted bacterium was able to reverse the stomatal defense response. Stomatal closure is now known to be one component of innate plant immunity, which can be overcome by pathogen virulence factors. This interaction is discussed in more detail in Chapter 9.

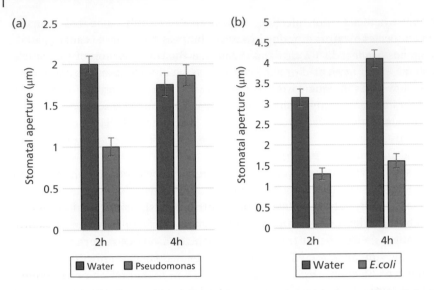

Figure 6.7 Response of *Arabidopsis* stomata to exposure to (a) a plant-pathogenic bacterium (*Pseudomonas syringae*) or (b) a human colonizing bacterium (*Escherichia coli*). Stomatal aperture (± standard error) was measured at two or four hours after inoculation of intact leaves (a) or epidermal peels (b). *Source:* Adapted from Melotto et al. (2006).

Lenticels, Hydathodes, and Nectaries

Lenticels allow gas exchange to occur through bark on woody stems and secondarily thick-ened roots. These loosely packed openings in the periderm are also abundant on potato tubers, where they provide suitable sites for the entry of a number of microorganisms, especially the common scab pathogen *Streptomyces scabies.*

Glandular tissues which have especially thin surface barriers, such as hydathodes and nectaries, are also exploited by pathogens. Bacterial lesions on leaves often develop at the margin, at sites where water exudes through hydathodes. *Erwinia amylovora*, the bacte-rium responsible for the destructive fireblight disease of pears and apples, enters through nectaries at the base of flowers. In this case, the sugary secretions of nectar, when diluted by rain, provide a favorable medium for multiplication of the pathogen prior to penetra-tion. Fireblight infections are also prevalent after severe thunderstorms, which suggests that the bacterium takes advantage of minor wounds caused by heavy rainstorms. Rainfall is an important predisposing factor in foliar infection by bacteria, as an external force suf-ficient to wash cells through natural openings into substomatal cavities and other internal tissues (see p.79).

Penetration Through Wounds

For many pathogens, especially bacteria and viruses which are incapable of penetrating plants directly, wounds are the most frequent or only avenue of entry. Wounds are caused by human activities, as well as by natural agencies, including wind, hail, extremes of temperature and light, and by pests. The external barriers of the host may also be broken temporarily as a natural consequence of plant growth and development.

Many agricultural and horticultural practices involve accidental or even deliberate wounding. Grafting, pruning, and picking may spread pathogens through a crop or create wounds which can be exploited by opportunist fungi and bacteria. Mechanical harvesters often increase the incidence of wounding of plant produce. Postharvest rots of apples (caused by *Penicillium expansum*) and citrus fruits (caused by *Penicillium digitatum* [Figure 6.8] and *P. italicum*) are only important if the fruits are mechanically wounded during harvesting, packing or transport. Many important forest pathogens also enter through wounds. *Heterobasidion annosum*, which is a destructive pathogen of conifers (see Chapter 13, Figure 13.8), normally colonizes wounds caused by high winds, snow or other natural agencies. It has become a particularly damaging pathogen in plantations where it takes advantage of the stumps left after felling as sites for entry (see p.348).

Leaf abscission provides opportunities for infection, as does any other point in the life cycle at which parts of the plant are detached. *Nectria galligena*, which causes apple canker, enters woody twigs through the vascular bundles which are exposed at leaf fall, and hence avoids the problem of penetrating intact bark. The vascular bundles in the leaf scar are, however, soon sealed off by the development of a cork layer and hence the pathogen must take immediate advantage of the infection sites created at leaf fall. Lateral roots emerge by breaking out through the cortex of the parent root. Lesions caused by root pathogens, such as soil-borne *Phytophthora* species, are often initiated at these sites. Soft rot pathogens of potatoes commonly enter tubers through the scar left during separation of the tuber from the stolon of the parent plant (Figure 6.8).

As well as providing entry sites, wounds may release solutions rich in carbohydrates and amino acids, which stimulate germination of spores, or attract motile bacteria and fungal zoospores. The crown gall bacterium *Agrobacterium* is dependent on wounds to initiate tumors; exudates leaking from wounded cells have been shown to contain phenolic compounds such as acetosyringone, which serve as molecular signals activating virulence genes on the Ti plasmid (see Chapter 8, Figure 8.15). This specific recognition–response system ensures that virulence functions are expressed only in the presence of susceptible host cells.

Senescent tissues which remain attached to plants also serve as an entry route, and facilitate the invasion of adjoining healthy tissues by opportunist pathogens. The gray mold pathogen *B. cinerea* colonizes vegetables such as courgettes (see Chapter 11, Figure 11.11) or tomatoes by vegetative growth from the senescing remains of flowers, causing a disease known as blossom end rot.

Disease lesions may themselves allow the entry of other pathogens. In these instances, the host is initially infected by a pathogen which may or may not itself cause serious damage. This pathogen, however, paves the way for more aggressive organisms. Potato late blight lesions in tubers may be exploited by soft rot bacteria such as *Pectobacterium carotovora* (see Chapter 8, Figure 8.3) which can destroy the tuber much more quickly than the blight pathogen itself. It is becoming increasingly clear from recent studies analyzing DNA sequences isolated directly from disease lesions (an example of **metagenomics** – the analysis of genetic material from environmental samples) that a mixed community of microorganisms is often present, some of which may act as secondary pathogens.

It was noted in Chapter 3 that many pests are also important vectors for plant pathogens. As well as dispersing the pathogen, their feeding activities cause wounds which serve as an entry route. The Dutch elm fungus, *Ophiostoma novo-ulmi*, is introduced directly into

(a) (b)

(c)

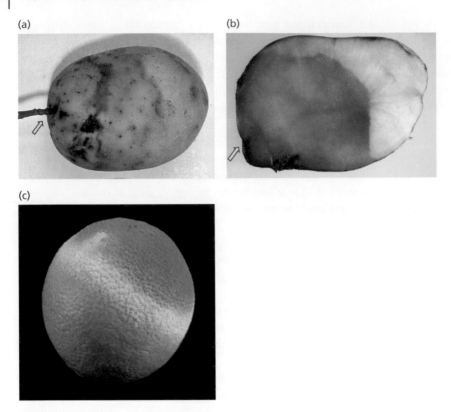

Figure 6.8 Pathogens that exploit wounds to infect the host. (a) Pink rot of potatoes caused by *Phytophthora erythroseptica*. Infected tuber showing external symptoms. The pathogen has gained entry through the join between the stolon and tuber (*arrow*). (b) Cut tuber with characteristic pink lesion. Infection has spread from the point of separation of the stolon (*arrow*). (c). Green mold of citrus fruit caused *by Penicillium digitatum*. The fungus enters fruit via wounds in the peel so careful handling is required to prevent postharvest spoilage by this and related *Penicillium* species. *Source:* Photos by Brian Case.

sapwood by its vector, the bark beetle *Scolytus*. In this example, the feeding tunnels not only breach the external protective layers of bark but also provide direct access to the vascular tissues in which the fungus can flourish. An even more elegant means of entry is provided by the aphid vectors of many viruses. The aphid stylet injects the virus into the sieve cells of the host with clinical efficiency (see Chapter 3, Figure 3.13) and subsequently the virus can spread freely via the phloem. Soil-borne viruses are often introduced via wounds caused by nematodes or fungal pathogens. The wide range of vectors exploited by viruses is paralleled by a similar variety of infection routes.

In some cases, wounds caused by animal pests can increase the incidence and severity of plant diseases, although the pest itself is not a vector. One of the best-known examples is the interaction between vascular wilt pathogens and nematodes. The fungus *Verticillium dahliae* causes a wilt disease in potatoes known as early dying, characterized by premature senescence of leaves and haulms. The pathogen survives in soil as microsclerotia, and there is a correlation between the number of pathogen propagules present in soil and the incidence of early-dying disease. If, however, the soil also contains significant numbers of nematodes capable of causing lesions on potato roots, the disease is much more severe. The most likely explanation for this synergistic effect is that feeding wounds caused by the nematodes provide enhanced access to the vascular tissues of host roots.

The Host–Pathogen Interface

Once inside the plant, pathogens exhibit a wide variety of modes of growth within host tissues (Table 6.1). The site of contact between a pathogen and host cells is known as the host–pathogen interface. This zone is vital in understanding the nature of different host–pathogen interactions, as it is the site at which nutrient uptake by the parasite occurs, and also where molecular communication between the two partners takes place. It is likely, for instance, that recognition events determining active resistance or susceptibility to infection

Table 6.1 Some modes of growth of pathogens in host plants

Type	Pathogen	Host
Subcuticular	*Rhynchosporium*	Barley
	Venturia	Apple
Intercellular	*Cladosporium*	Tomato
	Zymoseptoria	Wheat
	Most bacteria	Various
Vascular		
Xylem elements	*Fusarium*	Various
	Verticillium	Various
	Ophiostoma	Elm
	Some bacteria	Various
Phloem elements	*Candidatus Liberibacter*	Citrus greening (huanglongbing)
	Phytoplasmas	Various
	Spiroplasmas	Various
Haustorial		
Epiphytic with haustoria	Powdery mildews	Various
Intercellular with haustoria	Rust fungi	Various
Intracellular vesicle with intercellular hyphae and haustoria	Oomycetes including:	
	Phytophthora	Potato
	Bremia	Lettuce
	Hyaloperonospora	*Arabidopsis*
Intracellular		
Intracellular vesicle with invasive hyphae	*Colletotrichum*	Legumes, *Arabidopsis*
	Magnaporthe	Rice and grasses
Wholly intracellular	*Plasmodiophora*	Brassicas
	Polymyxa	Cereals and sugarbeet
	Viruses	Various

are often initiated at this interface. Three main types of interface can be distinguished: intercellular, where the pathogen grows outside host cells; partly intracellular, where limited penetration of cells by parasitic structures occurs; and intracellular, where growth and development take place entirely within host cells (Table 6.1). These categories are not absolute as many pathogens which initially grow in the apoplast between host cells subsequently invade them once tissues become moribund.

Intercellular relationships are characteristic of bacteria and some fungi which grow between cell walls and through intercellular spaces without penetrating host cells. Soluble nutrients such as sugars and amino acids are scavenged from the apoplast or released from cell walls by secreted hydrolytic enzymes (see p.188). Hence, there is no intimate contact with living host protoplasts. With necrotrophic pathogens, such as *Sclerotinia sclerotiorum* and *B. cinerea*, host cells are often killed in advance of invasion, through the action of enzymes or toxins. A structurally defined interface is therefore short-lived, as host cells rapidly die and release their cell contents. In other cases, such as *Cladosporium fulvum* on tomato and *Zymoseptoria tritici* on wheat, growth in the apoplast takes place for several days before any signs of host cell damage are evident (see Chapter 7, Figure 7.12). Subsequently, cells collapse and necrotic lesions appear. A different infection strategy is seen with the barley leaf blotch fungus *Rhynchosporium commune* and apple scab *Venturia inaequalis* that penetrate the leaf surface but then grow beneath the cuticle for an extended period without penetrating epidermal cells, until eventually sporulation occurs (see Chapter 2, Figure 2.3).

Intracellular relationships typically involve a more permanent contact between the partners, and penetrated host cells may remain viable for an extended period of time. In these cases, the host–pathogen interface is a living and dynamic zone, often involving the formation of modifed membranes or specialized parasitic structures such as infection vesicles and haustoria.

Structure and Function of Haustoria

Many biotrophic fungi form modified hyphae, known as haustoria, which enter host cells. Haustoria typically develop from intercellular hyphae as narrow branches which penetrate through the plant cell wall and then expand inside the cell (Figure 6.9b,c). They are diverse in morphology, ranging from small, club-shaped extensions to much larger, lobed or branched structures (see Chapter 3, Figure 3.3). Other fungi, such as hemibiotrophic species of *Colletotrichum* and the rice blast pathogen *M. grisea*, form intracellular vesicles (Figure 6.9a) and hyphae within initially penetrated cells, which have some similarities with haustoria, as the host–pathogen interface is a fungal cell in intimate contact with the host protoplast.

Although haustoria and equivalent structures are formed within plant cells, the host plasma membrane is not penetrated and remains intact as an invagination surrounding the fungal cell (Figure 6.10a). The interface between host and pathogen is therefore a complex zone comprising the fungal plasma membrane, the fungal cell wall, and an extrahaustorial

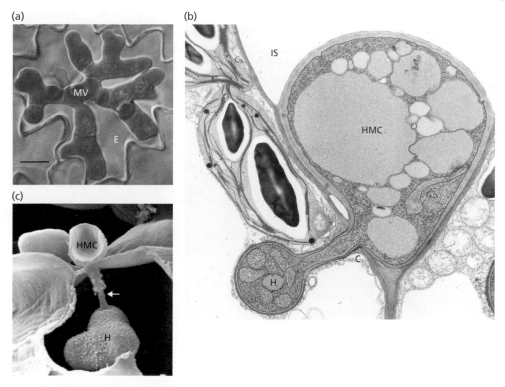

Figure 6.9 Intracellular structures formed by biotrophic fungi and oomycetes. (a) Multilobed vesicle (MV) of *Colletotrichum destructivum* inside epidermal cell (E) of the host plant alfalfa. Scale bar = 10 μm. *Source:* Latunde-Dada et al. (1997). (b) White blister rust *Albugo candida* in mesophyll tissues of cabbage, showing haustorial mother cell (HMC) in intercellular space (IS) and club-shaped haustorium (H) adjacent to chloroplast with dark starch grains. A collar (C) surrounds the penetration site. ×9600. *Source:* Coffey (1975). (c) Scanning electron micrograph of coffee leaf tissue infected by rust, *Hemileia vastatrix*. The tissue has been frozen and fractured to reveal a haustorium (H) within a mesophyll cell. Note a slight swelling (*arrow*) in the haustorial neck at the position of the neck band and the haustorial mother cell (HMC) external to the penetrated host cell. ×5000. *Source:* Courtesy of Rosemarie Honneger.

membrane, or EHM (Figure 6.10b). In addition, there is often an amorphous matrix, probably secreted by the host, between the EHM and the fungal cell wall. Typically, where the haustorial neck breaches the host cell wall, a collar of callose-like material is deposited (Figure 6.9b). In an incompatible host–pathogen combination, this may extend to form a sheath completely encasing the haustorium. Finally, in the majority of haustoria, a discrete, electron-dense ring is visible in the fungal cell wall in the neck region (Figure 6.10a,b). This is not observed in haustoria formed by oomycete pathogens such as the downy mildews and *Albugo* (Figure 6.9b).

The structure and location of haustoria, which provide an enlarged surface area of the parasite directly adjacent to nutrient sources such as chloroplasts (Figure 6.9b), suggest

(a) (b)

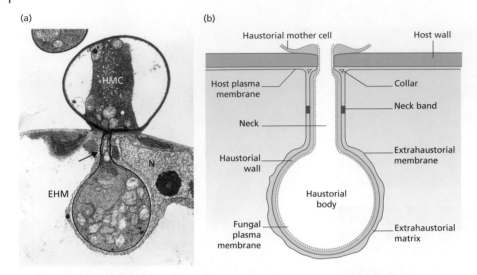

Figure 6.10 The structure of haustoria. (a) Haustorium of flax rust *Melampsora lini* penetrating host cell from haustorial mother cell (HMC). Note invaginated host membrane (EHM) and host cell nucleus (N) adjacent to haustorium and electron-dense neck band (*arrow*) (×8000). *Source:* Coffey (1976). (b) Diagrammatic interpretation of haustorial structure showing main interfacial components.

that they play a role in nutrient uptake. Obtaining direct physiological evidence to confirm this idea proved difficult. Initially, most of the work on haustorial function was done with powdery mildew fungi, as these epiphytic parasites form haustoria only in epidermal cells and are therefore a convenient experimental system for analysis (see Chapter 3, Figure 3.3). Most of the fungal biomass can be stripped off the leaf and separated from the host tissues. If plants infected by powdery mildew are fed radiolabeled carbon as $^{14}CO_2$, a proportion of the carbon fixed in photosynthesis travels to the epiphytic hyphae and spores of the fungus. No significant uptake of radiolabeled solute occurs until after formation of the first haustoria.

An important technical advance for analyzing haustorial function was the development of methods for isolating them from host cells. This approach was pioneered with powdery mildews by peeling off the infected epidermal layer and then macerating the tissue. Intact haustoria were recovered comprising the haustorial neck and body with the EHM still attached. Experiments with labeled sugars and amino acids showed that solutes cross the EHM, and that the epidermal cell cytoplasm plays an essential role in transporting assimilates into haustoria. The main compound initially moving from the host to the fungus is glucose. Thus, with powdery mildew fungi, the pathway of carbon flow is from the source (chloroplasts in leaf mesophyll cells) to the primary sink (epidermal cells which lack chloroplasts), and then into a secondary sink, the fungus, via haustoria in epidermal cells. Sucrose, the main transport sugar, is hydrolyzed prior to the uptake of glucose into the fungus, where it is converted into fungal metabolites such as mannitol and glycogen.

Subsequently, methods were developed to isolate haustoria from rust-infected bean leaves using affinity chromatography to separate them from plant organelles such as chloroplasts. cDNA libraries were then prepared from the purified fractions to identify fungal genes expressed in haustoria. Among those identified were genes encoding putative transporter proteins for sugars and amino acids; one of these proteins from the bean rust *Uromyces fabae* was confirmed to be able to transport glucose and fructose by expression in a culturable fungus, yeast. Hence haustoria have the molecular machinery necessary to take up plant nutrients from the host.

What is the actual mechanism by which solutes are removed from host cells? Electron micrographs of stained or freeze-fractured haustoria suggest that the invaginated region of the host membrane, the EHM, is altered in structure and composition by comparison with the rest of the host plasma membrane. In particular, the EHM lacks intramembrane particles and ATPase, an enzyme involved in the active transport of solutes. ATPase activity can be detected in the host membrane where it lines the plant cell wall, and also in the fungal plasma membrane inside the haustorium, but not in the EHM. It appears, therefore, that both the host cell protoplast and the fungal protoplast are actively importing solutes, while the membrane enclosing the haustorium has diminished control of solute transport, and leaks nutrients into the extrahaustorial matrix, from where they are scavenged by the fungus. Transporter proteins in the haustorial plasma membrane actively import sugars and amino acids into the fungus, thereby maintaining a concentration gradient across the host–pathogen interface (Figure 6.11).

One further feature of this model is the electron-dense band of impermeable material (Figure 6.10b) where the EHM contacts the haustorial neck. This is presumed to prevent solutes diffusing along the haustorial cell wall in the neck region. The extrahaustorial matrix and the haustorial wall are therefore a sealed compartment, and any solutes leaking across the EHM can only enter the fungus via active transport through the haustorial plasma membrane. In biotrophs lacking a haustorial neckband, such as the downy mildews, the plasma membrane of the penetrated cell still appears to comprise two functional domains, so there may be parallel processes of nutrient acquisition.

These experiments confirmed the importance of haustoria in nutrient uptake by biotrophic fungi. Calculations suggest, however, that with many biotrophs, a significant proportion of nutrients can also be acquired from the host apoplast via intercellular hyphae. A detailed three-dimensional analysis of colonies of brown rust, *Puccinia hordei*, on barley estimated that the total length of intercellular hyphae in a colony is approximately 1 m. Haustoria occur at a frequency of one every 70 µm, giving a total number of more than 10 000 per colony. However, haustoria accounted for less than 20% of the total colony surface area, and the major area of contact between host and pathogen was therefore between intercellular hyphae and host cell walls.

Some pathogens appear to grow preferentially at sites in the plant where nutrient transfer is occurring. A good example of this type of relationship is the ergot fungus *Claviceps purpurea*, which colonizes ovary tissues and diverts nutrients passing from transfer cells to the developing embryo. The pathogen competes with the embryo and ultimately replaces it with a fungal structure, the sclerotium. It has also been noted that many biotrophic fungi, such as rusts, invade host tissues adjacent to vascular elements, where loading or unloading of sugars into or from phloem cells is occurring. The term **transfer-intercept infection** has been used to describe such behavior.

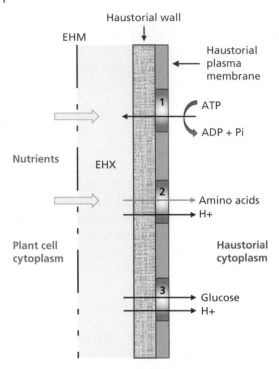

Figure 6.11 Proton symport model for nutrient transport across the haustorium–plant interface. Nutrients from the host cell leak across the modified extrahaustorial membrane (EHM) into the extrahaustorial matrix (EHX). A membrane H+ ATPase (1) generates a proton gradient across the haustorial plasma membrane, providing energy for the transport of nutrients. Transporter proteins for amino acids (2) and glucose (3) actively import host nutrients into the haustorial cytoplasm. ADP, adenosine diphosphate; ATP, adenosine triphosphate. *Source:* Adapted from Szabo and Bushnell (2001).

Other Roles for Haustoria

While initial studies focused on the role of haustoria in nutrient uptake from the host, it was presumed that they might also play other roles in host–pathogen interaction. The intracellular location of these structures and their proximity to host organelles such as nuclei (Figure 6.10a) suggested that they are a prime site for molecular communication between the partners. Could haustoria play an important role in subverting the host cell during maintenance of the biotrophic relationship? This idea has now been confirmed with the discovery that haustoria can act as delivery sites for fungal effector molecules targeting host defense pathways. Further details of this role are presented in Chapter 10.

Intracellular Pathogens

Intracellular relationships are typical of some mutualistic associations, for example the root nodule bacterium *Rhizobium* and some types of mycorrhizal fungi. In this context, it is interesting to note the theory that the chloroplasts and mitochondria of eukaryotic cells may have arisen from endosymbiotic microorganisms. A few pathogenic fungi also live inside host cells. The clubroot pathogen *Plasmodiophora* exists in the form of a naked cell or **plasmodium**, and the interface consists simply of the plasmodial cell membrane surrounded by a second membrane which is presumed to originate from the host. An even more intimate contact is found in the parasitic chytrids such as *Olpidium*, where the fungal cell is not surrounded by a host membrane and is therefore in direct contact with the host cytoplasm. A similar type of relationship has been found in cells infected by phytoplasmas.

The ultimate examples of intracellular pathogens are the viruses and viroids. Virus particles occur within the cytoplasm, plastids, and nuclei. Hybridization techniques used to locate and visualize specific nucleic acid sequences have recently shown that viroids accumulate in the nucleolus. Due to their unique properties, viruses and viroids are not comparable to cellular pathogens regarding the nature of the host–pathogen interface. Successful replication of viruses requires removal of the coat protein, so that the interface during multiplication is between a nucleic acid molecule and the synthetic machinery of the host cell.

Development Following Infection

Following entry, there are wide variations in the extent and pattern of colonization of host tissues (Figure 6.12). Further development is related both to the nature of the parasitic relationship between the two partners, and to the relative success of host resistance mechanisms in limiting pathogen invasion (see Chapter 9). Broadly speaking, two main patterns of colonization occur: **localized** or **systemic**. In localized infection, the pathogen multiplies or grows within a particular tissue or organ to give discrete lesions. In a systemic

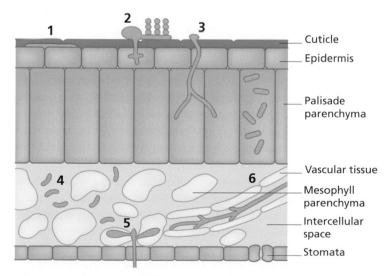

Figure 6.12 Some patterns of pathogen invasion of plant tissues, based on cross-section of leaf. 1. Subcuticular. 2. Epiphytic with haustoria. 3. Intracellular. 4. Intercellular. 5. Intercellular with haustoria. 6. Vascular.

infection, the pathogen spreads widely throughout the plant, and in extreme cases occurs in every part of the root and shoot system. Complete systemic colonization, in which literally every cell is infected, probably never occurs, as even viruses which spread efficiently from cell to cell are usually absent from meristems and from gametophyte tissue (see later in this chapter).

Within these two broad categories, there are numerous subtle variations in pathogen behavior, and in the ultimate extent of damage to the host. Many pathogens exhibit **tissue specificity**, in other words they grow preferentially in certain host tissues. Vascular pathogens, for instance, including both bacterial and fungal examples, grow within the xylem, while phytoplasmas are usually confined to the phloem. The reasons for such behavior are poorly understood, although the mode of nutrition of the pathogen is clearly important. The less specialized necrotrophic pathogens tend to spread indiscriminately through plant organs, while biotrophs, in keeping with their more benign form of parasitism, grow selectively within certain well-defined host tissues. A further special case is where a pathogen induces major changes in the organization and morphology of host tissues, for instance tumors, in which it subsequently lives; the classic example of this mode of colonization is provided by *Agrobacterium* (see Chapter 8, Figure 8.14).

One might assume that there is a correlation between the extent of host colonization by a pathogen and the eventual severity of disease, but this is by no means always the case. For instance, a localized pathogen may disrupt an essential physiological function, such as water transport, or produce a diffusible toxin which can act at a distance from the lesion itself. In an extreme example of this type of behavior, the pathogen causing choke disease of grasses, *Epichloe typhina*, is restricted to a short section of leaf sheath tissue but its growth in this strategic position prevents the emergence of the flowering axis and hence its effect on the life cycle of the host is dramatic. Conversely, it is quite common to encounter systemic virus or viroid infections in which the host is asymptomatic. Such cryptic infections pose a particular problem when attempting to eradicate a pathogen from a crop.

The pattern of colonization can be determined by the infection route. Downy mildews, such as *Peronospora* and *Plasmopara* species, typically infect leaf tissues, where growth is restricted by large veins, resulting in angular, localized lesions. Infection of the stem apex of young seedlings, however, leads to a more systemic mode of growth below the dividing meristem, causing severe stunting of the host. Some modern pea cultivars are more susceptible to this type of infection by *Peronospora pisi*, simply because they lack large stipules, leaf-like structures which normally enclose and protect the stem apex.

Only a few fungi, notably the smuts, are truly systemic. *Ustilago nuda*, causing loose smut of barley, occurs as a dormant mycelium in the cotyledon of infected grain. During germination of the seed, the pathogen also resumes activity and grows intercellularly within the young seedling. As the plant matures the pathogen keeps pace just behind the apical meristem, and eventually invades the developing flower head to form a mass of black teliospores which replace the grain. The older mycelium in the stem may break down as the host matures, but it often persists in the nodes. One interesting feature of smut diseases is that visible symptoms are not obviously manifest until the pathogen sporulates. Infected plants may, however, be stunted or in some cases, such as loose smut, slightly taller than normal.

Among those fungi which are specific to particular host tissues, the vascular wilt pathogens have a particularly interesting mode of spread within the host. Fungi such as *Fusarium oxysporum* and *Verticillium albo-atrum* enter in the apical zone of the root. In this region, the endodermis is not fully differentiated and the fungi are able to grow through it and reach the developing xylem. Further colonization of the living host is restricted to the vascular tissues. Hyphae grow through the vessels and tracheids and pass from cell to cell via pit pairs. In addition, long-distance movement is accomplished by the production of microconidia which are carried in the transpiration stream. This mode of spread is much more rapid than would be possible by mycelial growth; the Panama disease wilt pathogen, *Fusarium oxysporum* f.sp. *cubense*, can migrate from the bottom to the top of an 8 m tall banana plant in less than two weeks. Because xylem tissues ramify throughout the plant, wilt pathogens can migrate into every part of their host. Thus, although they are tissue specific, these pathogens can become virtually systemic.

Colonization by Bacteria

The morphology of bacterial cells limits their capacity for invasive growth through compact tissues. Instead, they colonize intercellular spaces or spread through natural channels, such as vascular elements. However, necrotrophic species can spread within the host by macerating and destroying tissues. For example, *Pectobacterium carotovorum* and the related *P. atrosepticum* cause common storage soft rots of potato tubers (see Chapter 8, Figure 8.3). The pathogen is unable to pass through intact periderm and therefore gains entry via lenticels or wounds, including those caused by other agents. Once inside the tuber, the bacteria spread rapidly through the parenchyma, giving rise to a soft, slimy lesion, and under favorable conditions can quickly destroy the tuber. At all stages of colonization, the bacteria occupy intercellular spaces, and spread is facilitated by the production of pectolytic enzymes which degrade middle lamellae and hence separate host cells.

Under some conditions, the bacteria can also move from infected tubers into stem tissues, causing a spreading necrotic lesion known as blackleg disease. The fire blight bacterium *E. amylovora* invades flowers, leaves, and stem tissues of rosaceous hosts such as apples and pears, spreading through vascular elements causing water-soaked lesions and a necrotic dieback. In contrast to *Pectobacterium*, this bacterium produces only low amounts of cell wall-degrading enzymes, and pathogenicity appears to be related instead to production of exopolysaccharides and some proteins delivered via the Type III secretion system (p.257).

Other bacterial pathogens grow biotrophically within plant tissues without separating or killing host cells. *P. syringae* pathovars proliferate in intercellular spaces causing a water-soaked lesion in which host cells initially remain alive. Evidence suggests that the bacteria are able to create an environment more favorable for growth by altering water relations within the lesion and stimulating efflux of sugars from host cells. This manipulation of plant metabolism is mediated by effector molecules produced by the pathogen (see p.255). In some hosts, spread over longer distances may then occur by invasion of xylem elements (Figure 6.13), with subsequent escape to form colonies in other parts of the leaf or stem tissues.

(a)　　　　　　　　　　　　　　　　(b)

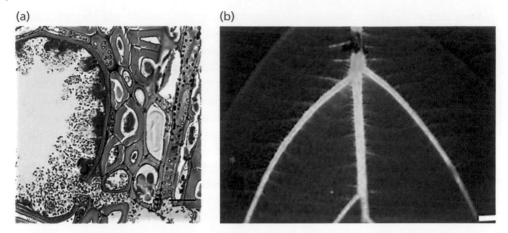

Figure 6.13 Invasion of vascular tissues of kiwifruit by *Pseudomonas syringae* pv. *actinidiae*. (a) Cross-section of vascular tissues showing large numbers of bacterial cells (Ba) within xylem elements. Bar = 20 μm. (b) Fluorescence image of leaf showing bacteria labeled with green fluorescent protein spreading through veins 14 days after inoculation. Chlorophyll in leaf lamina fluoresces red. Bar = 2 mm. *Source:* Gao et al. (2016).

Bacteria can also cause serious vascular wilt diseases, in which the symptoms parallel those caused by fungal pathogens. Their unicellular morphology then makes them ideally suited for transport within xylem vessels. Vascular wilt bacteria are found in several genera, including *Ralstonia*, *Erwinia*, *Xanthomonas*, *Clavibacter*, and *Pantoea*, causing a variety of important diseases in both tropical and temperate regions. *Ralstonia solanacearum* is one of the most destructive bacterial plant pathogens known, due to its worldwide distribution and very wide host range which includes crops such as potato, tomato, tobacco, and banana (Moko disease). The bacterium is soil-borne and infects via wounds, root tips, or sites of lateral root emergence, and then invades xylem vessels where it multiplies to high cell densities, leading to wilting and eventual death of the plant. Bacterial wilt of cucurbits, *Erwinia tracheiphila*, was one of the first bacterial diseases of plants described when, in the late nineteenth century, fields of cucumbers and squashes in the USA were devastated by a wilt disease. In this case, disease transmission is aided by beetles that feed on leaves and nectaries, creating wounds via which the bacteria gain entry to xylem vessels, where they multiply and produce extracellular polysaccharides that block the flow of water.

Several important and emerging diseases, such as Pierce's disease of grapevine, citrus variegated chlorosis, and olive quick decline syndrome, that is now invading southern Europe, are caused by the xylem-limited bacterium *Xylella fastidiosa*. The pathogen is introduced into xylem tissues by insect vectors (sharp-shooter leafhoppers) and then multiplies, forming microfilms and dense colonies that block water transport. This pathogen is adapted to survive and spread in the nutrient-poor environment of xylem vessels.

Other tissue-specific bacterial pathogens invade the phloem. Huanglongbing (HLB, citrus greening), now a worldwide threat to citrus crops, is caused by strains of the nonculturable bacterium *Candidatus Liberibacter*. The pathogen is spread by, and introduced into, host phloem cells by psyllid insect vectors. The bacteria also colonize and multiply in the hemolymph of the vector, and there is speculation that this pathosystem might be of recent origin, with an insect endosymbiont invading the nutrient-transporting cells of a plant host.

A unique form of host colonization occurs with *Agrobacterium* species that initially colonize wounds but then transform host cells to differentiate into tumor tissues (*A. tumefaciens*) or abnormal hairy roots (*Agrobacterium rhizogenes*). The bacteria multiply within these tissues, utilizing nutrients produced by the transformed cells. This interaction is described in more detail on p.205.

Colonization by Viruses

The spread of viruses within their hosts is distinct in that they can only multiply within cells and are small enough to behave as subcellular particles. Plant cells are connected by plasmodesmata, small channels which provide a pathway for communication and trafficking of macromolecules such as proteins and nucleic acids between cells. Plant viruses have evolved to exploit these channels as gateways to transmit infection from cell to cell.

Early studies using electron microscopy observed virus particles within plasmodesmata. The transport of viruses within cells and through plasmodesmata is mediated by "movement proteins" encoded by the virus itself, for instance the P30 protein of tobacco mosaic virus (TMV) (see p.67). When labeled with fluorescent markers, movement proteins localize to the endoplasmic reticulum (ER) and accumulate at plasmodesmata. It is now known that the endomembrane system is continuous between plant cells via these channels. Virus movement proteins therefore appear to play several roles, associating with virus replication sites on membranes, aiding movement of the virus within the cell, and modifying the exclusion size of plasmodesmata to allow passage of the virus genome or nucleoprotein complexes containing the virus to adjacent cells.

As plasmodesmata provide a network of channels connecting cells, it would be possible for infection to eventually spread throughout the plant via this route. However, this would be a very slow process as trafficking from one cell to the next takes several hours. Much faster long-distance spread takes place via the vascular system in phloem tissues where the rate of movement has been estimated to be as high as several centimeters per hour. How exactly the virus is loaded into, or unloaded from, phloem sieve cells is not yet clearly understood. The virus may be transported as intact virions or nucleic acid complexes. Phloem transport plays an important part in the development of systemic virus infections. Literally every cell in the plant may become infected, although the small numbers of infected seeds and pollen grains in most virus diseases suggests that movement into gametophyte tissue of the developing embryo is restricted. Often, meristematic tissues are also virus free; this fact has been put to good use in the production of virus-free plants by meristem culture.

Further Reading

Books

Bailey, M.J., Lilley, A.K., Timms-Wilson, T.M. et al. (eds.) (2006). *Microbial Ecology of Aerial Plant Surfaces. Wallingford.* CABI.

Lindow, S.E., Hecht-Poinar, E.I., and Elliott, V.J. (eds.) (2002). *Phyllosphere Microbiology.* St Paul, Minnesota: APS Press.

Reviews and Papers

Bendix, C. and Lewis, J.D. (2018). The enemy within: phloem-limited pathogens. *Molecular Plant Pathology* 19 (1): 238–254.

Chen, L.Q., Hou, B.H., Lalonde, S. et al. (2010). Sugar transporters for intercellular exchange and nutrition of pathogens. *Nature* 468: 527–532.

Danhorn, T. and Fuqua, C. (2007). Biofilm formation by plant-associated bacteria. *Annual Review of Microbiology* 61: 401–422.

Dodds, P.N., Lawrence, G.J., Catanzariti, A.M. et al. (2004). The *Melampsora lini AvrL567* avirulence genes are expressed in haustoria and their products are recognized inside plant cells. *Plant Cell* 16: 755–768.

Folimonova, S.Y. and Tilsner, J. (2018). Hitchhikers, highway tolls and roadworks: the interactions of plant viruses with the phloem. *Current Opinion in Plant Biology* 43: 82–88.

Hamer, J.E., Howard, R.J., Chumley, F.G. et al. (1988). A mechanism for surface attachment in spores of a plant pathogenic fungus. *Science* 239: 288–290.

Martin-Urdiroz, M., Oses-Ruiz, M., Ryder, L.S. et al. (2016). Investigating the biology of plant infection by the rice blast fungus *Magnaporthe oryzae*. *Fungal Genetics and Biology* 90: 61–68.

Melotto, M., Underwood, W., Koczan, J. et al. (2006). Plant stomata function in innate immunity against bacterial invasion. *Cell* 126 (5): 969–980.

Read, N.D., Kellock, L.J., Collins, T.J. et al. (1997). Role of topography sensing for infection-structure differentiation in cereal rust fungi. *Planta* 202 (2): 163–170.

Serrano, M., Coluccia, F., Torres, M. et al. (2014). The cuticle and plant defense to pathogens. *Frontiers in Plant Science* 5 (274) https://doi.org/10.3389/fpls.2014.00274.

Voegele, R.T., Struck, C., Hahn, M. et al. (2001). The role of haustoria in sugar supply during infection of broad bean by the rust fungus *Uromyces fabae*. *Proceedings of the National Academy of Sciences USA* 98: 8133–8138.

Xin, X.F., Nomura, K., Aung, K. et al. (2016). Bacteria establish an aqueous living space in plants crucial for virulence. *Nature* 539: 524–529.

Yi, M. and Valent, B. (2013). Communication between filamentous pathogens and plants at the biotrophic interface. *Annual Review of Phytopathology* 51: 587–611.

7

The Physiology of Plant Disease

It is of the first importance to understand that disease is a condition of abnormal physiology, and that boundary lines between health and ill health are vague and difficult to define.

(H. Marshall Ward, 1854–1906)

The invasion of the host by a foreign organism leads, sooner or later, to changes in host physiology. If the organism is a pathogen, these changes will eventually prove deleterious to the host. Alternatively, where the pathogen fails to establish itself, these changes may be important in preventing the pathogen from gaining a foothold. In practice, it is often difficult to distinguish between postinfectional changes which are linked with plant defense and those which are related to pathogenesis, that is, disease development. Plants infected by quite different types of pathogens often exhibit very similar physiological symptoms. These similarities at first sight suggest that plants employ a common pathway of response to infection. However, many of these physiological effects also result if plants are subject to other forms of stress, such as mechanical or chemical injury. Hence, some of the gross changes in the physiology of diseased plants represent a nonspecific response to cellular damage inflicted by physical, chemical or microbial agents. An analysis of postinfectional changes in host physiology may, nevertheless, clarify the ways in which pathogens cause disease, and help to elucidate the mechanisms by which plants resist pathogenic attack.

Postinfectional Changes in Host Physiology

Respiration

As parasitism involves a nutritional relationship, much of the work on the physiology of diseased plants has been concerned with energy metabolism. One of the most prominent changes which occurs following infection is an increase in respiration rate. This is equally true for diseases involving fungi, bacteria, and viruses, although most of the available information has been obtained from plants infected by biotrophic fungi. Figure 7.1 shows the rate of oxygen uptake in barley leaves infected by two different fungi: *Pyrenophora teres*, a nectrotroph, and *Blumeria graminis*, a biotroph. In both instances, oxygen consumption is

Plant Pathology and Plant Pathogens, Fourth Edition. John A. Lucas.
© 2020 John Wiley & Sons Ltd. Published 2020 by John Wiley & Sons Ltd.
Companion website: www.wiley.com/go/Lucas_PlantPathology4

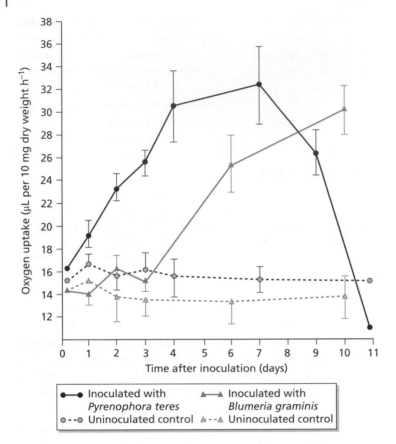

Figure 7.1 Time course of respiration in two susceptible barley cultivars, Wing and Sultan, inoculated respectively with the net blotch pathogen *Pyrenophora teres* (a necrotroph) and the powdery mildew pathogen *Blumeria (Erysiphe) graminis* f.sp. *hordei* (a biotroph). Vertical bars represent standard errors. *Source:* After Smedegaard-Petersen (1984).

increased but with the nectrotroph, the respiration rate peaks earlier and then declines as the host tissues are progressively destroyed. With the biotroph, the rate increases steadily as the pathogen sporulates and remains high until the host tissues senesce.

This pattern inevitably raises the question as to whether the increase is simply due to the additional respiration of the pathogen itself, rather than to a genuine host response. Several lines of evidence suggest that while the pathogen makes some contribution to the increase, the greater part of it cannot be explained on this basis.

Powdery mildew fungi are only in intimate contact with their host where the haustoria enter epidermal cells. It is possible to peel off the epiphytic mycelium of the pathogen and measure the respiration rate of the host leaf with only a small portion of the fungus (viz. the haustoria) remaining. Experiments like this have shown that the increase in respiration is maintained even after removal of the pathogen. There are alternative ways of approaching this problem; for instance, one can measure the respiratory rate of uninfected tissues adjacent to lesions or, in the case of nectrotrophic pathogens, one can examine the effects of toxic factors produced in culture on the respiratory metabolism of host cells. Using both approaches, enhanced rates of host respiration have been found.

Perhaps the most convincing argument in support of the idea that increased postinfectional respiration is due to a stimulation of host metabolism comes, however, from studies on virus diseases. Viruses, being noncellular, possess no respiratory apparatus of their own, and yet a similar stimulation of respiration rate is found in a variety of viral infections. For example, the development of necrotic local lesions in *Nicotiana glutinosa* inoculated with tobacco mosaic virus (TMV) is accompanied by a pronounced increase in respiration. In systemically infected hosts, an increase may accompany symptom development but the change is less marked.

All in all, the rise in respiration rates following infection would seem mainly to represent a response by host tissues. This response bears similarities to the transitory increase in respiration observed in plants subjected to mechanical injury.

Mechanism of Respiratory Increase

There are several theories which seek to explain the enhanced respiration rate in diseased plants (Figure 7.2).

In healthy cells, respiration is regulated by a number of factors, the most important of which is the availability of adenosine diphosphate (ADP). Agents such as 2,4-dinitrophenol (DNP) stimulate the respiration rate by "uncoupling" electron transfer from oxidative phosphorylation. In essence, this means that while electron flow continues, the regeneration of adenosine triphosphate (ATP) from ADP and inorganic phosphate no longer takes place. Due to the continued consumption of ATP in cellular metabolism, the pool of ADP is replenished and the usual feedback mechanism based on the availability of ADP no longer operates. It has been suggested that pathogens may uncouple host respiration (**1** in Figure 7.2). This hypothesis is based on evidence that respiration in diseased tissues is no longer stimulated by treatment with DNP. In addition, the level of ATP is often lower in infected tissues.

The major metabolic pathway for the degradation of glucose to pyruvate is glycolysis, otherwise known as the Embden–Meyerhof pathway. An alternative route for the production of pyruvate, the pentose phosphate pathway, is generally considered to be of less importance although it does provide intermediates for the biosynthesis of many vital cellular materials, including nucleic acids. Assessment of the relative contribution of each pathway relies largely on data from labeling studies in which radioactive carbon is incorporated into either the C_6 or the C_1 position of the glucose molecule. Subsequent measurement of the ratio of labeled carbon dioxide released from each source during respiration indicates which pathway is predominant; activation of the pentose phosphate pathway leads to an increased contribution from the C_1 position, and hence lowers the C_6/C_1 ratio. In infected plants, the C_6/C_1 ratio is typically lower than the values obtained from healthy tissues, suggesting that there is increased participation of the pentose route (**2** in Figure 7.2). Assays of several pentose phosphate pathway enzymes support this idea, their activity being higher in diseased tissues. Alterations in the pathway of glucose degradation would not necessarily result in a rise in gross respiration rate, although the pentose phosphate pathway is marginally less efficient in generating ATP. It now seems likely that the real

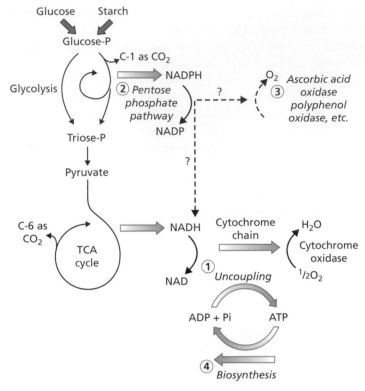

Figure 7.2 Theories concerning stimulation of respiration in infected plants. ADP, adenosine diphosphate; ATP, adenosine triphosphate; NAD, nicotinamide adenine dinucleotide; NADH, reduced NAD; NADP, nicotinamide adenine dinucleotide diphosphate; NADPH, reduced NADP; Pi, inorganic phosphate; TCA, tricarboxylic acid. See text for further details of 1–4.

significance of the switch to the pentose pathway is linked to its role in the biosynthesis of various compounds.

As well as providing pentoses for the biosynthesis of nucleic acids, pentose phosphate intermediates are involved in the production of numerous aromatic compounds, notably phenols and their derivatives. Many of these are antimicrobial and associated with host defense, a topic further discussed in Chapter 9.

In addition to changes in respiratory pathways, alternative terminal oxidation systems may operate in diseased tissues (**3** in Figure 7.2). Apart from the usual cytochrome system terminating with cytochrome oxidase, systems involving phenol oxidases and ascorbic acid oxidase have been detected in plants. Both of these enzymes appear to be activated in diseased tissues. The precise mechanisms and significance of such oxidation systems are not clear, but phenol oxidases play a part in the production of phenolic compounds, an observation consistent with the general pattern of postinfectional metabolism. It also seems likely that the reduced coenzyme nicotinamide adenosine dinucleotide phosphate (NADPH) generated by the pentose pathway is oxidized by one of these enzymes, rather than by the cytochrome system. As far as is known, these alternative oxidases are unable to participate in the formation of ATP during oxygen uptake.

While evidence exists in support of each of the above theories regarding respiratory changes in the infected host, the most satisfactory explanation for the increased rate of metabolism in infected plants is perhaps the most obvious. With biotrophic pathogens, the increased respiration is associated with enhanced biosynthetic, rather than degradative, metabolism (**4** in Figure 7.2). This increase in turn requires more rapid utilization of ATP and thereby removes the restraints imposed by the availability of ADP. It is significant that in many diseases caused by fungi, the major increase in respiration coincides with the onset of sporulation by the pathogen. This is precisely the time when the fungus will be exerting the maximum drain on host nutrients due to the considerable energy requirement for the production of spores or other propagules. The switch to active biosynthesis of defense compounds by the host plant will also consume energy in the form of ATP.

Photosynthesis

Photosynthesis is the most distinctive physiological activity of green plants. The capture of solar energy by chlorophyll and its subsequent utilization to fix carbon dioxide into organic compounds is the key process underlying the productivity of plants, and is the ultimate source of food for all consumers in the ecosystem. Several parameters determine the efficiency of photosynthesis, including the amount of radiation initially captured by green tissue and its rate of conversion into sugars. Together, these are often expressed as **radiation use efficiency** (RUE) that relates the amount of dry matter produced (biomass) to the photosynthetically active radiation that is intercepted. Plant disease can have both direct and indirect effects on RUE. Pathogens that attack aerial plant tissues causing necrotic lesions directly reduce the area available for the interception of solar radiation, while others have more subtle effects, for instance by reducing the amount of chlorophyll or the rate of carbon fixation. From a practical viewpoint, one vital factor is the area of photosynthetically active crop canopy (described as green leaf area [GLA]), as well as its longevity at critical points in the season during which the production and assimilation of carbohydrates determine final yield. Any reduction in GLA, either by destruction of tissues or accelerated senescence due to disease, will impact on the final productivity of the crop.

In many cases, the harmful effects of a pathogen can be directly attributed to the loss of photosynthetic tissues. A serious outbreak of potato blight can quickly defoliate an entire field, while *Botrytis fabae* causes necrotic patches which can occupy over 50% of the leaf area of broad beans (see Chapter 2, Figure 2.2). In the latter case, initial measurements of the relative growth rate of diseased plants showed that it is similar to the growth rate of healthy plants, even when 30–40% of the leaf area is removed, suggesting that the remaining leaf tissues might compensate for the loss in area. However, there is eventually an impact on yield, due to a reduction in the number of pods formed per plant.

Chlorosis is one of the most common symptoms of plant disease, and is indicative of a reduction in the chlorophyll content of green tissues. A reduced chlorophyll content could be due to the breakdown of chlorophyll, inhibition of chlorophyll synthesis, or a reduction in the number of chloroplasts. In chlorosis associated with some virus infections, higher levels of the enzyme chlorophyllase have been detected, suggesting that chlorophyll is being degraded by the enzymic reaction:

$$\text{Chlorophyll} \xrightarrow{\text{Chlorophyllase}} \text{Chlorophyllide} + \text{Phytol}$$

Symptoms of virus infection in leaves often include characteristic mosaics of green and yellow areas in which chloroplasts are reduced in number, or show ultrastructural abnormalities such as swelling and fewer lamellae, which are the sites of photochemical reactions. The chlorotic areas are often rich in starch, suggesting that virus infection affects both photosynthetic capacity and carbon partitioning within diseased leaves.

In leaves infected by biotrophic fungi, there is a progressive loss in overall photosynthetic activity. Figure 7.3 shows the net photosynthetic rate of oak leaves infected by the powdery mildew pathogen *Microsphaera alphitoides*. There is a slight initial stimulation of photosynthesis in inoculated leaves but this is followed by a gradual decline. The rate of $^{14}CO_2$ uptake by leaf discs from sugar beet infected by powdery mildew is also reduced compared with healthy tissues (Figure 7.4). Chloroplasts isolated from mildewed beet leaves show a reduced capacity to form ATP by noncyclic photophosphorylation. Other work on powdery mildew-infected barley leaves has shown that there is also a progressive reduction in the activity of several key enzymes of the Calvin cycle. This downregulation may be linked to changes in the concentration of soluble carbohydrates in infected tissues (see later in this chapter), which in turn affects the rate of photosynthetic CO_2 fixation. In *Arabidopsis* leaves infected by the biotrophic pathogen *Albugo candida*, reductions in the rate of photosynthesis are paralleled by decreases in the amounts of ribulose biphosphate carboxylase (RuBisCo), the major enzyme involved in carbon assimilation.

These examples show that effects of pathogens on photosynthesis are related not only to loss of green tissue but also to changes in the performance of the photosynthetic apparatus itself. Chlorophyll fluorescence imaging is a noninvasive technique that can measure photosynthetic electron transport and related parameters at the cellular level, for

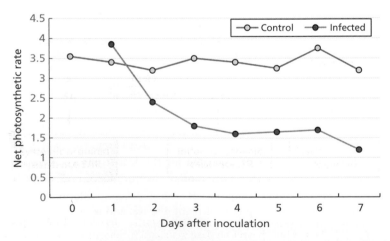

Figure 7.3 Changes in photosynthesis of oak leaves following infection by the powdery mildew fungus *Microsphaera alphitoides*. *Source:* Data from Hewitt and Ayres (1975).

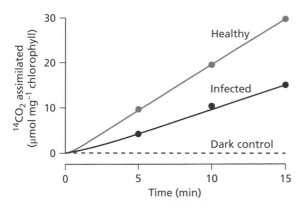

Figure 7.4 Effect of *Erysiphe polygoni* on the rate of photosynthetic $^{14}CO_2$ assimilation by sugar beet leaf discs. *Source:* Data from Magyarosy et al. (1976).

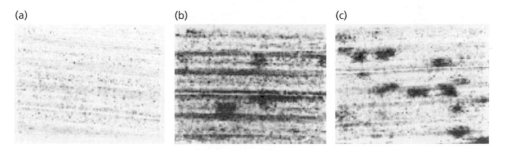

Figure 7.5 Chlorophyll fluorescence images of barley leaf tissues infected with the powdery mildew fungus *Blumeria graminis*, three days after inoculation, showing effects on the rate of photosynthetic electron transport (ETR). (a) Uninfected control leaf. (b) Reduction of ETR associated with colony development on susceptible leaf. (c) Incompatible reaction on *mlo* barley genotype showing reduction in ETR associated with resistance response to attempted infection. *Source:* Scholes and Rolfe (2009).

instance in and around disease lesions or in resistant tissues mounting a defense response. Such studies have confirmed the rapid loss of photosynthetic capacity during disease development with necrotrophic pathogens, or the transition from asymptomatic to symptomatic infection with hemibiotrophic fungi such as *Zymoseptoria* and *Pyricularia* on cereal hosts. With biotrophs, such as the powdery mildew fungus *B. graminis*, reductions in photosynthesis can be detected not only in cells directly beneath fungal colonies but also in adjacent uncolonized cells, and resistant leaf cells responding to attempted penetration by the pathogen (Figure 7.5).

An apparent exception to the predominantly negative impacts of disease on photosynthesis is the so-called "green island" effect observed on leaves infected by various parasitic agents, including biotrophic fungi and leaf-mining insects. Tissues surrounding colonized areas remain green, while the intervening parts of the leaf become chlorotic and eventually senesce (Figure 7.6). There has been much discussion of the significance of these green islands because the selective retention of chlorophyll around infection sites implies that the pathogen exerts some degree of control over host physiology. The similarities between

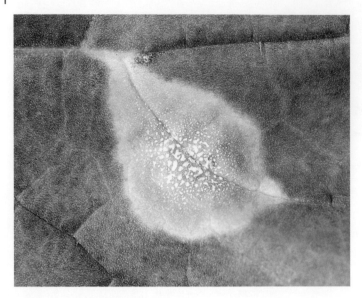

Figure 7.6 Green island surrounding a lesion caused by the light leaf spot pathogen *Pyrenopeziza brassicae* on a leaf of oilseed rape (*Brassica napus*). The infected area in which the fungus is sporulating (*white spots*) is associated with green tissues in which chlorophyll is retained while the rest of the leaf is senescent.

this and the effects of cytokinins in delaying senescence have suggested that these hormones might play a role in green island formation. It is now believed that green islands are mainly a consequence of the overall redirection of host nutrients associated with colonization by parasites (see later). They are not exclusive to biotrophs, and can occur in some circumstances with more destructive pathogens. The hemibiotrophic fungus *Colletotrichum graminicola* infects maize leaves, causing chlorotic spots that transition to necrotic lesions. On aging leaves, however, characteristic green islands are formed in which tissues remain intact for an extended period. Chlorophyll fluorescence imaging showed that photosynthesis is transiently maintained at these sites, while the expression of some plant genes associated with senescence is reduced.

Translocation and Assimilation of Nutrients

The damage caused by biotrophic pathogens is due, to some extent at least, to their ability to redirect host nutrients for their own use. The idea that fungal colonies act as "metabolic sinks" in their hosts was initially supported by radioisotope tracer experiments in which labeled carbon accumulated preferentially in disease lesions (Figure 7.7). In this way, photosynthate originally destined for developing host tissues, such as new shoots or roots, is instead utilized by the pathogen. The reduced root growth and grain yield of cereals infected by rusts and powdery mildews are due to this disturbance in the nutrient balance of the plant.

As shown in Figure 7.7, a large part of the imbalance seems to be caused by the retention of sugars and amino acids in the diseased older leaves. This has been confirmed by feeding

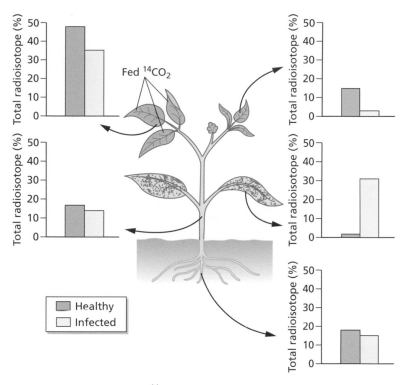

Figure 7.7 Translocation of ^{14}C in healthy bean plants and plants infected by the rust *Uromyces phaseoli* after feeding $^{14}CO_2$ to uninfected trifoliate leaves. Stippling indicates rust pustules on unifoliate leaves of infected plants and histograms show percentage of total radioisotope in each organ in healthy versus infected plants. *Source:* Data from Livne and Daly (1966).

a "pulse" of $^{14}CO_2$ to rust-infected leaves and then comparing the amount of carbon remaining with that retained by healthy leaves (Figure 7.8). Efflux of fixed carbon from the rusted leaf is substantially reduced. Tracer experiments have also shown that carbon originally present in host sugars, such as sucrose, can be subsequently detected in typical fungal metabolites like the sugar alcohols, mannitol and sorbitol. The conversion of host sugars to fungal metabolites could in fact serve to maintain a concentration gradient and ensure a continued flow from host to fungus.

The enzyme invertase plays a central role in carbohydrate metabolism and assimilation through changes in the relative levels of hexose sugars, such as glucose, and the principal translocated sugar, sucrose. Plant tissues infected by biotrophic fungi show substantial increases in invertase activity (Figure 7.9). Increased hydrolysis of sucrose may have effects on both photosynthesis and carbon partitioning in diseased leaves. The accumulation of glucose and fructose is one possible explanation for the downregulation of the photosynthetic Calvin cycle noted earlier, due to end-product inhibition. It will also affect the range of soluble sugars available to the invading pathogen.

The movement of sugars and amino acids in and out of plant cells is mediated by carrier proteins that transport them across cell membranes. It is now clear that pathogens have evolved mechanisms to hijack such nutrient efflux or uptake. As described in the

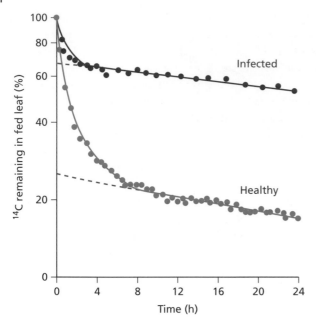

Figure 7.8 Kinetics of efflux of ^{14}C from healthy and rusted first leaves of barley, following a pulse of labeled CO_2. *Source:* After Owera et al. (1983).

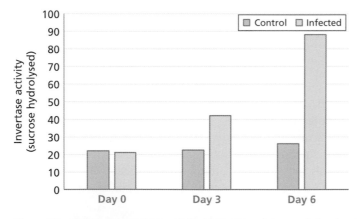

Figure 7.9 Cell wall invertase activity in *Arabidopsis* leaves following inoculation with or without (control) the powdery mildew fungus *Erysiphe cichoracearum*. *Source:* Fotopoulos et al. (2003).

previous chapter, many biotrophic fungi and oomycetes produce haustoria that enter plant cells, disrupt host membrane function, and import hexose sugars and amino acids via their own transporter proteins. The maize smut fungus, *Ustilago maydis*, which does not produce haustoria, has recently been shown to possess a high-affinity sucrose transporter that enables it to compete effectively with plant cells for this major transport sugar. It also turns out that some plant-pathogenic bacteria target sugar export by host cells as a means to acquire nutrients. Recently, a class of sugar transporters, known as SWEETs (Sugars Will Eventually be Exported Transporters), were identified that facilitate efflux from parenchyma cells. This is a crucial process during loading of sucrose into phloem tissues for long-distance translocation. Some *Xanthomonas* species, including *X. oryzae* and *X. axonopodis*, causal agents of destructive bacterial blight disease of rice and cassava respectively, produce effector proteins that increase expression of host SWEET genes, thereby promoting sugar export into the apoplast. Bacterial strains with mutated versions of these effectors display reduced virulence, suggesting that intervention in sugar transport is a key process in pathogenicity.

The multiplication of viruses in plant cells is entirely at the expense of the host. Sequestration of host metabolites to make more virus particles may explain some of the deleterious effects of virus infection. In severe infections, the virus particles themselves come to represent about 10% of the dry weight of leaf tissues, and multiplication on this scale must severely tax the synthetic capacities of the host. Virus replication requires the synthesis of two components, nucleic acid and coat protein, and one major side effect of infection may therefore be a deficiency of inorganic nutrients, in particular phosphorus and nitrogen, available to the host. There is evidence in virus-infected plants for a shift in photosynthetic products from sugars to amino acids, perhaps reflecting the biosynthetic demand for coat protein.

Global Changes in the Metabolism of Infected Plants

The development of technologies to analyze gene expression profiles (**transcriptomics**), the protein complements of cells and tissues (**proteomics**), and simultaneous changes in different classes of metabolites (**metabolomics**) means that it is now possible, in theory at least, to analyze the effects of disease on host plant metabolism in great detail at different stages during infection. The large amounts of data generated by these approaches require careful interpretation, but such studies have confirmed that infection leads to profound changes in patterns of plant gene expression and a corresponding reprogramming of metabolism. Analysis of groups of genes, proteins, and metabolites affected indicates alterations in primary metabolic pathways, source–sink relations, downregulation of photosynthetic genes, and increased expression of some genes involved in the synthesis of secondary metabolites, most likely related to plant defense. Changes also occur in genes involved in hormone signaling pathways known to be induced in response to pathogen attack. This is further discussed in Chapter 9.

Translocation of Water

The nutrient stress imposed by a redirection of host nutrients to satisfy the energy and biosynthetic needs of the pathogen is very different from that caused by pathogens which actually colonize the transporting tissues of the plant. Vascular wilt pathogens impair the flow of water and mineral salts through the xylem. Translocation of sugars through phloem tissues is also disrupted by some pathogens; a number of virus infections cause necrosis of phloem elements, resulting in nutrient imbalances in the host. For instance, potato plants infected by leafroll virus have higher than normal carbohydrate levels in their leaves, while that of the tubers is reduced. Two possible explanations for this symptom are the inhibition of translocation of sugars or the breakdown of the phloem tissue. High concentrations of phytoplasmas often build up in the phloem elements of diseased plants but in this case the symptoms, such as wilting and stunting, are thought to be due to production of a diffusible toxin rather than to occlusion of the sieve tubes.

The interdependence of physiological processes such as ion uptake and the translocation of water and nutrients should be emphasized. In take-all disease of cereals, invasion of the root cortex by the pathogen does not significantly affect ion uptake or translocation. Instead, the crucial stage appears to be the subsequent colonization of phloem tissues by the fungus. This reduces the translocation of nutrients to the apical meristems, with the result that the distal portions of the root cease to function and ion uptake is impaired.

Wilt Syndrome

The most pronounced effect of vascular wilt pathogens is on the water economy of the host. In tomatoes infected by *Fusarium oxysporum* f.sp. *lycopersici*, the resistance to water flow through the xylem is substantially increased compared to the resistance of uninfected stems. This effect can be partially explained on the basis of physical obstruction of the vessels by hyphae, but the vascular wilt syndrome is complex and involves host responses to infection as well as direct effects of the pathogen and its products (Figure 7.10). Blockages caused by the growth of the pathogen are compounded by its secretion of polysaccharides and pectolytic enzymes; in turn, the host responds by producing gums and mucilages, and by forming tyloses, bubble-like outgrowths from xylem

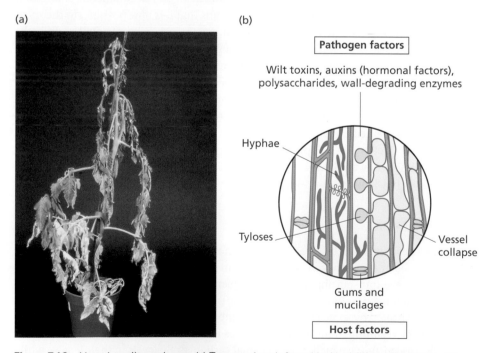

(a)

(b)

Figure 7.10 Vascular wilt syndrome. (a) Tomato plant infected by *Verticillium albo-atrum*, with wilting, downward angle of petioles (epinasty) and chlorosis followed by necrosis of leaves. *Source:* Photo by Brian Case. (b) Diagrammatic longitudinal section of vascular tissue from an infected stem.

parenchyma cells, in the vessels. The end-result is that water flow may be reduced to less than 5% of that in healthy plants. A further factor interfering with xylem function may be gas bubbles, or embolisms, breaking the water column; this has been observed in sap-wood colonized by the Dutch elm pathogen, *Ophiostoma novo-ulmi*. As well as causing severe water stress, infection by wilt pathogens also reduces the passage of essential mineral ions to the leaves.

Many plant-pathogenic bacteria can enter the vascular system through wounds, and spread and multiply in xylem vessels. Proliferation of bacterial cells may occlude vessels, while secretion of extracellular polysaccharide or other high molecular weight materials further reduces flow by plugging pit membranes.

Transpiration

Wilting is one of the most common disease symptoms in plants, but the physiological basis of the symptom is not the same in all cases. The water economy can be disrupted through reduced water uptake, reduced flow rate, or increased water loss through transpiration. Root-rot pathogens destroy root tissues and therefore reduce the surface area available for uptake, and disrupt the transport of water through the root system. The vascular wilt fungi reduce flow and at the same time reduce the transpiration rate. This second effect can be explained on the basis of water stress in the leaves, coupled with stomatal closure. Many other pathogens increase the transpiration rate (Figure 7.11). This effect is predictable inasmuch as any pathogen which damages the surface layers of the leaf will increase cuticular transpiration and this may, in fact, be the major source of the increased rate of water loss. Damage of this sort is often restricted to the reproductive phase of the pathogen's life cycle; the rust fungi form pustules which tear through the host epidermis prior to release of the spores. At this point, wilting often occurs for the first time. The physical damage inflicted on the host during sporulation is a good example of the indirect and harmful effects that biotrophic fungi have on the plant.

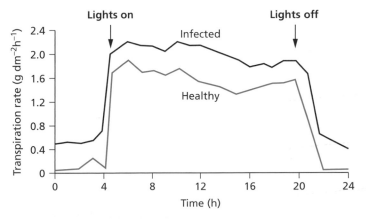

Figure 7.11 Transpiration rates over 24 hours of healthy barley plants and barley plants infected by *Rhynchosporium commune*. *Source:* Data from Ayres and Jones (1975).

It is often difficult, however, to assess the reasons for increased transpiration in diseased plants. The data shown in Figure 7.11 are from a relatively early stage of barley infection by *Rhynchosporium* when the host cuticle is still intact. A complicating factor is the stomatal behavior of diseased leaves. In this example, a higher proportion of stomata remain open in the dark in infected plants, which no doubt contributes to the increased level of water loss. Abnormal opening of stomata also occurs in potato leaves infected by *Phytophthora infestans*. Here, the effect is confined to a zone surrounding the necrotic lesions. Stomata within this zone open abnormally wide and remain open in the dark; they therefore do not present an obstacle to the developing fungal sporangiophores, which characteristically emerge through the stomata at night.

The few studies which have been made on the water relations of virus-infected plants indicate that the transpiration rate is reduced and the total water content of the host is lower. This is especially true in severely diseased plants.

Cell Water Relations

In addition to effects on the water economy of the whole plant, pathogens may also influence the water relations of individual cells. It is generally accepted that cell water relations are controlled by the plasma membrane, although in plant cells the rigid wall is also important in maintaining turgor. Due to its semipermeable properties, the plasma membrane is of central importance in regulating the passage of ions and organic molecules into and out of the cell. This membrane is a complex and dynamic structure which maintains a suitable intracellular environment for metabolism.

One of the most common effects of pathogens on plant cells is to increase their permeability. The membrane apparently loses its semipermeable properties and mineral ions and other electrolytes leak out into the external medium. This effect is pronounced in soft-rot diseases caused by necrotrophic pathogens, in which host cell necrosis is a major feature. Changes in cell permeability also take place during the transition from the asymptomatic to symptomatic phase of infection with hemibiotrophs. Figure 7.12 shows the progressive development of symptoms on wheat leaves infected by the leaf blotch pathogen *Zymoseptoria tritici*, along with data on electrolyte leakage from leaf tissues. At the point where the first chlorotic symptoms appear on leaves, there is an efflux of electrolytes into the apoplast. Following this change in cell permeability, the growth rate of the fungus begins to accelerate. Further analysis of this switch based on the expression of fungal genes indicates that the increased leakage of electrolytes into intercellular spaces alters the growth environment for the fungus from nutrient poor to nutrient rich. The greater availability of soluble nutrients fuels the more rapid growth and eventual reproduction of the pathogen within host tissues.

It has been shown that biotrophic pathogens also increase the permeability of host tissues. Cells penetrated by haustoria can still be plasmolyzed, so the integrity of the host membrane is maintained. Nevertheless, there is an increase in leakage of electrolytes from diseased tissues, related to the uptake of nutrients by the pathogen. The extrahaustorial membrane is modified and loses control of nutrient transport, so that sugars pass more freely to the fungus (see p.159).

The importance of endomembrane systems in the compartmentalization and regulation of cell metabolism is now clearly understood. In addition to the plasma membrane itself, the

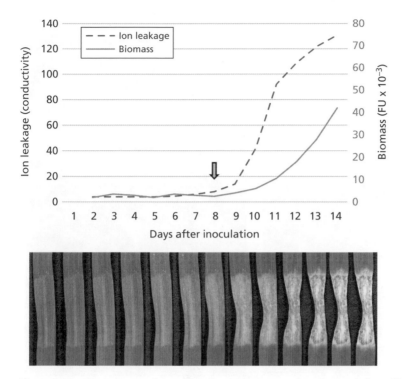

Figure 7.12 Time course of symptom development on wheat leaves infected by *Zymoseptoria tritici.* A spore suspension was applied to the central area of detached leaves and photos were taken at daily intervals showing the initial asymptomatic phase followed by chlorosis and necrosis of tissues. The graph above shows corresponding changes in ion leakage and the biomass of the fungus measured by real-time PCR. The arrow indicates the point at which the membrane integrity of host cells begins to be lost followed by accelerating fungal growth as leaking nutrients become available in the apoplast. *Source:* Adapted from Keon et al. (2007).

cell also contains membrane-bound organelles, such as plastids, mitochondria, peroxisomes and lysosomes, and alterations in one or several of these types of organelle are involved in many aspects of disease physiology.

Growth Regulation

All plant pathogens affect the growth and development of their hosts to a greater or lesser extent. The diversion of nutrients or the destruction of host tissues will inevitably lead to reduced performance, and in some cases may severely stunt the plant. These effects on plant growth are, however, essentially indirect and therefore different from the specific growth abnormalities induced by a variety of pathogens (see Table 8.3). Symptoms such as galls and tumors, excessive branching (witch's brooms), leaf epinasty, phyllody (replacement of floral organs by leaf-like structures), misplaced induction of adventitious roots, and premature leaf abscission are all associated with changes in the control of plant growth and differentiation. Such abnormal growth is characteristic of many diseases involving microbial pathogens. Two examples are shown in Figure 7.13.

(a) (b)

Figure 7.13 (a) Maize cob infected by smut, *Ustilago maydis*. Infection induces tumors on aerial plant organs. *Source:* Science Photo Library. (b) Peach leaf curl caused by the yeast-like fungus *Taphrina deformans*. Infected leaves are distorted and accumulate red carotenoid pigments. Symptoms in both cases are associated with production of hormones such as auxins and cytokinins by the pathogen. *Source:* Photo by John Lucas.

Although plant morphogenesis is influenced by environmental conditions, control of the basic processes of cell division and differentiation is mediated by hormonal compounds such as indoleacetic acid, gibberellins, cytokinins, abscisic acid and ethylene. Changes in the concentration or distribution of these hormones have widespread effects on the physiology of plants. As alterations in growth regulation can often be attributed to the production of hormonally active compounds by the pathogen, further discussion of diseases involving growth abnormalities will be deferred until the next chapter. Plant hormones such as salicylic acid, jasmonic acid, and ethylene also play key roles as signal molecules in plant defense, a topic further explored in Chapter 9.

Conclusion

It should be apparent from this brief review that many of the measurable changes in the physiology of infected plants are common to a variety of diseases. In the search for a unifying concept in host–pathogen interaction, it is tempting to interpret these similarities as evidence for a common pathway or sequence of biochemical events following infection. However, many of these gross alterations, such as increased respiration and permeability changes, are also characteristic of plants damaged by nonmicrobial agents. In view of this, the explanation for pathogenesis and host resistance may be more to do with changes unique to diseased plants rather than to these general responses to stress conditions. This explanation must ultimately be sought in terms of molecular interactions occurring in the host–pathogen complex.

Further Reading

Reviews and Papers

Baron, M., Pineda, M., and Perez-Bueno, M.L. (2016). Picturing pathogen infection in plants. *Zeitschrift Fur Naturforschung Section C: A Journal of Biosciences* 71 (9–10): 355–367.

Behr, M., Humbeck, K., Hause, G. et al. (2010). The hemibiotroph *Colletotrichum graminicola* locally induces photosynthetically active green islands but globally accelerates senescence on aging maize leaves. *Molecular Plant-Microbe Interactions* 23 (7): 879–892.

Berger, S., Sinha, A.K., and Roitsch, T. (2007). Plant physiology meets phytopathology: plant primary metabolism and plant–pathogen interactions. *Journal of Experimental Botany* 58 (15–16): 4019–4026.

Chanclud, E., Kisiala, A., Emery, N.R.J. et al. (2016). Cytokinin production by the rice blast fungus is a pivotal requirement for full virulence. *PLoS Pathogens* 12 (2): e1005457. https://doi.org/10.1371/journal.ppat.1005457.

Fernandez, J., Marroquin-Guzman, M., and Wilson, R.A. (2014). Mechanisms of nutrient acquisition and utilization during fungal infections of leaves. *Annual Review of Phytopathology* 52: 155–174.

Horst, R.J., Doehlemann, G., Wahl, R. et al. (2010). *Ustilago maydis* infection strongly alters organic nitrogen allocation in maize and stimulates productivity of systemic source leaves. *Plant Physiology* 152 (1): 293–308.

Hulsmans, S., Rodriguez, M., de Coninck, B. et al. (2016). The SnRK1 energy sensor in plant biotic interactions. *Trends in Plant Science* 21 (8): 648–661.

Jiang, Z.H., He, F., and Zhang, Z.D. (2017). Large-scale transcriptome analysis reveals Arabidopsis metabolic pathways are frequently influenced by different pathogens. *Plant Molecular Biology* 94 (4–5): 453–467.

Liu, M., Gao, J., Yin, F.Q. et al. (2015). Transcriptome analysis of maize leaf systemic symptoms infected by *Bipolaris zeicola*. *PLoS One* 10 (3): e0119858. https://doi.org/10.1371/journal.pone.0119858.

Malinowski, R., Novak, O., Borhan, M.H. et al. (2016). The role of cytokinins in clubroot disease. *European Journal of Plant Pathology* 145 (3): 543–557.

Moscatello, S., Proietti, S., Buonaurio, R. et al. (2017). Peach leaf curl disease shifts sugar metabolism in severely infected leaves from source to sink. *Plant Physiology and Biochemistry* 112: 9–18.

Proels, R.K. and Huckelhoven, R. (2014). Cell-wall invertases, key enzymes in the modulation of plant metabolism during defence responses. *Molecular Plant Pathology* 15 (8): 858–864.

Robert-Seilaniantz, A., Navarro, L., Bari, R. et al. (2007). Pathological hormone imbalances. *Current Opinion in Plant Biology* 10 (4): 372–379.

Rolfe, S.A. and Scholes, J.D. (2010). Chlorophyll fluorescence imaging of plant-pathogen interactions. *Protoplasma* 247 (3–4): 163–175.

Schierenbeck, M., Fleitas, M.C., Miralles, D.J. et al. (2016). Does radiation interception or radiation use efficiency limit the growth of wheat inoculated with tan spot or leaf rust? *Field Crops Research* 199: 65–76.

Schirawski, J. (2015). Invasion is sweet. *New Phytologist* 206 (3): 892–894.

Struck, C. (2015). Amino acid uptake in rust fungi. *Frontiers in Plant Science* 6: 40. https://doi.org/10.3389/fpls.2015.00040.

Walters, D.R. and McRoberts, N. (2006). Plants and biotrophs: a pivotal role for cytokinins? *Trends in Plant Science* 11 (12): 581–586.

8

Microbial Pathogenicity

We have little exact knowledge of the chemico-physiological processes in the life of the parasitic fungi because the symbiotic relation puts great complications and diffficulties in the way of their precise investigation.

(Anton de Bary, 1831–1888)

The majority of microorganisms, if inoculated into a plant, fail to grow or to cause disease. Pathogenicity is the exception rather than the rule. The successful pathogen must possess special properties enabling growth and multiplication in the host to the extent that disease develops.

Originally, identifying the factors determining microbial pathogenicity relied upon a combination of microscopy and painstaking biochemical analyses of microbial cultures or extracts from infected plants. Observation of pathogens on or in plants, especially by electron microscopy, provided clues as to how host defenses are breached, or tissues colonized and damaged. Evidence obtained in this way is useful but usually inconclusive; for instance, changes in the ultrastructure of cell walls or membranes might suggest the action of enzymes or toxins, but the factor(s) responsible remain unidentified. Alternatively, the pathogen may be grown in culture and the filtrate analyzed for the presence of substances which reproduce disease symptoms when introduced into the host. Obviously, this approach is limited to culturable microorganisms.

Quite apart from the difficulty of purifying molecules of interest from complex mixtures, there is the additional problem of proving that a factor produced *in vitro* is also produced in the host plant. There are many examples of biologically active compounds produced by pathogens in culture which have never been detected in infected plants, and which probably play little or no part in pathogenesis.

Even greater problems are encountered when trying to identify key factors in the host–pathogen complex itself. The crucial events may be localized to only a few cells, and the molecules of interest may be unstable or present in low concentrations. Improvements in the sensitivity and accuracy of analytical methods, for instance by using antibodies to detect molecules *in situ*, have extended the scope of such studies but many difficulties remain.

A Molecular Genetic Approach

The development of molecular techniques for isolating DNA from cells, cutting it into smaller pieces with restriction enzymes and cloning these pieces in a suitable vector (usually a bacterial plasmid or a bacteriophage), provided a new way of identifying the factors responsible for microbial pathogenicity.

The basic idea in this approach is straightforward (Figure 8.1). Starting with a wild-type pathogenic strain of the microorganism, one generates mutants which have lost the ability to infect the host, using a chemical mutagen or ultraviolet (UV) light. These nonpathogenic mutants are identified by inoculation tests onto a normally susceptible plant. At the same time, a genomic library of DNA from the pathogenic wild type is prepared; this consists of pieces of the pathogen genome isolated and cloned in a phage or bacterium to produce multiple copies of each piece. Each piece or clone is now transferred individually back into the nonpathogenic mutant and the transformed strains are tested on plants. Provided the library contains clones representing the whole genome of the pathogen, sooner or later a DNA sequence complementing the mutation, and thereby restoring pathogenicity, will be introduced. This clone should contain a gene, or genes, encoding a product essential for pathogenicity. Once the specific piece of DNA is found, the gene can then be sequenced and the gene product identified.

While specific functions can be inactivated or restored by this approach, the process of testing all the clones is laborious, and also depends on the library representing the complete genome of the pathogen. An alternative and potentially more efficient procedure is to generate mutations in such a way that the site of the inactivated gene can be readily identified. This can be achieved in a number of ways, for instance by using transposons, restriction enzymes or the bacterial genetic engineer *Agrobacterium* (see p.205). Each method generates a large number of mutants by inserting a specific piece of DNA at random sites in the genome; if such a sequence inserts into a gene essential for pathogenicity, then the gene is disrupted and pathogenicity is reduced or lost. The inserted DNA carries a recognizable genetic marker, such as resistance to an antibiotic, and the disrupted gene is therefore "tagged" and can be amplified by sequence-specific primers and subsequently cloned. Such **random insertional mutagenesis** has been widely used to identify candidate genes involved in pathogenicity or host specificity. It proved a powerful tool, especially in bacteria, which have relatively small, haploid genomes. With viruses and viroids, which have even smaller genomes, it is possible to manipulate part or all of the nucleic acid, and to define gene functions by inserting mutations at strategic points in the genome. This approach was used, for instanc,e to identify the role of the TMV P30 protein in virus movement between cells (see p.56).

Stage-Specific Gene Expression

An alternative but complementary way to gain molecular insights into pathogenicity is to study the expression of genes at different stages in the infection process. This has already been touched upon in Chapter 6 when describing penetration and entry to the host plant, and interactions taking place at the host–pathogen interface. For instance,

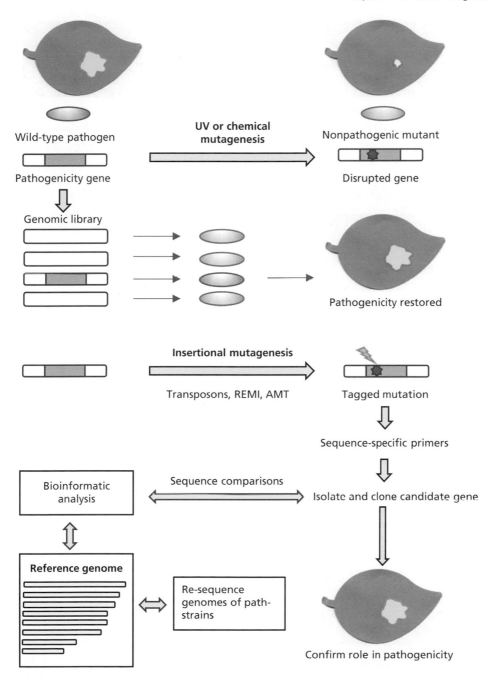

Figure 8.1 Experimental approaches to identifying pathogenicity genes. The example shown is based on an infection assay inoculating host leaves. Further explanation is in the main text. AMT, *Agrobacterium*-mediated transformation; REMI, restriction enzyme-mediated integration.

gene libraries prepared from specific infection structures such as haustoria have provided important clues to the processes of nutrient uptake, as well as delivery of effectors into host cells.

RNA-sequencing (RNA-seq) is a relatively recent method for revealing the presence and quantity of RNA in a biological sample at a given moment in time. The technique provides a snapshot of all the genes being expressed in cells or tissues, otherwise known as the transcriptome. If repeated throughout a particular process, such as the infection of a plant organ by a pathogen, one can compile a transcript profile indicating which genes are up- or downregulated at critical points such as penetration, invasion of tissues, and symptom development. One can identify patterns of gene expression associated with each stage, and groups of genes that might therefore play a role in pathogenicity. The output from this type of analysis can be presented graphically or as a heat map (Figure 8.2).

Comparative Genomics and Pathogenicity

Recent advances in genome sequencing technology and bioinformatic analysis are now superseding the molecular genetic approaches described above. With the number of complete genome sequences of plant-pathogenic microorganisms increasing year on year, the scope for such analysis is constantly expanding and in many cases, a reference genome may already be available (Figure 8.1). The increased speed and decreasing cost of sequencing mean that it is now possible to resequence variant strains to detect changes, such as single nucleotide polymorphisms (SNPs), in genes that are linked to particular traits, including pathogenicity. There are also specialist databases such as PHI-base (www.phi-base.org), which contains curated molecular and biological information on genes proven to affect the outcome of pathogen–host interactions. Comparisons between different types of pathogens are of particular value to identify common features and conserved pathways involved in pathogenesis, as well as genes associated with particular pathogen lifestyles. Conclusive proof of a key role in disease may, however, still require gene disruption, cloning, and complementation in a pathogenicity assay.

Enzymes and Microbial Pathogenicity

Free-living bacteria and fungi produce a wide variety of enzymes which are secreted into the external environment and play an important role in the utilization of nutrient substrates. Many plant-pathogenic species also produce extracellular enzymes (Table 8.1). Whether at least some of these are determinants of pathogenicity was debated for many years but with the application of molecular genetic techniques, the diverse roles played by enzymes in host penetration, colonization, and countering plant defense are now being clarified. Enzymes are involved in breaching the external protective layers of the plant, modifying cell walls and releasing vital nutrients for use by the pathogen. They may degrade inhibitors present in plant tissues and inactivate proteins required for active host defense. Enzymes that kill host cells may also be important in suppressing defense responses which depend on active metabolism.

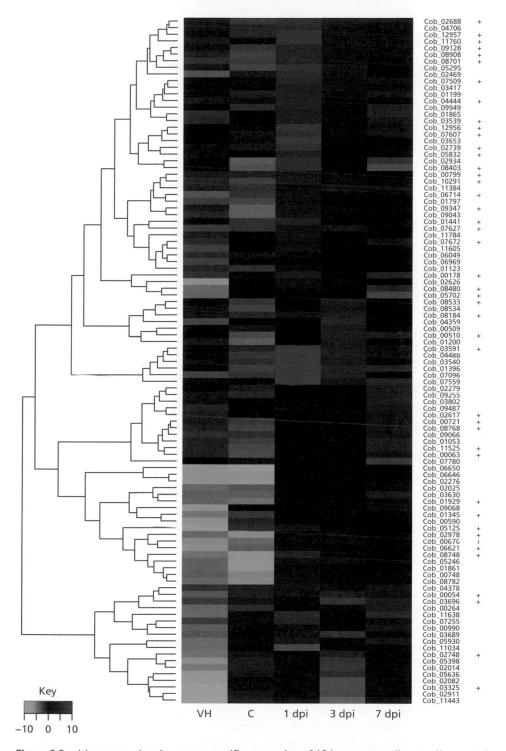

Figure 8.2 A heat map showing stage-specific expression of 104 genes encoding small secreted proteins of *Collectotrichum orbiculare* during infection of tobacco leaves. Samples shown are one, three, or seven days post inoculation (dpi) compared with vegetative hyphae (VH) or conidia (C) of the fungus. The color key indicates levels of expression from green (low) to red (high). Note many of the genes are more highly expressed *in planta*, especially during early host invasion. Clustering (*left*) indicates genes with similar expression patterns, individual gene codes listed on right. *Source:* Gan et al. (2013).

Table 8.1 Some extracellular enzymes produced by plant pathogens

Enzyme	Substrate attacked	Comments
Cutinase	Cutin	Implicated in host penetration
Pectolytic enzymes		
• Pectin esterase	Pectin	Cell-separating activities in soft-rot diseases
• Polygalacturonase	Pectate	
• Pectate lyase	Pectate	
Hemicellulases		
• Xylanase	Xylan	Predominant polymers found in monocot cell walls
• Arabanase	Araban	
Cellulase	Cellulose	Often produced in later stages of infection
Ligninase	Lignin	Produced by timber decay fungi
Phospholipase	Phospholids	Breaks down membranes
Protease	Protein	May target host proteins involved in defense
β-glucosidase	Glucosides	Degrades plant inhibitors

Cell Wall-Degrading Enzymes (CWDE)

One group of plant pathogens in which enzymes have been extensively studied are the soft-rot bacteria and fungi which typically attack storage tissues and cause spreading, necrotic lesions (Figure 8.3). More than 100 years ago, de Bary demonstrated that extracts of such rotted tissue can macerate firm, healthy tissues. Further work in the early part of the last century showed that the maceration of host tissues is due to the action of cell-separating enzymes produced by the pathogen. These enzymes degrade the pectic substances in the middle lamella between cells, thereby facilitating the colonization of host tissues. The ability to produce pectic enzymes is widely distributed among fungi and bacteria and is characteristic of necrotrophic fungal pathogens such as *Botrytis* and *Sclerotinia* and soft-rot bacteria like *Erwinia* and *Pectobacterium*.

The soft-rot syndrome involves two processes: host cells within the lesion are separated through the action of enzymes and the cells subsequently die. There was a long-running debate about whether the enzymes themselves are responsible for killing cells or whether some other lethal factor is involved. Resolving this question proved difficult, because soft-rot pathogens produce not one but a whole series of CWDEs, each with a different mode of action. For instance, pectin methyl esterase removes the methyl groups of pectin to yield pectic acid, which is then more susceptible to chain-splitting enzymes such as pectate lyase and polygalacturonase, which cleave the polymer into fragments. The complete digestion of pectic polymers is hence a multienzyme process. It is now known that certain pectic enzymes, when purified, can in isolation kill host cells as well as macerating tissues. The way in which the enzyme actually kills cells is not entirely clear although evidence favors the idea that cell walls in plant tissues treated with pectic enzymes lose their ability to support the plasma membrane, particularly under osmotic stress. Direct effects of reaction products of the enzyme on the cells do not appear to be responsible for cell death, although such products may be important as signal molecules triggering host defense responses.

Figure 8.3 Soft-rot symptoms caused by *Pectobacterium carotovora* on potato (*left*) and the brown-rot fungus *Monilinia* infecting apple (*right*). *Source:* Photos by Brian Case.

As well as pectolytic activities, there are enzymes which attack other wall polymers, such as cellulases, hemicellulases, and ligninases (Table 8.1). The latter are particularly important in diseases involving wood decay, such as heart-rots of trees. In combination, these enzymes can degrade all the polymers present in higher plant cell walls.

The relative importance of different pectic enzymes for the virulence of soft-rot pathogens has been intensively studied in the bacterium *Erwinia* which, as a close relative of *Escherichia coli*, is a convenient model for genetic manipulation. Genes encoding individual enzymes were isolated from *Erwinia* and cloned in *E.coli*, thereby allowing analysis of their structure and regulation. Mutants deficient in particular enzymes were also produced. Several interesting discoveries emerged from this work. First, some pectic enzymes are coded for by several genes; in *E. chrysanthemi*, for instance, there are at least five pectate lyase genes (designated *pelA–E*), each producing a different pectate lyase isoenzyme. The genes are grouped into two clusters (Figure 8.4). Second, not all of the genes are essential for tissue maceration and host colonization. Mutations in *pelE* have the greatest effect on pathogenicity. Third, nonpathogenic mutants were identified which still produce the full repertoire of pectic enzymes, but fail to secrete them to the external environment. These Out⁻ bacteria have mutations in genes encoding proteins which are essential components of a secretion pathway exporting enzymes and other proteins across the outer membrane (see p.251). Hence, pathogenicity requires not only the synthesis of certain CWDEs, but also effective export out of the cell.

A general conclusion of this and subsequent work on soft-rot pathogens is that they possess an arsenal of different CWDEs that work synergistically to enable colonization of host tissues. Other necrotrophs and hemibiotrophs also produce a range of CWDEs, but the relative importance of the different types of enzyme varies depending on the chemistry of the host cell wall. Monocotyledons, for instance, have cell walls rich in hemicellulose, so pathogens colonizing these hosts, such as the stem base-infecting fungi *Oculimacula* spp., *Fusarium culmorum*, and *Rhizoctonia cerealis*, produce significant amounts of xylanase and other hemicellulases. Enzymes degrading host cell walls are usually produced by pathogens in a specific sequence – first pectic enzymes, followed by hemicellulases, then cellulases. This is consistent with a stepwise digestion of the cell wall, pectic matrix polymers

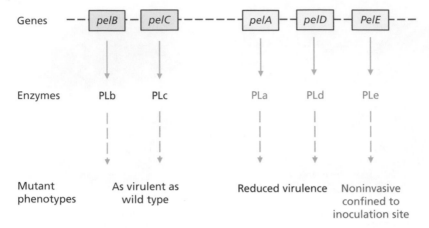

Figure 8.4 Diagrammatic version of the organization of pectate lyase (*pel*) genes in *Erwinia chrysanthemi*, showing two gene clusters producing different PL isoenzymes and the effect of mutations disrupting each gene on the virulence of the pathogen.

being removed first to expose the microfibrillar skeleton (Figure 8.5). Recent studies on the wheat pathogen *Fusarium graminearum* have shown that the concerted action of polygalacturonase, xylanase, and cellulase is necessary for full virulence.

Further insights into the potential roles of enzymes in pathogenicity have been gained from comparative genomic studies of microorganisms with different lifestyles, including saprotrophs, necrotrophs, hemibiotrophs, and biotrophs. Generally, necrotrophs and hemibiotrophs possess larger numbers of genes encoding CWDEs and other hydrolytic enzymes than biotrophs. This is consistent with the relative extent of tissue damage associated with parasitism by these contrasting types. Figure 8.6 compares the number of genes encoding cellulase or hemicellulase activities in the rice blast fungus *Magnaporthe oryzae*, a hemibiotroph with an invasive, destructive second phase of host colonization, with the powdery mildew *Blumeria graminis*, a strict biotroph that only penetrates the epidermis and establishes a sustained relationship with living host cells. This graph also includes a comparison of certain classes of secondary metabolism genes such as polyketide synthases (PKS) and nonribosomal peptide synthetases (NRPS) implicated in toxin production. As expected, the powdery mildew has far fewer genes for CWDEs and potential toxin biosynthesis.

The ability to degrade lignin is rare, and restricted mainly to specialized parasites of woody hosts, such as the white-rot fungi. The mode of action of ligninase is unusual, involving the oxidative production of highly reactive free radicals which break the diverse chemical bonds in the lignin molecule, a process described as "enzymic combustion." The lack of ligninase activity in most bacteria and fungi explains why lignified cell walls are a highly effective barrier to microbial penetration.

Enzymes Degrading Host Inhibitors

A further group of microbial enzymes which may play an important role in pathogenicity are those with activity toward plant antibiotics, such as preformed inhibitors,

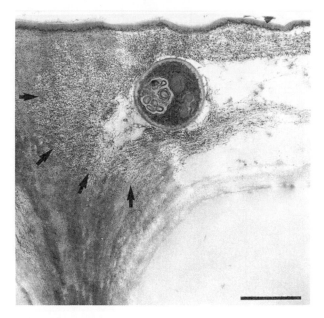

Figure 8.5 Intramural growth of hypha of the cereal eyespot fungus *Oculimacula* showing degradation of host cell wall adjacent to hypha and unmasking of wall polymers (*arrows*). Scale bar – 2 μm. *Source:* Courtesy of Alison Daniels.

and phytoalexins produced in response to infection. For instance, β-glucosidases can cleave toxic glucosides present in plants, thereby inactivating such compounds and allowing the pathogen to establish itself in the host. Detailed discussion of these interactions is postponed until after the plant defense factors involved have been described (see p.212).

Toxins

The idea that pathogenesis might be due to the production of poisons by the pathogen is by no means new, as the potent bacterial toxins involved in human diseases such as diphtheria and tetanus were discovered more than 100 years ago. The importance of toxins in plant pathology took longer to establish, but there are now many examples of bacterial and fungal products known to play at least some role in symptom development in plants. The term **toxin** can refer to any pathogen product which is harmful to the host, but is usually restricted to secondary metabolites or peptides which do not attack the structural integrity of plant tissues but instead affect metabolism in some other, more subtle fashion. Two important properties of toxins are (i) that they are active at low concentrations, and (ii) that they are mobile within the plant and may therefore act at a distance from the site of infection.

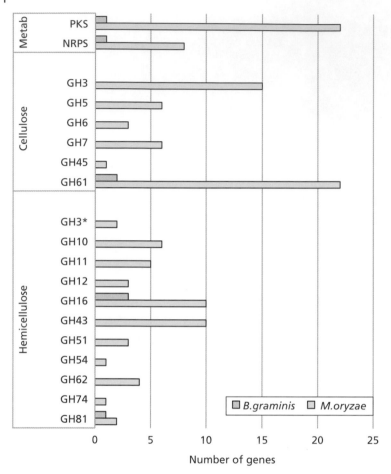

Figure 8.6 Comparison of numbers of genes encoding cellulose- and hemicellulose-degrading enzymes (glycosyl hydrolases – GH) in genomes of the powdery mildew *Blumeria graminis* and the rice blast pathogen *Magnaporthe oryzae*, along with two classes of genes (PKS and NRPS) potentially involved in the biosynthesis of toxic secondary metabolites. The powdery mildew appears to have lost many of these genes during the evolution of extreme biotrophy. *Source:* Adapted from Spanu et al. (2010).

Several schemes have been proposed for the classification of toxins produced by plant pathogens, based on biological activity rather than the chemical structure of the compounds. Two main groups are usually distinguished: **host-specific (selective) toxins** (HSTs), which only affect plants susceptible to the pathogen producing them, and **non-host-specific (nonselective)** toxins which also affect other plants as well (Table 8.2). Pathogen compounds which act directly as plant growth regulators are often treated separately, although it should be noted that certain toxins possess hormone-like properties.

Host-Specific Toxins

In 1946, oat crops in the USA were affected by a seedling blight caused by *Cochliobolus (Helminthosporium) victoriae*. The fungus was especially damaging on an oat cultivar known as Victoria. The pathogen itself was typically localized to the basal part of infected

Table 8.2 Some toxins involved in plant disease

Toxin	Pathogen	Host	Mode of action
Host-specific (host-selective) toxins			
Fungal			
Victorin	*Cochliobolus victoriae*	Oats	Induces programmed cell death (PCD)
HC-toxin	*Cochliobolus carbonum*	Maize	Suppression of host defense
HmT toxin	*Cochliobolus heterostrophus*	Maize	Disruption of mitochondrial function
PC-toxin	*Periconia circinata*	Sorghum	Induces PCD
PtrTox A and B	*Pyrenophora tritici-repentis*	Wheat	Induces PCD
SnTox A	*Parastagonospora nodorum*	Wheat	Induces PCD
AK-toxin	*Alternaria alternata*	Japanese pear	Targets membrane proteins
Nonhost-specific (nonselective) toxins			
Fungal			
Tentoxin	*Alternaria* spp.	Various	Inhibits adenosine triphosphate synthesis in chloroplast
Cercosporin	*Cercospora* spp.	Soybean, tobacco, and others	Production of reactive oxygen species (ROS)
Rubellin	*Ramularia collo-cygni*	Barley	Production of ROS
Sirodesmin PL	*Leptosphaeria maculans*	Oilseed rape	Production of ROS
Oxalic acid	*Sclerotinia sclerotiorum*	Numerous	Induces PCD
Bacterial			
Tabtoxin	*Pseudomonas syringae* pv. *tabaci*	Tobacco	Inhibition of amino acid synthesis
Phaseolotoxin	*Ps. syringae* pv. *phaseolicola*	Beans	Inhibition of amino acid synthesis
Coronatine	*Ps. syringae* pathovars	Various	Hormone mimic (jasmonic acid)
Syringomycin	*Ps. syringae* pathovars	Various	Membrane disruption

plants but symptoms extended into the leaves, which often collapsed. Suspicions that a mobile toxin might be involved were confirmed when fungal culture filtrates were shown to cause the same symptoms, and an active compound was isolated and given the name victorin (HV-toxin). The most interesting property of this toxin, subsequently characterized as a cyclic pentapeptide (Figure 8.8), was that oat cultivars resistant to the fungus were unaffected by it, while susceptible cultivars such as Victoria were affected by very low concentrations. Sensitivity to the toxin therefore showed the same specificity as the pathogen itself. A classic demonstration of this genotype sensitivity to victorin is shown in Figure 8.7. Furthermore, there was a direct correlation between toxin production and pathogenicity; genetic analysis showed that a single gene in the fungus determines toxin production, and

Figure 8.7 Original demonstration of effects of victorin on resistant (R) and susceptible (S) oat seedlings. Toxin was added to the nutrient solution where indicated three days before the photo was taken. *Source:* Scheffer and Yoder (1972).

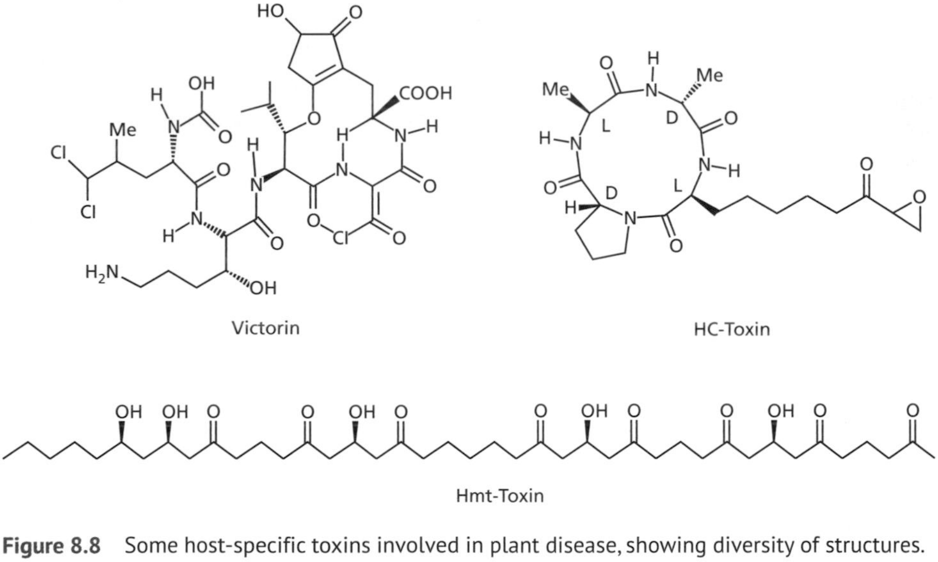

Victorin

HC-Toxin

Hmt-Toxin

Figure 8.8 Some host-specific toxins involved in plant disease, showing diversity of structures.

in crosses between toxin-producing and nonproducing strains, progeny lacking the toxin gene were only weakly pathogenic. All this evidence pointed to victorin as the major determinant of the disease.

Victorin has widespread physiological effects in the host. It stimulates respiration, the increase being proportional to the concentration of the toxin applied to the plant. Toxin-treated tissues leak electrolytes, and electron micrographs of treated cells revealed damage to the plasma membrane. In sensitive plants, victorin triggers responses typical of programmed cell death (PCD). The pathogen might therefore hijack a host defense pathway to kill cells and colonize host tissues.

Sensitivity to victorin in oats was shown to be controlled by a single gene, designated *Vb*. Interestingly, sensitivity to the toxin is also correlated with resistance to an unrelated pathogen, the rust fungus *Puccinia coronata*, conferred by the gene *Pc-2*. The genes responsible for toxin sensitivity and rust resistance are therefore closely linked or even identical. This raises the possibility that a gene conferring resistance to a biotroph is also involved in susceptibility to a necrotrophic pathogen.

Several other *Cochliobolus (Helminthosporium)* species have been shown to produce HSTs, including *C. heterostrophus*, the causal agent of southern corn leaf blight, and *C. carbonum*, responsible for leaf spot and ear mold of maize (Table 8.2). These toxins differ in chemical structure and apparent mode of action. HmT-toxin, for instance, consists of several linear polyketols (Figure 8.8) which interfere with mitochondrial function, while HC-toxin is a cyclic tetrapeptide which inhibits enzymes that modify histones, the proteins associated with DNA in chromatin. The primary role of HC-toxin in disease is therefore believed to be in repressing the expression of host defense genes, thereby making the plant more susceptible to infection.

Host resistance to HC-toxin is conferred by a dominant gene, *Hm1*, which encodes a reductase enzyme that chemically modifies and inactivates the toxin. Susceptible maize genotypes lack the enzyme and are therefore unable to degrade the toxin. This discovery was significant as it confirmed the central role played by an HST in disease, and also assigned a specific biochemical function to a plant resistance gene (see Chapter 10).

Tan spot, caused by the fungus *Pyrenophora tritici-repentis*, is an economically important disease of wheat worldwide. Virulent isolates of the pathogen produce a protein toxin known as Ptr ToxA encoded by a single fungal gene. The toxin induces necrotic lesions surrounded by chlorotic tissue on susceptible wheat varieties. Sensitivity to the toxin is determined by a dominant gene, *Tsn1*, present in such wheat genotypes. Disease therefore results from the interaction of a protein HST with the product of a susceptibility gene in the host plant. The related fungus *Parastagonospora nodorum* causes a similar necrotic leaf blotch disease in wheat.

While analyzing the complete genome sequence of *P. nodorum*, researchers came across a gene with a high degree of similarity to the *P. tritici-repentis* gene encoding the ToxA HST. Regions flanking the gene also showed a close match with the locus surrounding the Ptr toxin gene (Figure 8.9), suggesting a common ancestry. The most likely explanation for this similarity is that at some point, the *ToxA* gene was transferred between the two pathogen species. Further analysis of the genetic diversity of *ToxA* variants in pathogen populations concluded that *P. tritici-repentis* acquired the gene by horizonal transfer from *P. nodorum* relatively recently, which might also explain the emergence of tan spot as a more damaging disease since the mid-twentieth century.

Many of the HSTs produced by necrotrophic fungi, such as the examples shown in Table 8.2 and described above, induce cell death or otherwise disrupt host defense, so they are now considered to be **necrotrophic effectors** that contribute to pathogen virulence. The diversity of microbial products that act as effectors in host–pathogen interactions is covered in more detail later (see p.255).

The discovery of host-specific toxins raised hopes that the biochemical basis of pathogenicity, and conversely host resistance, might turn out to be simple, determined by a single compound of major effect. Overall, these hopes have not been fully realized, as relatively few of the toxic compounds investigated to date have proved to be host specific. Instead, most toxins appear to be nonspecific in effect, and play a less prominent role in the disease process.

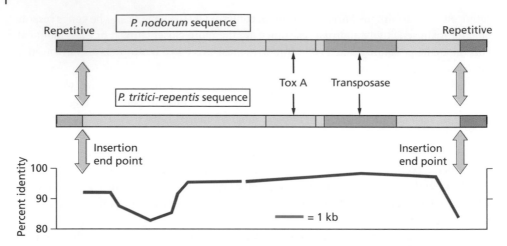

Figure 8.9 Evidence for lateral transfer of a fungal virulence gene. An 11 kb region of the *Parastagonspora nodorum* genome is collinear with a sequence from *Pyrenophora tritici-repentis* containing the Tox A gene and a transposase enzyme gene often associated with transposons. Sequence similarity (shown in lower diagram) ranges from 80–90% at the periphery of the region to 98–100% in the central part. Flanking sequences are repetitive in *P. nodorum* while those in *P. tritici-repentis* show some similarity to unlinked parts of the *P. nodorum* genome. *Source:* Adapted from Friesen et al. (2006).

Nonhost-Specific Toxins

A wide variety of secondary metabolites from plant-pathogenic fungi and bacteria have been implicated in disease, occurring in the infected host where they account for at least some of the disease symptoms. These toxins, which do not seem to be the sole determinants of pathogenicity, can instead be regarded as factors contributing to the virulence of a pathogen.

Nonhost-specific toxins have effects on plant species other than the natural host. Some examples are shown in Table 8.2, and diversity of chemical structures in Figure 8.10. Tentoxin, a cyclic tetrapeptide produced by the fungus *Alternaria*, induces chlorosis in a wide range of green plants, including pteridophytes, mosses and green algae, as well as seed plants. Its primary effect appears to be on energy transfer and adenosine triphosphate (ATP) synthesis in the chloroplast. Cercosporin, produced by several plant-pathogenic species of *Cercospora*, is a pigmented toxin that in the presence of light generates reactive oxygen species (ROS) that damage membrane lipids and have other nonspecific toxic effects on cells. The recently emerged barley pathogen *Ramularia collo-cygni* produces several potentially toxic secondary metabolites, including rubellin D which is also light-activated and generates ROS leading to chlorosis and necrosis of leaf tissues. The role rubellin plays in host colonization and pathogenesis has not yet been fully determined.

Sirodesmin PL is a secondary metabolite toxin produced by *Leptosphaeria maculans*, cause of stem canker of oilseed rape. Mutants of *L. maculans* that are unable to synthesize sirodesmin are still pathogenic on seedling leaves but less virulent on stems, with smaller

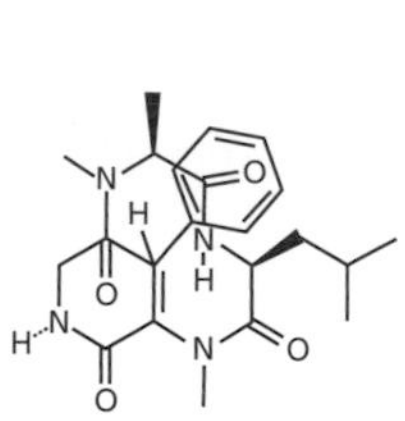

| Tentoxin | Cercosporin | Oxalic acid | Coronatine |

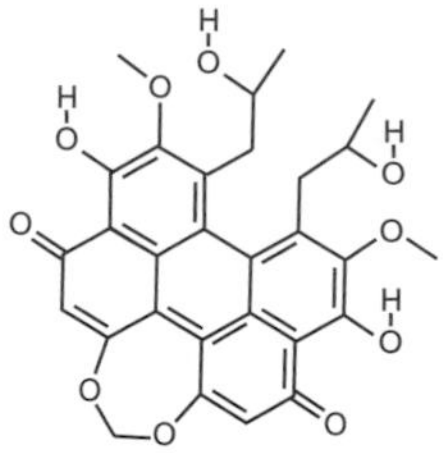

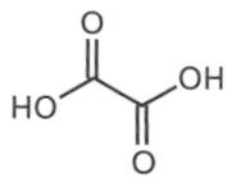

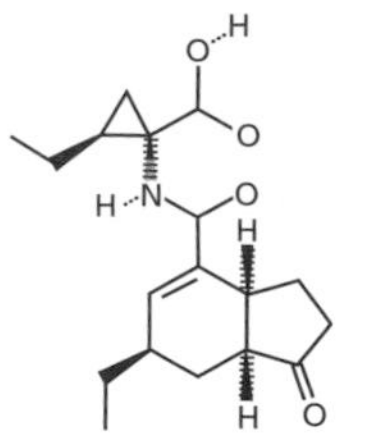

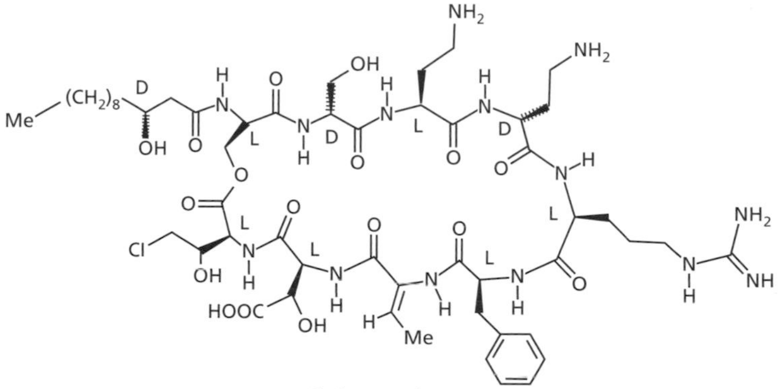

Syringomycin

Figure 8.10 Structures of some nonhost-specific toxins involved in plant disease.

lesions and reduced biomass. The toxin is therefore implicated as a virulence factor in this disease. The necrotrophic fungus *Sclerotinia sclerotiorum* has a very broad host range, infecting more than 400 different plant species. During infection, the fungus produces a simple organic compound, oxalic acid, which reduces the pH but also induces a programmed cell death response as well as increased ROS levels in the host. Oxalic acid may therefore have both direct and indirect toxic effects by manipulating the redox (oxidation versus reduction) status of host tissues.

One interesting revelation from fungal genome sequencing is that genes encoding toxin biosynthesis can be found in some biotrophic species. For instance, *Cladosporium fulvum*, which has a biotrophic lifestyle growing in the apoplast of tomato leaves, possesses several gene clusters with the potential to produce toxic secondary metabolites, including cercosporin. However, these are not expressed during infection of the host. Downregulation of pathways producing harmful metabolites might therefore be another strategy for maintaining a biotrophic relationship.

Several bacterial plant pathogens also produce toxic secondary metabolites that play some role in disease. The examples listed in Table 8.2 are all from *Pseudomonas syringae*, a well-studied and economically important species which exists as a series of pathovars infecting different host plants. Wildfire disease of tobacco, caused by *Ps.* pv. *tabaci,* is characterized by lesions on leaves which are surrounded by a conspicuous chlorotic halo. Formation of this chlorotic zone has been attributed to a dipeptide, tabtoxin, which inhibits the enzyme

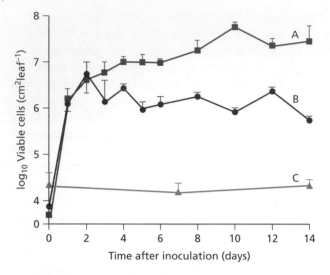

Figure 8.11   Multiplication of the wildfire disease bacterium, *Pseudomonas syringae* pv. *tabaci* in tobacco leaves. A is a toxin-producing wild-type strain, while B is a nontoxin-producing mutant. A saprophytic bacterium, *P. putida* (C), is included for comparison. *Source:* Turner (1984).

glutamine synthetase, leading to the accumulation of toxic concentrations of ammonia. Bacterial mutants not producing tabtoxin (Tox-) do not induce chlorotic lesions, but are still able to grow in host leaves, albeit less well than wild-type toxigenic strains (Figure 8.11). Strains of *Ps. syringae* infecting beans (*Ps.* pv. *phaseolicola*) produce a tripeptide, phaseolo- toxin, which also inhibits amino acid synthesis, in this case arginine. The molecule is con- verted into a smaller active toxin by plant enzymes that cleave peptides. Many strains of *Ps. syringae* also produce another nonhost specific toxin, coronatine (Figure 8.10) which is of particular interest as it is a molecular mimic of the plant hormone jasmonic acid. The toxin interferes with plant defense signaling in several ways, including reversing the closure of stomata which would otherwise prevent entry into the leaf (see p.151). The final example listed, syringomycin, consists of a ring of amino acids bonded to a fatty acid hydrocarbon tail (Figure 8.10); aggregates of the toxin can insert into cell membranes to form small pores, allowing leakage of ions and electrolytes. This mode of action might therefore increase the amount of nutrients available to the pathogen as it colonizes intercellular spaces.

A general conclusion from the above is that some nonhost-specific toxins play a poten- tially important role in disease development. In *Ps. syringae*, genes for toxin biosynthesis are often clustered on plasmids and conserved among groups of strains, suggesting that they are important for survival of the bacteria in the plant environment. Where more than one toxin is produced, they may act at different stages during infection, thereby aiding host colonization. It should also be noted that many of the toxins listed in Table 8.2 cause sig- nificant metabolic disruption, often resulting in cell death. This may be an advantage for necrotrophic pathogens in terms of releasing nutrients or preventing an active defense response by the plant, but is clearly not an option for biotrophic pathogens which require living host cells to survive.

High Molecular Weight Compounds and Pathogenesis

Not all toxic products of pathogens are small molecules. Several compounds implicated in vascular wilt diseases have been shown to be high molecular weight polysaccharides or glycoproteins. These substances may impair water flow simply by virtue of their size. Experiments with synthetic polymers have shown that if their molecular weight is less than 50 kDa, they can be transported through the xylem and act in the leaves; if greater than 50 kDa, the polymers remain in the xylem and contribute to physical plugging of the vessels. Strictly speaking, such disruption of the water economy of the host, while damaging to the plant, is not equivalent to a direct toxic effect on metabolism.

The role played by toxins in vascular wilt diseases is a controversial research topic. A diverse array of substances with toxic effects has been isolated from the culture filtrates of wilt pathogens, ranging from simple organic acids to complex macromolecules. After further investigation, most have been relegated to a subsidiary role in disease causation. For instance, toxic peptides from *Verticillium dahliae* evoke wilt symptoms when introduced into hosts such as cotton, tomato and potato, but a central role in disease has not been confirmed. *Ophiostoma novo-ulmi*, the fungus responsible for the devastating Dutch elm disease, produces a small hydrophobin protein, cerato-ulmin, which causes morphological and physiological symptoms similar to those seen in the disease. The protein has also been detected in elm wood in the early stages of disease development. In some studies, the amount of toxin produced by different isolates of the fungus correlated with their virulence. However, targeted disruption of cerato-ulmin synthesis did not result in a reduced ability to cause disease. The significance of this protein in pathogenicity is therefore still unresolved.

More convincing evidence for a key role in the pathogenesis of wilt diseases has been obtained for the exopolysaccharides (EPS) produced by many bacterial pathogens. These polymers are constituents of the protective bacterial capsule as well as the ooze formed during infection. The fireblight pathogen *Erwinia amylovora* spreads within xylem vessels, producing an EPS called amylovoran. This aids the formation of biofilms whereby the bacteria are able to stick to surfaces and also to each other (Figure 8.12). Together, the proliferating bacterial cells and EPS block the passage of water. *E. amylovora* strains that cannot produce amylovoran have been shown to be nonpathogenic. A structurally similar EPS that is essential in pathogenesis is produced by the related bacterium *Pantoea stewartii*, causing Stewart's wilt of maize. A key role for EPS has also been implicated in *Ralstonia solanacearum*, cause of a lethal wilt of crops such as potato and tomato. Figure 8.13 shows that mutant strains compromised in EPS production cause less severe wilting than the EPS+ wild type. However, mutants defective in other properties, such as pectolytic enzyme production or export of extracellular proteins, are also reduced in virulence (Figure 8.13), indicating that pathogenesis in such wilt diseases is a complex process.

The xylem-limited bacterium *Xylella fastidiosa* infects a very wide range of economically important hosts. The pathogen is transmitted and introduced into the vascular system of the plant by sap-feeding leafhoppers. Analysis of the sequenced genome of the bacterium revealed a cluster of genes involved in the biosynthesis of an EPS similar to xanthan gum, a polysaccharide produced by *Xanthomonas* species. Disruption of two of these genes

(a)   (b)

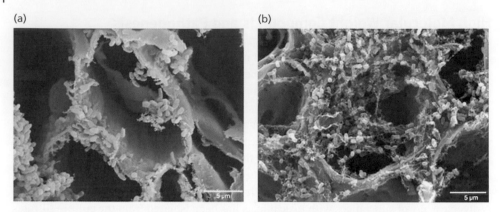

Figure 8.12 Colonization and biofilm formation by the fireblight pathogen *Erwinia amylovora* in xylem vessels of apple shoots. (a) Attachment of bacterial cells to the inner walls of xylem vessels, aggregation and initial growth into the vessel lumen. (b) Further multiplication and partial occlusion of the vessel by the biofilm. Note fibrillar matrix in which the cells are embedded. Bars = 5 μm. *Source:* Scanning electron micrographs from Koczan et al. (2009).

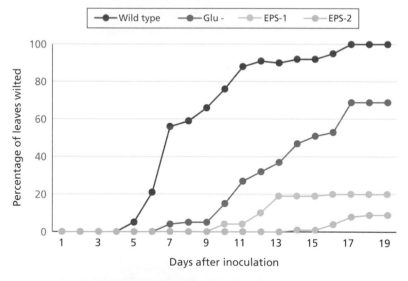

Figure 8.13 Relationship between extracellular polysaccharide (EPS) production and pathogenicity of *Ralstonia solanacearum* to tomatoes. Time course of wilting caused by wild-type EPS+ strain, and two mutants lacking EPS (EPS−) or the pectolytic enzyme β-1,4 endoglucanase (GLU-). *Source:* Saile et al. (1997).

reduced the ability of the pathogen to spread and cause symptoms in grapevines. Interestingly, transmission of these EPS− mutant strains by the insect vector was also reduced, suggesting that the polysaccharide is important for both pathogenicity and dispersal of the bacterium. It is clear from this study and the other cases described above that EPS plays a variety of roles in the biology and ecology of plant-infecting bacteria.

Hormones and Pathogenesis

Table 8.3 lists some examples of diseases in which invasion by the pathogen leads to abnormal growth of the plant. Such symptoms are consistent with an alteration in the hormonal status of the host. Several classes of compound are known to be involved: auxins, such as indole-3-acetic acid (IAA), gibberellins (GAs), cytokinins (CKs), abscisic acid (ABA), jasmonic acid (JA), and the volatile hormone ethylene (ET). Evaluation of the role played by individual growth regulators in pathogenesis is complex because their physiological effects in many cases overlap, and normal plant growth and development are regulated by the integrated action of several types of hormone rather than by changes in the concentration of a single compound. There is also the difficulty of determining whether any change in hormone levels is due to the accumulation or loss of hormones produced by the plant itself, or to the production of similar compounds by the pathogen.

Recent work analyzing the transcriptome and metabolome of diseased tissues in some of these pathosystems indicates that major changes occur in plant hormone metabolism. During gall formation in clubroot (see Chapter 1, Figure 1.3), host genes associated with cytokinin biosynthesis, signaling, and breakdown are repressed, while there is an increase in auxin levels. Plant tumors induced by *Ustilago maydis* (see Chapter 7, Figure 7.13) contain much higher concentrations of auxin than uninfected tissue, and there are also altered levels of ABA and cytokinins. The fungus *Moniliophthora perniciosa* invades young shoots of cocoa trees and activates multiple meristems, leading to repeated branching and formation of a characteristic "witch's broom"; increased expression of auxin- and

Table 8.3 Some pathogens which cause abnormal growth in the host plant

Pathogen	Host plant	Disease
Fungi		
Plasmodiophora brassicae[a]	Crucifers	Clubroot
Synchytrium endobioticum	Potato	Wart disease
Gibberella (Fusarium) fujikuroi	Rice	Bakanae (foolish seedling)
Moniliophthora (Crinipellis) perniciosa	Cocoa	Witch's broom
Taphrina deformans	Peach	Leaf curl
Ustilago maydis	Maize	Smut
Bacteria/Phytoplasmas		
Agrobacterium tumefaciens	Various	Crown gall
Pseudomonas syringae pv. *savastanoi*	Olive	Knot disease
Phytoplasmas	Various	Phyllody, witch's broom, dwarfism
Virus		
Cocoa swollen shoot virus (CSSV)	Cocoa	Swollen shoot
Potato leaf roll virus (PLRV)	Potato	Leaf roll
Beet necrotic yellow vein virus (BNYVV)	Sugar beet	Rhizomania

[a] *Plasmodiophora* is now classified as a protozoan.

gibberellin-responsive genes can be detected in developing brooms. These results confirm that infection by these pathogens leads to a reprogramming of plant hormone metabolism.

Genes for the biosynthesis of hormonally active compounds are present in the genomes of several of the fungi listed in Table 8.3, but the precise role played by hormones produced by the pathogen in these diseases is not yet fully resolved. The ergot fungus *Claviceps purpurea* is a specialized floral pathogen that invades the ovaries of grasses and cereals and converts them to fungal sclerotia (see Chapter 4, Figure 4.16). Analysis of the *C. purpurea* genome identified genes for cytokinin production by two different routes, and strains mutated in both pathways are depleted in CKs and also compromised in pathogenicity. This suggests that CKs act as virulence factors during host invasion by this biotrophic parasite.

Gibberellins are to some extent synonymous with plant disease as this class of physiologically active compounds was first discovered in culture filtrates of the fungus *Gibberella fujikuroi*, the causal agent of "foolish seedling" disease of rice. Infected plants outgrow normal plants and become weak and spindly, often collapsing. This symptom is linked with the production of GAs by the pathogen in infected seedlings. Conversely, the stunting caused by some viruses can be reversed by applying exogenous gibberellins, suggesting that the reduced growth of these plants might be due to a decreased gibberellin content.

Growth Disorders Caused by Bacterial Pathogens

Infection by bacteria and phytoplasmas can also lead to abnormal development in the host plant (Table 8.3). Phytoplasmas, in particular, are typically associated with a range of growth disorders including stunting, increased branching (witch's brooms and bunchy top) and phyllody, in which floral organs are converted into leaf-like structures. Such symptoms again suggest a profound disturbance to plant hormone balance. Changes in the levels of key plant growth regulators have been detected in such altered tissues, but the extent to which hormone production by the phytoplasma itself is involved has not been resolved.

Bacteria belonging to the *Pseudomonas syringae* complex are known to synthesize IAA, and recent work has implicated auxin as a potential virulence factor suppressing host defense responses. A more defined role for IAA produced by the pathogen has been shown in some pathovars causing growth disorders. *Ps. syringae* pv. *savastanoi* infects woody hosts such as olives, oleander, and privet, causing tumor-like overgrowths in which cell division, cell enlargement, and abnormal differentiation of vascular elements take place. Hormones produced by the bacterium are the main determinants of these symptoms. The biosynthetic pathway for conversion of the amino acid tryptophan into IAA involves two enzymes, tryptophan monooxygenase and indoleacetamide hydrolase, and genes encoding these two enzymes have been identified in the pathogen. Interestingly, in some bacterial strains these genes are located on a plasmid, rather than the bacterial chromosome. Loss of this plasmid, so that the bacterium is unable to synthesize IAA, prevents gall formation, although some limited invasion of host tissues can still take place. IAA-producing strains cause tissue proliferation and hence create a niche in which the pathogen can multiply to high population levels.

The classic example of a plant tumor is crown gall disease caused by *Agrobacterium tumefaciens*. The closely related *A. rhizogenes* causes hairy root disease, characterized by the formation of abnormal, root-like processes on aerial parts of the plant (Figure 8.14). Both pathogens have a very wide host range, although under natural conditions crown gall is mainly a problem in woody plants such as fruit trees or roses.

Crown gall has long attracted interest as, following initial infection via a wound, secondary galls are formed which do not contain the pathogen. These tumors are therefore autonomous and bear some resemblance to animal cancers. The identity of the factor which transforms healthy cells into rapidly dividing tumor cells was originally unknown, and described only as the "tumor-inducing principle." Subsequently, it was shown that virulent *A. tumefaciens* and *A. rhizogenes* strains contain large plasmids, described as either tumor-inducing (Ti) or root-inducing (Ri) plasmids respectively. During infection, part of the plasmid is transferred to the plant cell, where it integrates into the host genome (Figure 8.15a). The transferred (or T) region of plasmid DNA carries several genes which, when expressed in the host cell, cause the characteristic tumor or hairy root symptoms. The *onc* (as in oncogenic – cancer-causing) genes specify the synthesis of auxins and cytokinins, leading to uncontrolled cell division, while other genes code for the production of unusual amino acids known as opines, which cannot be utililized by the plant but are instead catabolized by the pathogen. *Agrobacterium* therefore acts as a natural "genetic engineer," reprogramming host cells to divide rapidly and then to produce amino acids which serve as a carbon and nitrogen source for the pathogen.

(a)   (b)

Figure 8.14 Crown gall *Agrobacterium tumefaciens* and hairy root disease *A. rhizogenes* on tomato, showing (a) tumor at stem base, and (b) hairy roots formed on stem. *Source:* Photos provided by M.R. Davey.

(a)

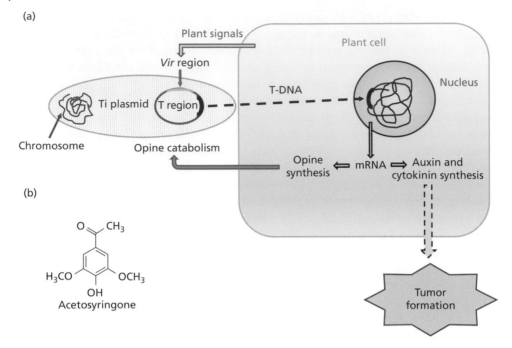

(b)

Figure 8.15 Molecular processes underlying tumor induction in crown gall disease. (a) The bacterium attaches to the plant cell wall, where plant signal molecules such as the phenolic compound acetosyringone (b) activate the Vir region of the Ti-plasmid. Transfer of the T-DNA and integration into the plant nuclear genome then take place. Expression of the T-DNA leads to the production of opines catabolized by the bacterium and plant hormones inducing tumor formation. *Source:* Modified from Melchers and Hooykaas (1987).

Further work identified other important components of the transformation process. Prior to transfer, the T-DNA must first be excised from the circular plasmid. This is regulated by another set of plasmid genes in the Vir (virulence) region. Activation of the *vir* genes takes place in response to certain wound chemicals, such as the phenolic compound acetosyringone (Figure 8.15b), leaking from damaged plant cells. Hence, the pathogen initiates mobilization of the T-DNA in response to a specific plant signal.

Because of its distinctive properties, the Ti plasmid has been widely used as a vector for the deliberate introduction of foreign genes into plant cells. Clearly, in any genetic engineering program designed to improve plants, it would be undesirable to introduce the *onc* genes responsible for hormone synthesis and disordered growth. These are therefore deleted prior to use, to give a "disarmed" plasmid which, while retaining the ability to stably integrate into the plant genome, produces none of the pathogenic effects of a natural infection with the bacterium.

Conclusion

Plant pathogens possess a variety of features which enable them to colonize and exploit host plants. Some of these are specific structures enabling entry to the host, such as appressoria and infection hyphae (Chapter 6), while others are extracellular secretions, enzymes, toxins or hormones, many of which contribute to pathogenesis. Export systems play an important role by transporting biologically active molecules across the host–pathogen interface. Disruption of genes encoding such putative pathogenicity factors has, in some cases, confirmed a role in disease but in others given inconclusive results. Loss of a particular product, for instance a single enzyme, does not necessarily reduce the capacity of the pathogen to cause disease. Instead, a suite of different genes and products contribute to successful infection and symptom development. Many virulence functions are only switched on in the presence of the host, often in response to a specific signal, which therefore serves as a host "signature." Host induction of microbial pathogenicity factors ensures that the biosynthetic investment required to produce them is expended only in situations where they will aid the growth and survival of the pathogen.

From a practical viewpoint, identification of such signal compounds and the pathways regulating pathogenicity should provide new opportunities for intervening in the disease process.

A Vital Role for Pathogen Effectors

In the introduction to Part II, a model was proposed for the interactions taking place between pathogens and their plant hosts during infection. Plants developed surveillance systems enabling them to detect potential invaders and trigger host defense mechanisms. In turn, the invaders deploy molecules, known as effectors, that interfere with plant defense. One of the most important revelations is that effectors are produced by a range of biotic agents attacking plants, including bacteria, fungi, oomycetes, nematodes, and aphids. They are chemically diverse, but most are small secreted proteins that act in the apoplast or are translocated into the host cell. Recent discoveries suggest that these proteins target common components of the plant immune system. For this reason, further discussion of effectors is deferred to Chapter 10, once the key elements of plant immunity have been described in Chapter 9.

Further Reading

Reviews and Papers

Arnold, D.L. and Jackson, R.W. (2011). Bacterial genomes: evolution of pathogenicity. *Current Opinion in Plant Biology* 14 (4): 385–391.

Collemare, J., Griffiths, S., Iida, Y. et al. (2014). Secondary metabolism and biotrophic lifestyle in the tomato pathogen *Cladosporium fulvum*. *PLoS One* 9 (1) https://doi.org/10.1371/journal.pone.0085877.

Friesen, T.L., Stukenbrock, E.H., Liu, Z.H. et al. (2006). Emergence of a new disease as a result of interspecific virulence gene transfer. *Nature Genetics* 38 (8): 953–956.

Gan, P., Ikeda, K., Irieda, H. et al. (2013). Comparative genomic and transcriptomic analyses reveal the hemibiotrophic stage shift of *Colletotrichum* fungi. *New Phytologist* 197 (4): 1236–1249.

Gonzalez-Mula, A., Lang, J.L., Grandclement, C. et al. (2018). Lifestyle of the biotroph *Agrobacterium tumefaciens* in the ecological niche constructed on its host plant. *New Phytologist* 219 (1): 350–362.

Howlett, B.J. (2006). Secondary metabolite toxins and nutrition of plant pathogenic fungi. *Current Opinion in Plant Biology* 9 (4): 371–375.

Jashni, M.K., Mehrabi, R., Collemare, J. et al. (2015). The battle in the apoplast: further insights into the roles of proteases and their inhibitors in plant-pathogen interactions. *Frontiers in Plant Science* 6: 584.

Kemen, A.C., Agler, M.T., and Kemen, E. (2015). Host-microbe and microbe-microbe interactions in the evolution of obligate plant parasitism. *New Phytologist* 206 (4): 1207–1228.

Killiny, N., Martinez, R.H., Dumenyo, C.K. et al. (2013). The exopolysaccharide of *Xylella fastidiosa* is essential for biofilm formation, plant virulence, and vector transmission. *Molecular Plant-Microbe Interactions* 26 (9): 1044–1053.

Koczan, J.M., McGrath, M.J., Zhao, Y.F. et al. (2009). Contribution of *Erwinia amylovora* exopolysaccharides amylovoran and Levan to biofilm formation: implications in pathogenicity. *Phytopathology* 99 (11): 1237–1244.

Kubicek, C.P., Starr, T.L., and Glass, N.L. (2014). Plant cell wall-degrading enzymes and their secretion in plant-pathogenic fungi. *Annual Review of Phytopathology* 52: 427–451.

Lanver, D., Muller, A.N., and Happel, P. (2018). The biotrophic development of *Ustilago maydis* studied by RNA-seq analysis. *Plant Cell* 30 (2): 300–323.

McClerklin, S.A., Lee, S.G., Harper, C.P. et al. (2018). Indole-3-acetaldehyde dehydrogenase-dependent auxin synthesis contributes to virulence of *Pseudomonas syringae* strain DC3000. *PLoS Pathogens* 14 (1) https://doi.org/10.1371/journal.ppat.1006811.

Nester, E.W. (2015). Agrobacterium: nature's genetic engineer. *Frontiers in Plant Science* 5: 730. https://doi.org/10.3389/fpls.2014.00730.

O'Connell, R.J., Thon, M.R., Hacquard, S. et al. (2012). Lifestyle transitions in plant pathogenic *Colletotrichum* fungi deciphered by genome and transcriptome analyses. *Nature Genetics* 44 (9): 1060–1065.

Paccanaro, M.C., Sella, L., Castiglioni, C. et al. (2017). Synergistic effect of different plant cell wall-degrading enzymes is important for virulence of *Fusarium graminearum*. *Molecular Plant-Microbe Interactions* 30 (11): 886–895.

Pusztahelyi, T., Holb, I.J., and Pocsi, I. (2015). Secondary metabolites in fungus-plant interactions. *Frontiers in Plant Science* 6: 573. https://doi.org/10.3389/fpls.2015.00573.

Spanu, P.D., Abbott, J.C., Amselem, J. et al. (2010). Genome expansion and gene loss in powdery mildew fungi reveal tradeoffs in extreme parasitism. *Science* 330 (6010): 1543–1546.

Stergiopoulos, I., Collemare, J., Mehrabi, R. et al. (2013). Phytotoxic secondary metabolites and peptides produced by plant pathogenic Dothideomycete fungi. *FEMS Microbiology Review* 37 (1): 67–93.

Toth, I.K. and Birch, P.R.J. (2005). Rotting softly and stealthily. *Current Opinion in Plant Biology* 8 (4): 424–429.

Vrancken, K., Holtappels, M., Schoofs, H. et al. (2013). Pathogenicity and infection strategies of the fire blight pathogen *Erwinia amylovora* in Rosaceae: state of the art. *Microbiology* 159: 823–832.

Williams, B., Kabbage, M., Kim, H.J. et al. (2011). Tipping the balance: *Sclerotinia sclerotiorum* secreted oxalic acid suppresses host defenses by manipulating the host redox environment. *PLoS Pathogens* 7 (6): e1002107.

Wolpert, T.J. and Lorang, J.M. (2016). Victoria blight, defense turned upside down. *Physiological and Molecular Plant Pathology* 95: 8–13.

9

Plant Defense

Plants have evolved with pathogens and insect pests for millions of years. It is therefore not surprising that a particular plant is resistant to most of them.

(Noel Keen, 1940–2002)

Plants are continually exposed to insects, nematodes, and other potentially damaging pests, as well as a wide variety of parasitic microorganisms. Yet the majority of plants remain healthy most of the time. This observation suggests that plants must possess highly effective mechanisms for preventing parasitism and predation, or at least limiting their effects. Plants, unlike animals, are sedentary organisms so cannot escape their enemies, and therefore have to rely on a series of defenses to ward off potential invaders. These include physical barriers, repellent or inhibitory chemicals, and an innate immune system mobilized once under attack. The combined action of these different layers of defense is usually sufficient to limit or prevent significant damage by pathogens and predators.

Types of Plant Defense

Plant defenses act at several points during the infection process. External structural barriers effectively exclude the majority of organisms; should these be breached then chemical inhibitors can slow the invasion process. At the same time, a sensitive surveillance system which can detect foreign cells triggers a rapid response to microbial attack. Plant defense systems can therefore be classified as either **passive (constitutive)** or **active (inducible)**, depending upon whether they are preexisting features of the plant or are switched on after challenge (Figure 9.1). These categories can be further subdivided into structural and chemical mechanisms, although such divisions are not mutually exclusive; for instance, a chemical inhibitor might be a component of a structural barrier. The picture of plant defense emerging from recent research is of a finely tuned and integrated system in which different components act together in a coordinated and complementary fashion to contain infection.

Passive anatomical features, such as the cuticle and bark, represent highly effective obstacles to penetration by most microorganisms (see p.143). Every plant cell is itself surrounded by a substantial obstacle, the cell wall. Where this is impregnated with chemicals such as suberin and lignin, it forms an even more effective barrier against pathogens.

Plant Pathology and Plant Pathogens, Fourth Edition. John A. Lucas.
© 2020 John Wiley & Sons Ltd. Published 2020 by John Wiley & Sons Ltd.
Companion website: www.wiley.com/go/Lucas_PlantPathology4

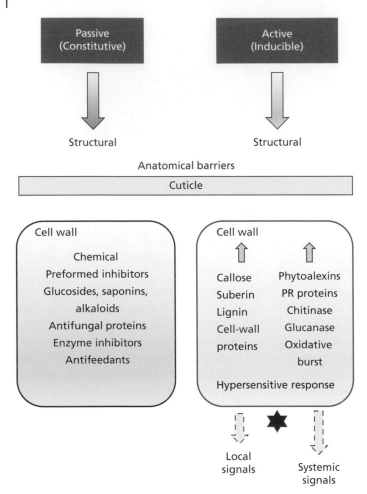

Figure 9.1   Types of plant defense, based on existing anatomical or structural features, or active changes induced after challenge by pathogens.

Preformed Inhibitors

Plants synthesize a vast array of secondary metabolites, many of which are toxic to potential pests and pathogens. Compounds such as phenols, alkaloids, glycosides, saponins, tannins, and resins all possess antibiotic properties and may therefore contribute to resistance. The remarkable chemical diversity of the plant kingdom has been partly linked to this role in defense, with each plant family evolving a different set of metabolites capable of repelling herbivorous animals or inhibiting pathogens. The well-known resistance of some plant materials, especially wood, to microbial decay is due in part to the presence of inhibitors such as terpenes, tropolones, and stilbenes (Figure 9.2) which act as natural preservatives.

   A direct correlation between the presence of a toxic compound in plant tissues and resistance to a pathogen has been established in relatively few cases. One of the first examples concerned smudge disease of onions, caused by *Colletotrichum circinans,* where the resistance of bulbs to infection was linked to two phenolic compounds, catechol and protocate-

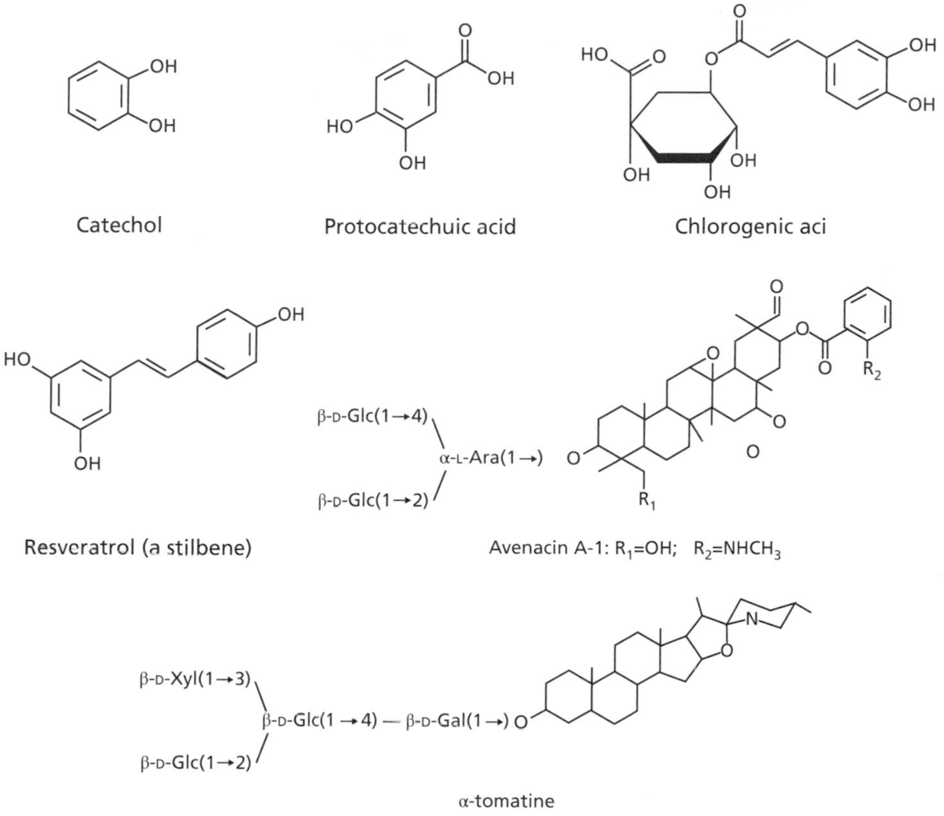

Figure 9.2 Some preformed antimicrobial compounds from plant tissues.

chuic acid (Figure 9.2), found mainly in the pigmented outer scales. Another preformed phenolic implicated in plant defense is chlorogenic acid, found in potato tubers resistant to common scab, *Streptomyces scabies*.

Evidence that preformed inhibitors may play a key role in host–pathogen interaction has been obtained from detailed studies on the host specificity of pathogens. The take-all fungus *Gaeumannomyces graminis* occurs as several pathogenic varieties which differ in their ability to attack different cereal hosts. *G. graminis* var. *tritici* (*Ggt*) is pathogenic to wheat but not oats, whereas var. *avenae* (*Gga*) can attack oats as well as wheat. Young oat roots, the normal site of infection by the take-all fungus, contain several chemically similar fluorescent saponin compounds known as avenacins (Figure 9.2), which are toxic to many fungi. Such saponins act to disrupt cells by complexing with sterols in membranes. Enzymatic removal of the sugar residues from avenacin reduces its toxicity. *Gga* produces significant amounts of the enzyme, avenacinase, which carries out this reaction, whereas *Ggt* produces none. The ability of the oat-attacking form of the fungus to colonize this host can therefore be linked to its ability to detoxify a preformed inhibitor.

Confirmation that production of avenacinase is a key determinant of the host range of the take-all fungus was obtained by isolating the gene encoding the enzyme from *Gga*, and then disrupting it to make a nonfunctional version which was reintroduced into the fungus. Pathogen strains in which the original functional avenacinase gene had been replaced

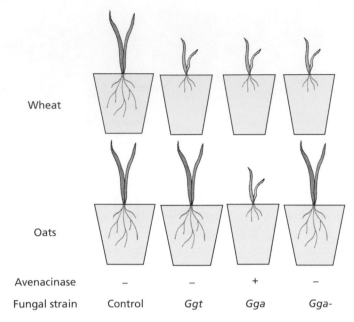

Wheat			
Oats			
Avenacinase	–	–	+	–
Fungal strain	Control	*Ggt*	*Gga*	*Gga-*

Figure 9.3 Relationship between production of the saponin-detoxifying enzyme avenacinase and host range in the take-all fungus *Gauemannomyces graminis*. Variety *tritici* (*Ggt*) does not produce the enzyme and is nonpathogenic to oats, which contain avenacin. Variety *avenae* (*Gga*) produces the enzyme and is able to infect oats. *Gga* strains in which the gene for avenacinase has been disrupted (*Gga-*) lose the ability to produce the enzyme and can no longer infects oats, but retain pathogenicity to wheat. *Source:* Based on Bowyer et al. (1995).

by a disrupted version no longer produced the enzyme, and had simultaneously lost the ability to infect oat roots (Figure 9.3). However, these mutants strains were still able to attack wheat roots, showing that their pathogenicity was unimpaired.

Interestingly, enzymes with structural, biochemical, and genetic sequence similarities to avenacinase have now been detected in some other groups of pathogenic fungi. For instance, *Septoria lycopersici*, a pathogen infecting tomato leaves, produces a related enzyme, tomatinase, active against the alkaloidal saponin tomatine (Figure 9.2) present in tomato leaf tissues. Thus, unrelated fungal pathogens appear to possess similar enzymatic mechanisms for dealing with the toxins they encounter when colonizing their particular hosts. It is possible that such enzymes might have diversified and developed distinct substrate specificities during evolution from a common ancestral form, encoded by a single original gene.

Inhibitors in plant tissues often occur as nontoxic precursors, which are converted to active forms following cell damage or exposure to enzymes. Apple and pear leaves contain glycosides such as arbutin, which are hydrolyzed by the enzyme β-glucosidase to form glucose, plus toxic quinones inhibitory to the fireblight bacterium *Erwinia amylovora*. Other plants including some legumes contain cyanogenic glycosides which release the familiar poison hydrogen cyanide as a breakdown product (Figure 9.4). Brassica crops such as cabbage are rich in sulfur-containing glucosinolates, which degrade to form volatile isothiocyanates inhibitory to several fungal pathogens. The cellular disruption caused by pathogens as they grow through host tissues may therefore initiate enzymatic and chemical reactions leading to a highly toxic local environment. These changes are induced by infection but as they involve conversion of preformed compounds, they are distinct from defense responses involving active host metabolism.

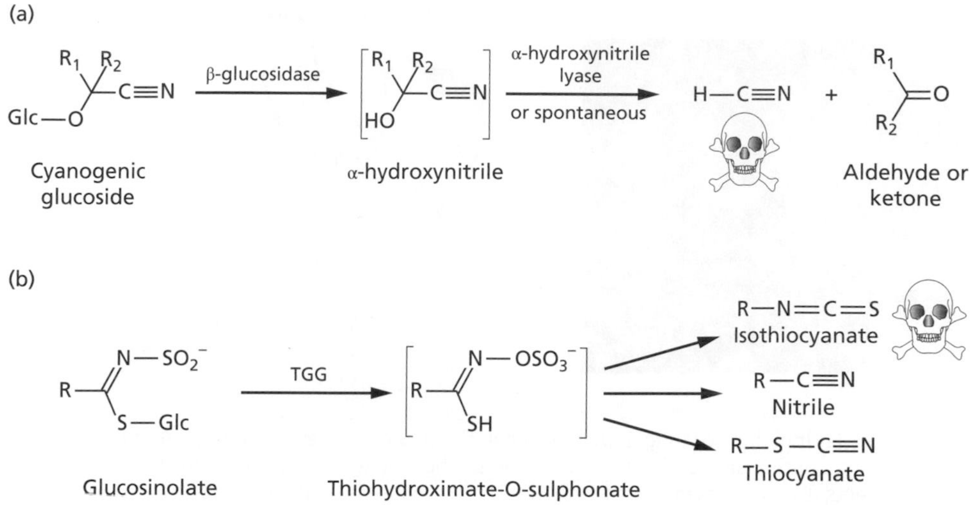

(a)

Cyanogenic glucoside → α-hydroxynitrile → Aldehyde or ketone

(b)

Glucosinolate → Thiohydroximate-O-sulphonate → Isothiocyanate / Nitrile / Thiocyanate

Figure 9.4 Examples of nontoxic glycosides converted to inhibitory breakdown products. (a) Cyanogenic glucoside breaks down to form the highly toxic hydrogen cyanide (*red*). (b) Glucosinolate converted to several end-products including toxic isothiocyanate (*red*). TGG is the enzyme β-thioglucoside glucohydrolase. *Source:* Adapted from Piasecka et al. (2015).

Antimicrobial Proteins

Extracts from seeds or other nutrient-rich plant storage tissues often prove to be inhibitory to the growth of microorganisms or invertebrates when tested in culture or added to insect diets. Figure 9.5 shows an example where exudate from germinating radish seeds prevents growth of a fungal pathogen in the zone surrounding the seed. This antibiotic effect helps to protect the emerging seedling during the vulnerable early stages of development. Analysis of such exudates has shown that the biological activity is due to the presence of potent antifungal peptides, known as defensins, which are released from the seed as it imbibes water. Defensins appear to act by inserting into the fungal cell membrane and increasing its permeability.

Plants contain several types of proteins which serve a defensive function (Table 9.1). They include the antimicrobial peptides found in seeds, toxic cell wall proteins known as thionins found in cereal leaves, hydrolytic enzymes such as glucanases, chitinases, and lysozyme, which can directly attack the cells walls of fungi and bacteria, and various classes of enzyme inhibitors. The latter are often abundant in storage tissues and show activity against digestive enzymes such as proteases occurring in the digestive tract of insects. For instance, cowpea seeds contain a trypsin inhibitor (CpTI) that may contribute to the resistance of such seeds to predation by insects. Other proteins may specifically inhibit microbial enzymes produced during tissue colonization.

The previous chapter discussed the role played by pectolytic enzymes such as polygalacturonase (PG) in the digestion of plant cell walls. Proteins present in the cell walls of many plants, for instance beans, can inhibit the action of PGs produced by fungi, thus interfering with the infection process. Such polygalacturonase-inhibiting proteins (PGIPs) not only enhance the resistance of plant cell walls to enzymic attack, but also influence the nature of any end-products released. Rather than being digested completely

(a)

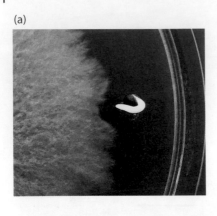

(b)

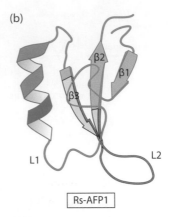

Rs-AFP1

Figure 9.5 (a) Inhibition of fungal growth by germinating radish seed and (b) three-dimensional structure of an antifungal peptide (AFP) from radish showing α-helix (*green*), loops (L1 and L2) and three β-sheets (β1–3). This conserved structure is believed to allow the peptide to insert into microbial membranes (b). *Source:* Lacerda et al. (2014).

Table 9.1 Some plant proteins that confer resistance to pests and pathogens

Type	Examples	Biological activity
Antimicrobial peptides	Defensins Thionins	Disrupt pathogen cell membranes
Hydrolase enzymes	Chitinase Glucanase	Digest fungal cell walls
	Lysozyme	Lyses bacterial cells
Enzyme inhibitors	Polygalacturonase-inhibiting proteins (PGIPs)	Inhibit fungal pectinases
	Protease inhibitors Amylase inhibitors	Inhibit insect digestive enzymes
Chitin-binding proteins	Hevein (in latex from rubber)	Inhibits growth of fungi
	Lectins	Inhibit growth of insects and fungi
Lipid transfer proteins	Various	Inhibit growth of fungi
Ribosome-inactivating proteins (RIPs)	Pokeweed antiviral protein (PAP) Dianthin from carnations	Inhibit viral replication and transmission

to monomers, pectic substrates are partially degraded to oligomers, many of which are active inducers of plant defense responses (see p.190). These inhibitory proteins therefore contribute to an early warning system whereby plant cells can detect the presence of an invading fungus.

Another well-characterized group with a possible role in plant defense are the lectins, many of which are secretory proteins that accumulate in cell walls or vacuoles. Lectins bind to different combinations of sugars present in polysaccharides, glycoproteins, and glycolipids. They attach to the surface of cells containing particular sugar residues, and may thus form part of a recognition system. For instance, one series (or family) of lectins binds to molecules containing the amino sugar N-acetyl glucosamine, the basic building block of chitin. Chitin-binding lectins are commonly found in cereal grains, as well as unrelated plants such as tomato and potato, and in the latex of rubber trees. These proteins, while widely distributed in the plant kingdom, share some sequence homology, suggesting that they may all have evolved from an ancestral gene with a common function. Chitin is found in the insect exoskeleton and is also a component of the cell walls of most fungi; these lectins may thus function in defense against both herbivorous insects and fungi. Binding of the protein to the fungus may slow hyphal growth, thereby creating time for other defense systems to act. If eaten by animals, other lectins bind to the digestive tract, preventing effective absorption of nutrients. Together, these proteins may protect plant tissues, and especially nutrient-rich seeds, against infection or predation.

Antiviral Proteins

It has been known for many years that protein extracts from certain plant species reduce the infectivity of plant viruses. For instance, when proteins from carnation and related *Dianthus* species are mixed with tobacco mosaic virus (TMV) prior to inoculation, local lesions are reduced or completely suppressed. Two theories have been proposed to explain this effect. One is that the proteins involved directly inactivate the virus; the second is that these inhibitors interfere with virus replication via effects on the host plant. Some evidence supports the latter view, as the molecules concerned appear to inhibit protein synthesis by affecting ribosomes in host cells and hence block virus replication. Consequently, these inhibitors are described as **ribosome-inactivating proteins** or RIPs.

The examples above illustrate some of the biochemical diversity of plant proteins with a proposed role in constitutive resistance to pathogens and pests. In fact, some of these inhibitors may also be produced in response to attempted infection or other forms of stress. Synthesis of antifungal thionins in cereal leaves, for instance, is triggered following inoculation with foliar pathogens such as powdery mildews. The distinction between constitutive and inducible resistance mechanisms is therefore not absolute, and the regulation of these defense proteins may vary from tissue to tissue. Some also serve as storage proteins in seeds, and hence perform more than one function in the life of the plant.

There has been considerable interest in plant defense proteins because many of them have been sequenced and the genes encoding them isolated, providing the opportunity to engineer transgenic plants expressing such proteins to improve resistance to both pathogens and pests. However, some types of antimicrobial proteins, such as the RIPs, also possess high toxicity to mammals which is likely to limit their practical use.

Active Defense Mechanisms

The various structural and chemical components described in the previous section serve as a first line of defense against microbial attack. These barriers are highly effective against the large majority of potential colonists. Nevertheless, in the constant evolutionary race between plant and pathogen, a proportion of microorganisms have developed pathogenicity factors able to breach or inactivate preformed plant defenses. Such pathogens then face a series of active defense responses induced once penetration of the host begins. The plant immune surveillance system based on pattern recognition receptors (PRRs) perceives the presence of chemical "signatures" characteristic of foreign invaders and triggers a complex of defense mechanisms.

Host Reactions to Penetration

The initial events during host–pathogen interaction appear to be similar irrespective of whether the combination is compatible or incompatible (see Chapter 2, Figure 2.6). Thus, in the case of a fungal pathogen penetrating a highly resistant host, spore germination, growth of the germ tube, appressorium formation, and subsequent entry of the infection hypha can take place on a comparable time scale to these events on a host lacking resistance. At this early stage in the interaction, it is often impossible to distinguish between the two host reaction types. Once the pathogen begins to breach the cell wall, however, events take a very different course, depending upon the degree of host resistance. If the combination is compatible, further hyphal growth and invasion of host tissues continue unrestricted. In resistant hosts, however, a number of changes take place in penetrated cells and adjacent tissues which ultimately halt the advance of the pathogen.

Changes in Host Cell Walls

The first detectable response of a plant cell to an invading microorganism is often an alteration in the appearance or properties of the cell wall. For instance, attempted penetration of cereal leaves by nonpathogenic or avirulent fungi is accompanied by the deposition of a plug of material, known as a papilla, directly beneath the penetration site (Figure 9.6). The epidermal cell wall surrounding the papilla may be changed to leave a disc-shaped zone or halo. Similar thickening and modification of host cell walls has been observed in other plant–fungus interactions; root cortical cells penetrated by root-infecting fungi may form characteristic protrusions of the wall known as lignitubers. Collectively, all these structures involving the accretion of new wall material are described as **wall appositions**.

Morphologically, wall deposits of various kinds appear similar, but what are they made of? Studies using histochemical stains, fluorescence microscopy, and X-ray microanalysis have shown that quite different types of material accumulate, ranging from polysaccharides and proteins to mineral elements such as calcium and silicon (Table 9.2). The β-1,3 glucan callose has been assumed to be the main constituent of papillae, but recent studies on the infection of barley epidermal cells by the powdery mildew fungus *Blumeria graminis* f.sp. *hordei* have shown that other cell wall polysaccharides such as cellulose and the hemicellulose arabinoxylan also accumulate and are important in reinforcing papillae against penetration by the fungus (Figure 9.6). Resistance of cell walls may be enhanced by impregnation with inhibitory phenolic compounds, some of which are themselves precursors of the complex polymer lignin involved in secondary thickening of cell walls in normal plant development. Induced lignification is a conserved defense mechanism against pathogens

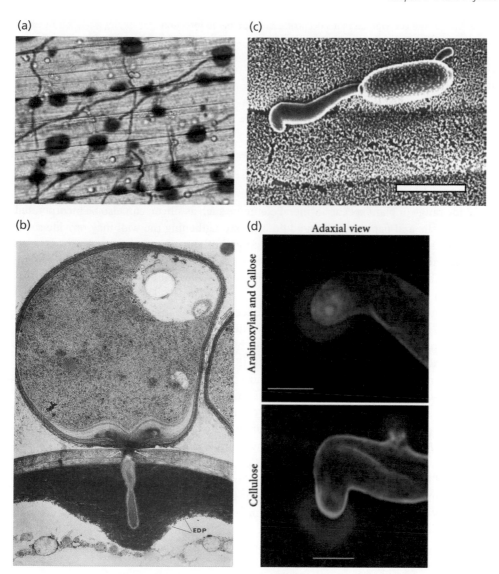

Figure 9.6 Cell wall reactions to penetration by fungal pathogens. (a) Stained epidermal strip from resistant cereal leaf inoculated with the leaf blotch pathogen *Parastagonospora nodorum* showing superficial hyphae and blue haloes of phenolic compounds at sites of attempted penetration. (b) Electron micrograph of appressorium of *Colletotrichum graminicola* on a resistant oat variety. The epidermal cell wall is penetrated but the infection peg is encased by an electron-dense papilla (EDP), preventing invasion of the cell (×16 000). *Source:* Politis (1976). (c) Scanning electron micrograph of germling of the powdery mildew fungus *Blumeria graminis* on surface of cereal leaf showing small primary germ tube, and appressorial germ tube with hooked tip. Bar = 25 μm. *Source:* Carver and Thomas (1990). (d) Fluorescent confocal images of the appressorium tip of *B. graminis* penetrating the epidermal cell wall with accumulation of hemicellulose (*purple*) and cellulose (*red*) in papillae formed in reaction to attempted invasion by the pathogen. Bars = 5 μm. *Source:* Chowdhury et al. (2014).

Table 9.2 Some cell wall changes occurring in response to infection

- Deposition of callose
- Deposition of suberin
- Addition of cellulose and hemicellulose
- Impregnation with phenolic compounds
- Accumulation of calcium, silicon or sulfur
- Lignification
- Changes in amount or type of cell wall proteins
- Oxidative cross-linking of cell wall proteins

in a wide range of plant species. Mineral elements such as silicon can also be incorporated into cell walls and papillae, and in addition to strengthening the wall may provide some protection against microbial enzymes digesting wall polymers. Surprisingly, it has even been shown that resistant cocoa (*Theobroma cacao*) varieties reacting to invasion by the vascular wilt fungus *Verticillium dahliae* accumulate elemental sulfur, a well-known fungicide, in cell walls adjacent to infected vessels.

Increases in the content of various classes of cell wall proteins also take place as an active response to invasion by pathogens. Typically, these structural proteins contain a high proportion of particular amino acids, for instance hydroxyproline and glycine, hence they are termed hydroxyproline-rich glycoproteins (HRGPs) and glycine-rich proteins (GRPs). Following infection, HRGPs accumulate in host cell walls and may contribute to resistance by binding to pathogen cells or acting as structural barriers and sites for lignin deposition. It has been shown that this response occurs earlier in incompatible host–pathogen combinations than compatible combinations. Figure 9.7 shows the time course of production of messenger RNA encoding HRGPs in bean hypocotyls inoculated with either a virulent or avirulent race of the anthracnose pathogen *Colletotrichum lindemuthianum*. Spore germination and formation of appressoria occur in both combinations, leading to host penetration at around 50 hours post inoculation. In the incompatible interaction, there is an early peak of HRGP mRNA, coinciding with induction of a resistance response by penetrated cells. An increase in HRGP mRNA also occurs in the compatible combination but later, coinciding with the onset of lesion formation. By this time, the pathogen is already invading the susceptible host.

During studies analyzing protein extracts of cell walls from plant cells exposed to microbial challenge, an unexpected observation was made. Certain types of HRGPs rapidly disappeared from the extracts following treatment of the cells with chemical inducers of plant defense. At first sight, this observation seemed to conflict with the notion that cell walls are reinforced by the addition of new protein and other polymers following infection. The mystery was solved when it was shown that the apparent disappearance of these proteins was in fact due to a sudden change in their properties. Cross-linking of amino acid chains in the proteins alters their solubility so that they remain bound to the wall during the extraction procedure, and hence are no longer detected. Insolubilization of these proteins appears to toughen the cell wall and also increases its resistance to digestive enzymes. The cross-linking reaction is extremely rapid, taking place within a few minutes of challenge, and hence is complete before any of the molecular changes based on gene activation occur. The reaction appeared to be triggered by sudden changes in oxidative metabolism, and in particular

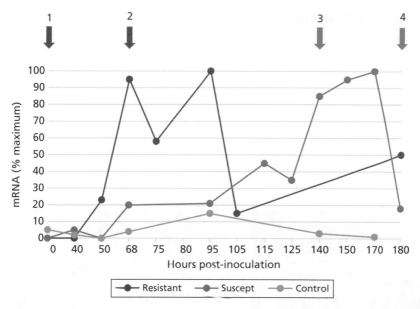

Figure 9.7   Time course of accumulation of mRNA for hydroxyproline-rich glycoprotein in *Phaseolus* bean hypocotyl tissues inoculated with different races of *Colletotrichum lindemuthianum*. Time points: 1. inoculation with spores; 2. hypersensitive flecks in R combination, no visible symptoms in S; 3. onset of symptoms in S combination; 4. water-soaked lesions in S. *Source:* Adapted from Showalter et al. (1985).

the generation of hydrogen peroxide at the cell surface (see p.223). Hence, transcription of host genes is not involved in this response. Similarly, the synthesis of callose (Table 9.2) appears to rely on enzyme activation rather than gene transcription.

The accumulation of reaction material in cell walls is one of the earliest and more distinctive plant responses to challenge by a pathogen, but is clearly part of a more complex cellular reaction. Genetic analysis of *Arabidopsis* mutants impaired in penetration resistance to powdery mildews from other host species (i.e., increased in susceptibility of what is usually a nonhost) has identified several genes playing different roles in the process, including a syntaxin protein involved in trafficking of secretory vesicles (*PEN1*), a glycoside hydrolase enzyme (*PEN2*) implicated in production of a toxic compound from a precursor (Figure 9.4), and an ABC transporter (*PEN3*) possibly exporting antimicrobial compounds. These genes may act together to focus defenses at sites of attempted penetration. The well-known and durable resistance of cereals to powdery mildews based on the *mlo* gene may also act via a protein complex targeting inhibitory compounds to such sites. Hence, papillae may vary in composition and function in several ways, for instance by increasing the strength of the cell wall, enhancing resistance to enzymic attack, or changing its permeability. The monomeric precursors of wall polymers such as lignin are themselves toxic to microorganisms, and further inhibitors may also be targeted at penetration sites.

What is not yet clear, however, is the extent to which any or all of these functions actually determine the outcome of a particular host–pathogen confrontation. For instance, some *Arabidopsis* mutants impaired in callose synthesis are still able to prevent penetration by powdery mildew fungi, while conversely, cell wall papillae are often successfully breached by penetrating hyphae, leaving a characteristic collar around the entry site.

The Hypersensitive Response

More than 100 years ago, the American plant pathologist Marshall Ward, working on rust fungi infecting grasses, noted that in resistant hosts the cells adjacent to the infection site rapidly became discolored, granular, and necrotic. A few years later the term "hypersensitive" was introduced to describe this type of reaction; host cells are apparently so sensitive to the pathogen that they collapse and die as soon as they are penetrated (Figure 9.8).

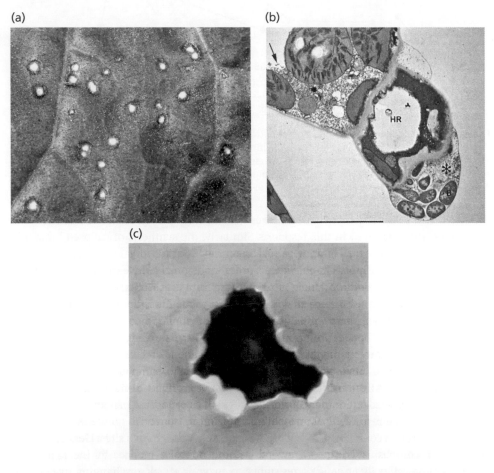

Figure 9.8   The hypersensitive response. (a) Local lesions formed on a tobacco leaf inoculated with TMV preventing systemic spread. (b) Electron micrograph of bean leaf tissues inoculated with an incompatible strain of the bacterium *Ps. syringae* pv. *phaseolicola*. A mesophyll cell has collapsed due to a hypersensitive response (HR) to adjacent colony of bacterial cells (B). The vacuolar membrane has ruptured (*arrow*) in the next host cell. Scale bar = 5 μm. *Source:* Brown and Mansfield (1988). (c) Single epidermal cell of *Brassica* cotyledon inoculated with an incompatible isolate of the downy mildew *Peronospora parasitica* undergoing HR at site of attempted penetration. Note granulation of cytoplasm, dark pigmentation, and deposition of fluorescent material on cell wall.

Initially, the hypersensitive response (HR) was thought to only be significant in interactions involving biotrophic fungi. Rapid death of host cells would be likely to prevent the fungus from establishing an effective relationship. For a while, this simple hypothesis seemed to be an adequate explanation for the basis of hypersensitive resistance. However, superficially similar and equally effective host reactions were found to occur with a wide range of different pathogens, including bacteria and viruses (Figure 9.8). The HR is now known to be a key feature of the plant innate immune system. In many respects, the local lesion reaction of plants to viruses bears similarities to an HR as necrosis of cells around the inoculation site prevents the virus from spreading systemically throughout the host.

The HR is a form of programmed cell death with some similarities to apoptosis, the tightly regulated death of cells occurring during the normal growth and development of multicellular organisms. It requires active metabolism and involves several organelles and subcellular compartments (Figure 9.9). The initial trigger for HR is recognition of molecules produced by pathogens, originally described as **elicitors**, which are now known to include pathogen-associated molecular patterns (PAMPS) and secreted effectors (see Chapter 10). The interaction of these pathogen products with either membrane-located PRRs or intracellular resistance (R) proteins sets in train a complex series of processes leading to changes in oxidative metabolism, the production of antimicrobial compounds known as phytoalexins and pathogenesis-related (PR) proteins, cell wall modification, membrane breakdown, and ultimately cell death. Early events include the influx of calcium (Ca^{2+}) ions and the production of reactive oxygen species (ROS) and nitric oxide (NO). After initial recognition of pathogen molecules, the signal is transduced by a cascade of kinase enzymes including mitogen-activated protein kinases (MAPKs) and calcium-dependent protein kinases (CDPKs). Signals entering the nucleus activate transcription factors such as the defense- and stress-related WRKY transcription factors which switch on genes required for the synthesis of antimicrobial proteins and secondary metabolites. At the same time, hormone signals such as salicylic acid (SA), jasmonic acid (JA), and ethylene are released, priming defense in neighboring cells or in more distant parts of the plant. The breakdown of the cell may itself release by-products, known as damage-associated molecular patterns (DAMPs), that serve as alarm signals for other cells.

There has been a long-standing debate about the role and significance of the HR in resistance to different types of plant pathogens. It is self-evident that cell death itself will inhibit the development of biotrophic parasites that depend on living host cells for their continued survival. With other types of pathogens, the generation of a hostile antimicrobial environment may be a more important factor in resistance. During cell death, cytoplasmic material and organelles may be recycled by an intracellular trafficking and degradation pathway known as **autophagy**. Spent structures are sequestered and digested in membrane-bound vesicles. It is possible that this process could remove pathogen products or even virus particles, adding another dimension to the complex of responses involved in HR. By contrast, in some interactions cell death may lead to disease susceptibility rather than resistance, for instance when a necrotrophic pathogen induces programmed cell death through production of a toxin (see p.196). Such pathogens hijack the HR machinery and trigger uncontrolled cell death at the infection site and beyond.

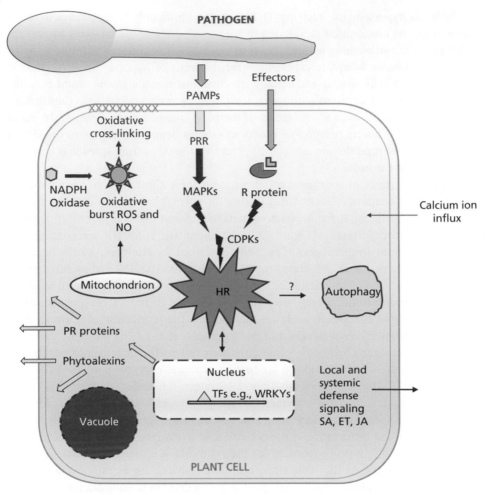

Figure 9.9 Diagrammatic overview of a plant cell undergoing the hypersensitive response. Pathogen-associated molecular patterns (PAMPS) and effector molecules are perceived by surface-located pattern recognition receptors (PRRs) or intracellular resistance R proteins, triggering a signal cascade via MAP and calcium-dependent kinases. Early responses include an influx of calcium ions and the generation of reactive oxygen species (ROS) and nitric oxide (NO) involving a plasma membrane NADPH oxidase and organelles including mitochondria and peroxisomes. The cell wall may be reinforced by oxidative cross-linking of proteins. Signals entering the nucleus interact with transcription factors such as WRKYs to switch on genes encoding PR proteins and phytoalexin biosynthetic enzymes. Hormones such as salicylic acid (SA), jasmonic acid (JA), and ethylene (ET) diffuse to neighboring cells and also generate systemic signals transmitted to more remote tissues. Programmed cell death may involve an autophagic pathway breaking down and recycling cellular components.

Hence, there is more than one outcome to a cell death response depending on the nature of the host–pathogen interaction. In some cases, a very rapid defense reaction may ensure early elimination of a pathogen without any visible necrosis, while at the other extreme a delayed response may lead to a spreading necrosis with some of the hallmarks of HR, but eventual susceptibility.

The Oxidative Burst

One of the earliest active responses of plant cells challenged by pathogens is the production of several ROS, including the superoxide anion (O_2^-), the hydroxyl radical ($\cdot OH$), and hydrogen peroxide H_2O_2. A similar process is seen in animal immunity, for instance during the killing of microbes by phagocytic cells. Several cellular compartments and enzymes contribute to this **oxidative burst** (Figure 9.9). For example, a membrane-located nicotinamide adenine dinucleotide phosphate (NADPH) oxidase enzyme produces the superoxide anion by the following reaction:

$$O_2 + NADPH \xrightarrow{\textit{NADPH oxidase}} O_2^- + NADP + H^+$$

with subsequent conversion to hydrogen peroxide by the enzyme superoxide dismutase:

$$2O_2^- + 2H^+ \xrightarrow{\textit{Superoxide dismutase}} H_2O_2 + O_2$$

Organelles such as mitochondria and peroxisomes are also a source of ROS contributing to the oxidative burst. ROS have multiple functions in plant defense, both directly as strong oxidants killing microbes and cross-linking cell wall proteins, or indirectly as signal molecules regulating other defense pathways.

Another bioactive molecule that plays a part in these early responses to pathogen attack is the gas NO. In mammalian macrophages, NO interacts with ROS to execute bacterial pathogens. It turns out that this gas may perform a similar role in plant innate immunity. NO is a small, highly diffusible molecule that can rapidly cross biological membranes and serve as a signal activating other components of the defense response. How NO is generated is not entirely clear although some studies have implicated the enzyme nitric oxide synthase. What is known is that NO production is associated with the early influx of calcium taking place in cells challenged by pathogens and that, together with ROS, NO is part of the signaling network integrating the different cellular reactions to invasion by foreign agents.

Phytoalexins

Evidence for the production of inhibitory chemicals during a resistant reaction to a pathogen was first obtained more than 70 years ago in experiments with potato blight *Phytophthora infestans* and potato tubers. When a compatible isolate of the blight fungus is applied to the cut surface of tubers, it invades the underlying tissues and forms a fluffy superficial mycelium. If an incompatible isolate is applied, the host tissues become discolored and necrotic and growth of the pathogen is restricted. Preinoculation of tuber tissues with an incompatible isolate, followed one or two days after by a compatible isolate, was shown to prevent infection by the usually virulent isolate. This second inoculation, which would normally lead to infection, failed to invade tissues and no mycelial growth was observed. The result was explained by postulating the production of inhibitors, termed phytoalexins, in response to the first inoculation, which subsequently prevented growth of the virulent isolate.

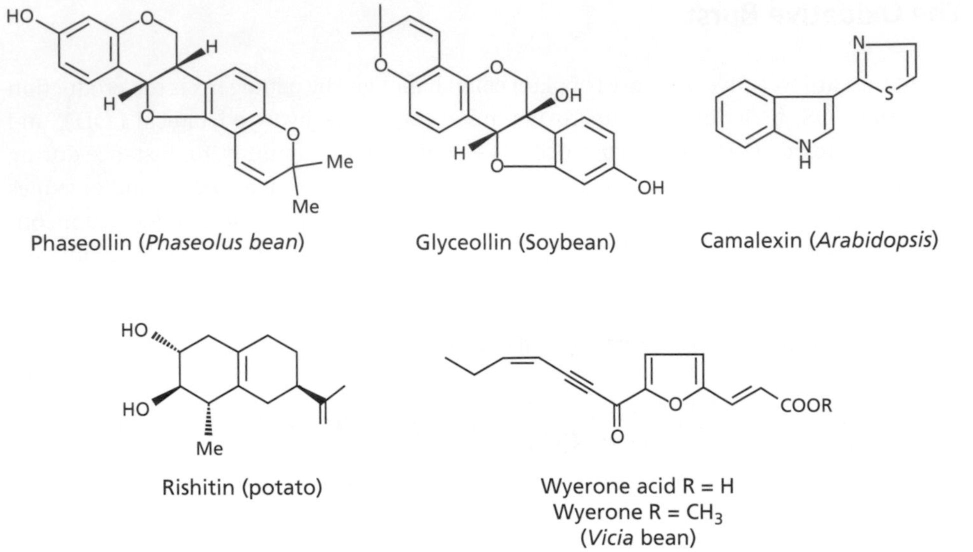

Phaseollin (*Phaseolus bean*)   Glyceollin (Soybean)   Camalexin (*Arabidopsis*)

Rishitin (potato)   Wyerone acid R = H
Wyerone R = CH_3
(*Vicia* bean)

Figure 9.10   Origin and structure of some phytoalexins.

Table 9.3   Some features of phytoalexins

- Low molecular weight antimicrobial compounds synthesized from remote precursors
- Broad-spectrum inhibitors active against a wide range of organisms (low specificity)
- Induced by microbial challenge, chemical or physical agents (low specificity)
- Chemically diverse, with different phytoalexins characteristic of particular plant species, genera or families, i.e., host specific rather than pathogen specific
- Induction associated with upregulation of genes encoding phytoalexin biosynthetic enzymes

Since this pioneering work, a large number of phytoalexins have been isolated from different plant tissues and chemically characterized (Figure 9.10). Phytoalexins are structurally diverse but share certain features in common (Table 9.3). A convenient working definition is "low molecular weight antibiotic compounds which are synthesized by and accumulate in plant tissues in response to microbial challenge or other types of stress." Unlike preformed inhibitors, phytoalexins are not present at significant levels in healthy tissues; instead, their production is induced by infection or injury. Detection of phytoalexins relies upon procedures for extracting and purifying the compounds, coupled with an appropriate bioassay to determine their antimicrobial activity. An example is shown in Figure 9.11, where extracts from broad bean (*Vicia faba*) leaves inoculated with *Botrytis cinerea* were separated by chromatography, and antifungal compounds visualized by spraying the plate with a colored fungus. Inhibitory zones show up as clear white areas of silica gel. This experiment demonstrates a further significant point: plants challenged with microorganisms often produce several phytoalexins which may be structurally related, even isomeric forms, or alternatively produced by quite different biosynthetic pathways.

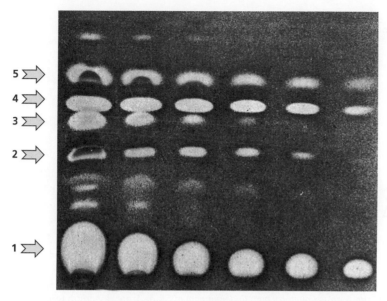

Figure 9.11 Bioassay of phytoalexins produced by broad bean (*Vicia faba*) leaves in response to inoculation with the fungus *Botrytis cinerea*. Extracts from leaf tissue were collected 3 days after inoculation and applied in decreasing amounts (left to right) to a silica gel plate and separated by thin-layer chromatography. Antifungal compounds were then detected by spraying the plates with a spore suspension of the dark-coloured fungus *Cladosporium herbarum*. Phytoalexins identified are 1) wyerone acid, 2) medicarpin, 3) wyerol, 4) wyerone epoxide, and 5) wyerone *Source:* Hargreaves et al. (1977).

Phytoalexins and Plant Defense

The role played by phytoalexins in defense proved difficult to determine for several reasons. First, phytoalexin induction is nonspecific. As well as microorganisms, treatment with a wide variety of chemical and physical agents can trigger their synthesis. Anything which injures plant cells is likely to induce phytoalexins, and they can therefore be regarded as stress metabolites rather than specific defense molecules.

The second problem is to establish that phytoalexins accumulate in the right place at the right time in sufficient concentrations to account for inhibition of pathogen growth. Bulk extracts of tissues may show the presence of inhibitory compounds, but often at low levels. Estimating the concentration of a phytoalexin at the actual site of infection is more difficult. Nonetheless, careful analysis of the growth of pathogens in resistant versus suscep-tible host plants and accumulation of a phytoalexin in tissues sampled around infection sites has in some cases shown that amounts are sufficient to account for the cessation of growth in the resistant combination. For instance, in the first eight hours after inoculation of soy-bean with *Phytophthora megasperma* f.sp. *sojae*, the growth rate of the pathogen is similar in both S and R genotypes but thereafter, growth in the R host is arrested, coinciding with the concentration of the phytoalexin glyceollin (Figure 9.10) reaching inhibitory levels.

The timing of phytoalexin production may be another critical factor. The relative rate at which a phytoalexin is produced is probably more important than the final concentration of compound present. Figure 9.12 shows that the french bean phytoalexin phaseollin

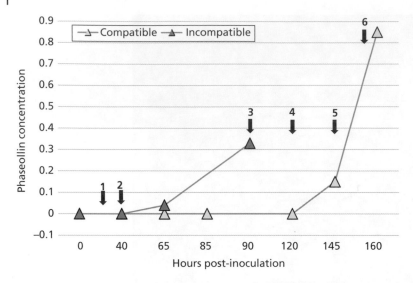

Figure 9.12   Accumulation of phaseollin (μg per inoculated site) in beans inoculated with compatible and incompatible races of the fungus *Colletotrichum lindemuthianum*. Stage 1, appressorium formation; 2, hypersensitive response visible; 3, HR complete; 4, 1% lesions in compatible combination; 5, 80% lesions; 6, 100% lesions. *Source:* Data from Bailey and Deverall (1971).

(Figure 9.10) accumulates earlier in plants reacting hypersensitively to the fungus *C. lindemuthianum* than in beans infected by a compatible isolate of the same pathogen. Although higher levels of phytoalexin eventually accumulate in the susceptible combination, by this time disease lesions are already well established and the fungus has, in effect, escaped.

Further insights into the role that phytoalexins play in plant defense have come from experiments with genetic mutants or inhibitors of the biosynthetic pathways leading to phytoalexin production. When phytoalexin synthesis is blocked, the resistance of tissues to pathogen attack is often reduced. In *Arabidopsis*, the major phytoalexin produced following microbial challenge is the indole derivative camalexin (Figure 9.10). Several phytoalexin-deficient (*pad*) mutants have been identified that produce reduced amounts of camalexin in response to inoculation with fungal, bacterial, and oomycete pathogens. When *pad* mutants impaired in phytoalexin synthesis were inoculated with virulent strains of the bacterium *Pseudomonas syringae*, the results varied; some proved to be more susceptible to infection while others, including the mutant most seriously compromised in camalexin production, showed no significant increase in susceptibility. This suggested that camalexin does not play a decisive role in defense against bacteria. Conversely, studies with the necrotrophic fungus *Alternaria brassicicola*, and the biotrophic oomycete *Hyaloperonospora parasitica* showed that susceptibility was increased in *pad* mutants. More recently, *Arabidopsis* mutants affected in other aspects of camalexin metabolism have been identified, including transporter proteins involved in the secretion of the phytoalexin to the leaf surface. Both phytoalexin synthesis and export may therefore be required for effective resistance to necrotrophic fungi such as *A. brassicicola*.

A further aspect concerns the ability of pathogens to tolerate exposure to phytoalexins. *Nectria haematococca* is a pathogen of peas that produces an enzyme, pisatin demethylase, which inactivates the major pea phytoalexin, pisatin (Figure 9.13). In crosses between fungus strains producing different amounts of the enzyme, only progeny strains producing

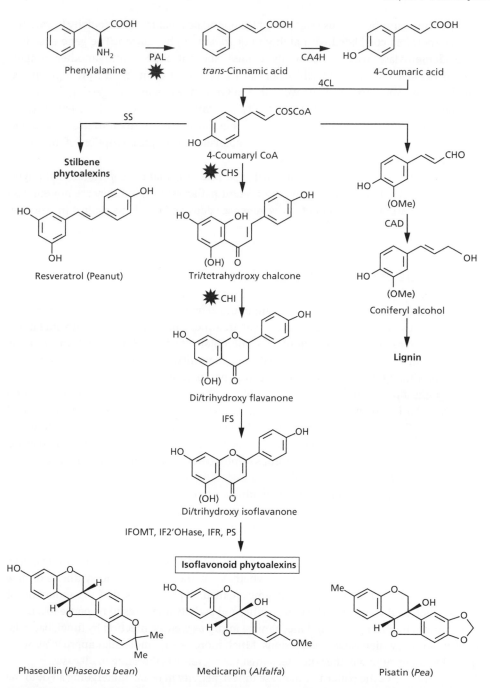

Figure 9.13 Pathway for biosynthesis of phenylpropanoid defense metabolites in legumes. Enzymes indicated: PAL, phenylalanine ammonia lyase; CA4H, cinnamic acid 4-hydroxylase; 4CL, 4-coumarate:CoA ligase; SS, stilbene synthase; CAD, coniferyl alcohol dehydrogenase; CHS, chalcone synthase; CHI, chalcone isomerase; IFS, isoflavone synthase; IFOMT, isoflavone-O-methyl transferase; IF2'OHase, isoflavone 2' hydroxylase; IFR, isoflavone reductase; PS, pterocarpan synthase. ✳ Enzymes for data shown in Figures 9.14 and 9.15. *Source:* After Dixon and Harrison (1990).

high amounts were able to cause significant lesions on pea plants. Pathogenicity toward the host was therefore correlated with ability to degrade the phytoalexin. The gene encoding pisatin demethylase was subsequently cloned and introduced into another fungus, *Cochliobolus heterostrophus*, which is normally a pathogen of maize. Surprisingly, transformed strains expressing the gene were able to grow in pea plants, suggesting that ability to degrade a host phytoalexin is a specificity determinant. However, there is still some debate over the importance of phytoalexin degradation, as some mutant strains of *N. haematococca*, in which enzyme production was abolished through disruption of the gene, apparently retain pathogenicity to peas.

An overall conclusion from these detailed experiments is that the contribution of phytoalexins to host resistance varies between different pathosystems, but they are nonetheless an important part of the armory of active defenses deployed by plants against invasion by pathogens.

Biosynthesis of Defense Metabolites

The participation of active host metabolism in the expression of induced defense has been known since early experiments showed that treatment of plant tissues with metabolic inhibitors increased their susceptibility to microbial infection. Subsequent analysis has defined the metabolic pathways involved, and identified some of the key biosynthetic enzymes producing defense compounds such as phytoalexins.

An important precursor for the synthesis of phenolic compounds is the aromatic amino acid phenylalanine; in healthy tissues, this and other amino acids are usually incorporated into proteins. Following infection or injury, however, phenylalanine is converted to trans-cinnamic acid and enters the pathway for biosynthesis of phenylpropanoid compounds (Figure 9.13). These include various groups of phytoalexins, as well as precursors of structural defense molecules such as lignin.

If one assays enzymes such as phenylalanine ammonia lyase (PAL), catalyzing the first step in the pathway, and chalcone synthase (CHS), catalyzing the branch point to isoflavonoid compounds (Figure 9.13), activity is usually found to be increased in infected plant tissues. This observation suggests that phenylpropanoid biosynthesis is accelerated as part of the plant response. Assays of activity alone, however, provide only limited information. It is not possible to decide, for instance, whether any increase is due to synthesis of new enzyme or activation of preformed enzyme. A further problem is that the induction of enzymes such as PAL seems to be nonspecific, and activity increases in response to a number of stimuli, including light and mechanical injury as well as infection by fungi, bacteria, and virus strains which cause local lesions. Much more precise molecular approaches were required to determine whether the mechanism of increase is the same in all cases.

Figure 9.14 shows the pattern of increase in PAL activity in beans undergoing a HR to the bacterium *Ps. syringae* pv. *phaseolicola*. Activity peaks around 20 hours after inoculation and then declines; the corresponding mRNA for PAL, detected using a complementary (cDNA) probe, peaks around 12 hours. Isoflavonoid phytoalexins begin to appear around 24 hours. This experiment suggests that following challenge with an incompatible bacterial isolate, the PAL gene is switched on, new message is transcribed for several hours, new enzyme

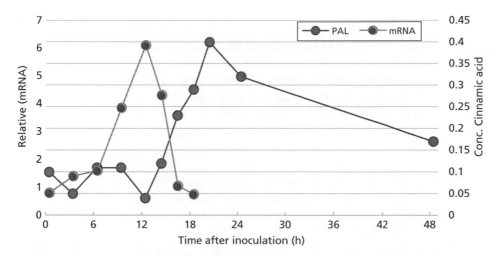

Figure 9.14 Timing of induction of phenylalanine ammonia lyase (PAL) mRNA and enzyme activity in the hypersensitive response of *Phaseolus* bean to an incompatible race of *Pseudomonas syringae* pv. *phaseolicola*. Relative mRNA concentration was determined using a cDNA probe for bean PAL. Chalcone synthase showed a similar pattern of induction. PAL enzyme activity increased in parallel with mRNA concentration but around six hours later. Antibacterial isoflavonoids were detectable by 24 hours after inoculation. *Source:* After Slusarenko et al. (1991), by permission of Oxford University Press.

protein is synthesized, and antibiotic secondary metabolites produced via the phenylpropanoid pathway accumulate. In the compatible (i.e., susceptible) bean plus bacterium combination, none of these changes occur within the first 24 hours after inoculation.

Even more rapid responses to microbial challenge can be demonstrated using cell cultures rather than tissues or intact plants. This approach permits simultaneous exposure of a whole population of cells to a pathogen, or more usually some biologically active fraction (or elicitor – see p.244) derived from the pathogen. Such studies show that defense genes encoding enzymes such as PAL, CHS, and cinnamyl-alcohol dehydrogenase (CAD), an enzyme on the branch to lignin synthesis (Figure 9.13), are activated within minutes of exposure to challenge. Figure 9.15 shows the rapid appearance of mRNA transcripts for several defense-related enzymes in bean suspension cells treated with an extract from fungal cell walls. This represents one of the most rapid gene activation responses known in plant cells. What is more, coordinated induction of these genes takes place; thus, in response to a microbial signal, genes specifying a series of enzymes in a biosynthetic pathway not operating in healthy plant cells are simultaneously activated, and the whole pattern of secondary metabolite production by the host is altered.

Pathogenesis-Related Proteins

If one prepares protein extracts from healthy plants and compares them with extracts from plants responding actively to microbial attack, it is clear that the pattern of protein synthesis is altered following challenge. Some of the new proteins detected are likely to be

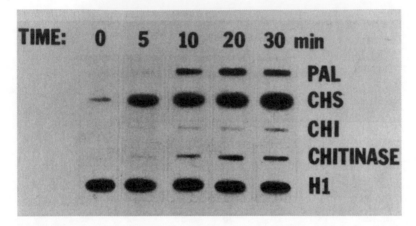

Figure 9.15 Induction of transcription of several defense-related enzymes in bean cells treated with an elicitor from cell walls of the fungal pathogen *Colletotrichum lindemuthianum*. Blot shows appearance of mRNA for phenylalanine ammonia lyase (PAL), chalcone synthase (CHS), chalcone isomerase (CHI), and chitinase within minutes of exposure to a fungal cell wall preparation. H1 is a constitutively expressed gene unaffected by microbial challenge. See Figure 9.13 for positions of PAL, CHS, and CHI in the phenylpropanoid pathway. *Source:* Hedrick et al. (1988). © American Society of Plant Physiologists. Reprinted by permission of the publishers.

enzymes involved in the synthesis of defense compounds such as phytoalexins, but others have quite different functions.

Inoculation of tobacco leaves with the virus TMV can lead to either systemic infection or the formation of local necrotic lesions, depending upon the presence of a dominant resistance gene (Figure 9.8). This hypersensitive-type reaction is accompanied by the production of new, host-encoded proteins described as pathogenesis-related proteins (usually abbreviated to PRs). PRs are relatively small, stable proteins which accumulate predominantly in the intercellular spaces of plant tissues. Their appearance is easily demonstrated by gel electrophoresis of tissue extracts (Figure 9.16). Similar proteins have now been detected in many other plant species responding to virus infection or challenge by microorganisms; PR production is also induced by various chemicals including hormones such as ethylene, as well as other types of stress.

PR proteins were initially classified on the basis of their physical properties, such as electrophoretic mobility, size, acidity, and so on, and grouped into "families" of related proteins (Table 9.4). The biological function of PRs was initially unknown, although their appearance in infected or stressed plants suggested some role in defense or protection from tissue damage. Work on the proteins from tobacco showed that some, such as PR2, PR3, and PR4, are enzymes active against polysaccharides containing β-1,3-glucans or another polymer, chitin. Interestingly, neither of these substrates occurs naturally in plants. Instead, they are the main polymers found in the cell walls of many fungi, while chitin is also a key component of the insect exoskeleton. PR1 was shown to inhibit the growth of oomycete pathogens such as *P. infestans*. A direct role in defense against pathogens and pests therefore seemed likely.

Figure 9.16 Electrophoretic separation of proteins extracted from healthy tobacco leaves (H) and leaves forming local lesions in response to inoculation (I) with tobacco mosaic virus. Note appearance of novel, PR proteins in infected leaves. *Source:* Photo supplied by Jon Antoniw.

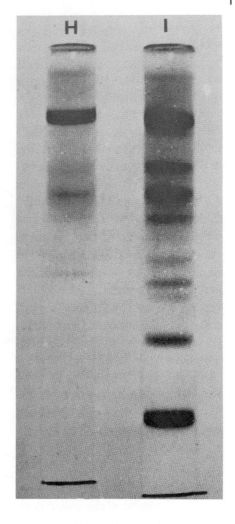

Purified chitinases from plants are able to inhibit the growth of at least some fungi *in vitro*, while mixtures of chitinase and β-1,3-glucanase are much more active. A direct test of the possible role of these enzymes in defense is to assess their effects in transgenic plants; when chitinase and glucanase genes were expressed in tobacco or oilseed rape plants (see p.327), the transformants showed enhanced resistance to certain fungal pathogens. These enzymes may have a direct role in slowing down fungal growth, thereby delaying invasion of the plant and allowing other host defense systems to operate, and also an indirect role by releasing fragments from fungal cell walls, which act as signal molecules which the plant can recognize. Such DAMPs are active inducers of plant immunity. Plants may therefore have evolved these enzymes both as a general type of resistance against fungi and as an early warning system.

The classification of PR proteins subsequently expanded to include other plant species and families as well as some smaller peptides with antimicrobial properties (Table 9.4).

Table 9.4 Some families of pathogenesis-related (PR) proteins

Family	Type member	Typical size (kDA)	Properties
PR1	Tobacco PR-1a	15	Antifungal/oomycete
PR2	Tobacco PR-2	30	β1,3-glucanase
PR3	Tobacco P,Q	25–30	Chitinase
PR4	Tobacco R	15–20	Chitinase
PR5	Tobacco S	25	Thaumatin-like
PR6	Tomato P_{69}	75	Endoproteinase
PR10	Parsley 'PR1'	17	"Ribonuclease-like"
PR12	Radish Rs-AFP3	5	Defensin
PR13	*Arabidopsis* TH12.1	5	Thionin
PR14	Barley LTP4	9	Lipid transfer protein

Original types found in tobacco highlighted in green, smaller PR peptides in yellow. At least 17 families have now been described.
Source: Adapted from Sels et al. (2008).

Some of these have already been described as preexisting inhibitors from seeds or other storage tissues but they may also be induced in response to infection. It is now known that some PR families, such as PR1, are widely distributed in the plant kingdom, and homologs have also been found in animals. They may therefore serve diverse roles in biological systems.

Can Plants Be Immunized?

Active defense in plants bears some similarity to the immune response of animals, and some molecular components of the response are conserved between the two kingdoms. Like animals, plants can recognize pathogens and respond by activating a series of defense systems, but can they remember the encounter so that they are protected against later exposure to a pathogen (Figure 9.17)?

During the early years of the last century, there were repeated attempts to demonstrate the existence in plants of an immune system producing specific antibodies against microorganisms. These experiments found no reliable evidence for plant antibodies or indeed defense cells similar to phagocytes. Nevertheless, the belief in plant "immunity" persisted, typified by Chester who in 1933 suggested that acquired resistance plays "an important role in the preservation of plants in nature." More recent work has partly validated this claim, as it has proved possible to protect plants against pathogens using immunization techniques similar to those employed so successfully in medicine.

Figure 9.18a diagrammatically shows a typical "immunization" experiment. Prior inoculation of a leaf with an incompatible pathogen, or pathogen products such as PAMPs, protects against infection by a live, virulent pathogen administered several days later. This phenomenon has been demonstrated with numerous host–pathogen combinations, and was instrumental in the discovery of phytoalexins. Such protection is, however, short-lived and localized to the treated tissues; developing leaves remain susceptible to subsequent challenge.

Figure 9.17 Plant and animal immunity compared.

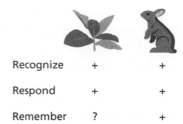

Recognize	+	+
Respond	+	+
Remember	?	+

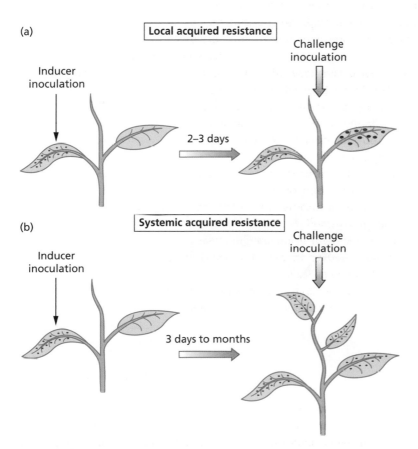

(a)

Local acquired resistance

Inducer inoculation

Challenge inoculation

2–3 days

(b)

Systemic acquired resistance

Inducer inoculation

Challenge inoculation

3 days to months

Figure 9.18 (a) Local and (b) systemic acquired resistance in plants.

A more intriguing experiment is shown in Figure 9.18b. In this case, exposure of leaf 1 to a pathogen leads to long-lasting protection of all subsequently developing tissues. Such **systemic acquired resistance** (SAR) was initially described in species from the cucumber family Cucurbitaceae inoculated with various fungi, bacteria or viruses, and tobacco plants exposed to local lesion-inducing strains of TMV or the blue mold oomycete *Peronospora hyoscyami*. Subsequently, SAR was demonstrated in the experimental model *Arabidopsis* and is now known to occur in a wide range of plants. Unlike animal immunization, this form of protection is nonspecific; the acquired resistance operates against a broad range of pathogens and can last for several weeks to months after the initial, inducing infection (Table 9.5). The degree of protection

Table 9.5 Some characteristics of systemic acquired resistance (SAR)

- Induced by agents/pathogens causing necrosis, e.g., local lesions
- Delay of several days between induction and full expression
- Protection conferred to tissues not exposed to inducer inoculation
- Expressed as reduction in lesion number, size, spore production, etc.
- Protection is long-lasting, often for weeks or even months
- Protection is nonspecific, i.e., effective against pathogens unrelated to inducing agent
- Development of SAR associated with expression of several gene families, e.g. PR proteins
- Signal for SAR is translocated and graft transmissible

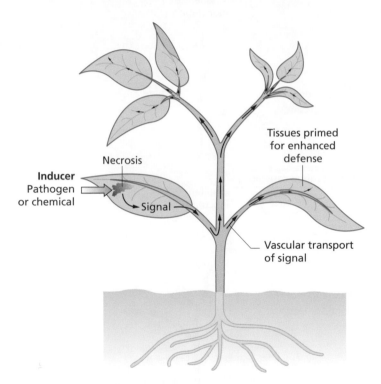

Figure 9.19 Basic model of systemic activation of plant defense. Inoculation of lower leaf with pathogen or chemical causing local necrosis induces production of mobile signal transmitted through the vascular system and potentiating distant tissues to respond more rapidly to infection.

is often high, and the resistance has been shown to be effective under field conditions. In the case of tobacco blue mold, protection compares favorably with that achieved using the best available fungicides.

Typically, in SAR, the primary inoculation involves a chemical or biological agent which induces cell death in the treated tissues. This triggers active defense systems in the surrounding cells. In addition, a signal must be produced and translocated to other parts of the plant, where it potentiates unexposed tissues to respond more quickly to a subsequent challenge (Figure 9.19). The identity of the translocated signal has been a topic of

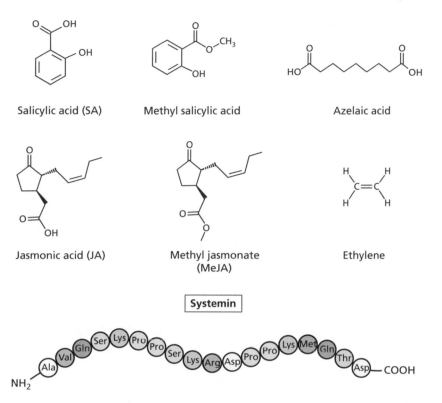

Salicylic acid (SA)   Methyl salicylic acid   Azelaic acid

Jasmonic acid (JA)   Methyl jasmonate   Ethylene
(MeJA)

Systemin

Figure 9.20   Some signal molecules involved in systemic resistance in plants.

intense interest and debate. Early experiments demonstrated that the signal is able to cross grafts between induced and uninduced parts of plants, most likely in the vascular tissues. It was also shown that SA, a chemical related to aspirin (Figure 9.20), is an active inducer of SAR and triggers the synthesis of PR proteins and defense-related enzymes in treated plants. Mutants defective in SA production or plants transformed with a bacterial enzyme that degrades SA lose their ability to express SAR. Hence SA plays an important role in SAR, and was initially regarded as a strong candidate for the translocated signal molecule. However, grafting experiments in which the plants transformed with a bacterial enzyme degrading SA were used as rootstocks and normal tobacco as the shoot (scion) showed that such grafted plants can still express SAR in their upper leaves following an inducer treatment of the rootstock. This result suggested that the signal moving from root to shoot cannot be SA.

An alternative candidate in tobacco plants turned out to be the methyl ester of SA, methyl salicylate (Figure 9.20), which may be transported and subsequently converted to SA in leaves developing SAR. To add to the story, methyl salicylate is volatile and could therefore act as an air-borne signal affecting distant parts of the plant, or even other plants. This initially fanciful idea received experimental support from work on tobacco reacting to inoculation with TMV. During local lesion formation, these plants emitted significant amounts of methyl salicylate, which was sufficient to induce PR protein expression in neighboring healthy plants and to boost their resistance to the virus. Generation of a volatile signal could therefore serve as an early warning system priming the defense pathways of plants not yet under attack.

More recently, several other signal candidates have been investigated, including azelaic acid (Figure 9.20) and glycerol-3-phosphate, acting downstream of the free radicals nitric oxide and ROS generated during programmed cell death. It now seems likely that there may be more than one signal pathway for SAR, and also the importance of the different signal molecules might vary between plant species.

Other Systemic Defense Responses

In addition to SAR, plants are capable of mounting other systemic responses to biological threats (Figure 9.21). Inoculation of roots with certain types of rhizosphere-colonizing bacteria can boost plant growth but also increases resistance to pathogens attacking aerial tissues. This phenomenon is described as **induced systemic resistance** (ISR). Phenotypically, ISR is similar to SAR but operates via a different signaling pathway regulated by the hormones JA and ethylene (Figure 9.21) rather than SA. While SAR is mainly effective against biotrophic pathogens, ISR confers increased resistance against necrotrophs and also some herbivorous insects. Some molecular cross-talk may occur between these two pathways, for

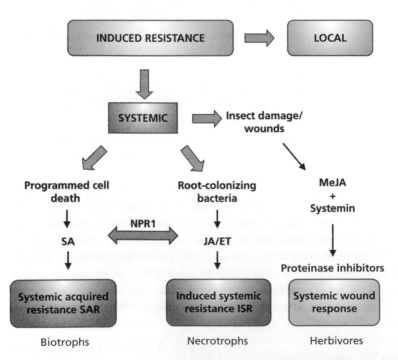

Figure 9.21 Induced resistance pathways in plants. Signal molecules implicated in these responses to pathogens, rhizosphere bacteria, insects, and wounding include salicylic acid (SA), jasmonic acid (JA), methyl jasmonate (JA), ethylene (ET), and the peptide systemin. NPR1 (nonexpressor of PR genes) is a regulatory protein involved in signal transduction.

instance involving the master regulator protein NPR1 which interacts with transcription factors controlling defense gene expression.

When plant tissues are damaged, for instance by mechanical wounding or insect feeding, breakdown products such as cell wall fragments induce the expression of defense-related genes in adjacent cells. Such injury can also activate defense in more remote parts of the plant. This **systemic wound response** (SWR) involves several potential signal molecules, including the hormones abscisic acid and methyl-jasmonate and a small peptide comprising 18 amino acids known as systemin (Figure 9.20). Systemin is derived from a much larger 200 amino acid precursor, prosystemin, from which it is released upon wounding. In this respect, systemin behaves much like certain polypeptide hormones found in animals. Wound-inducible genes encode products such as proteinase inhibitors which are active against digestive enzymes, and hence interfere with the gut function of herbivorous insects. Some wound responses occur so quickly that signaling via chemical messengers transported in the vascular system might not be the primary mechanism, and instead the rapid propagation of electrical signals has been implicated.

Overall, it is evident that rather than being controlled by individual chemical transmitters, systemic plant defense most likely depends upon an intricate network of hormones and other active components.

Defense Priming and the Immune Memory of Plants

One notable feature of systemic resistance as exemplified by SAR and ISR is the establishment of a heightened state of readiness to react to a biological threat. This has been described as defense **priming** and is typified by a more rapid and stronger response to potential invaders. An example is shown in Figure 9.22 in which the activity of two enzymes involved in phytoalexin synthesis increases more quickly and to a higher level in primed cowpea tissues than unprimed controls in response to inoculation with a fungal pathogen. Similar kinetics are seen with PR proteins and other defense-related molecules.

A further fascinating feature is that the primed state can last for a long time, as if the plant is able to "remember" the initial priming stimulus. Information from the first encounter must somehow be stored for a significant period, enabling the primed cells to react rapidly to any further challenge. A number of molecular changes have been noted in primed plants, including elevated levels of pattern recognition receptors, accumulation of protein kinase enzymes, and higher expression of some defense-related transcription factors, as well as alterations in histone proteins and DNA methylation, both of which may affect patterns of gene expression. While priming potentiates a more rapid defense response, it does not incur a high biosynthetic cost to the plant as the response is only triggered in the presence of a biotic challenge. Hence, priming represents an adaptive trait that enables plants to cope with diverse biotic threats in an unpredictable environment.

Recently, it has been experimentally demonstrated that in some cases priming can be passed on via seed to progeny, so that an enhanced immune capability can persist in the next generation. Such transgenerational defense priming is most likely based on epigenetic factors, in other words phenotypic changes not involving alterations in DNA sequences.

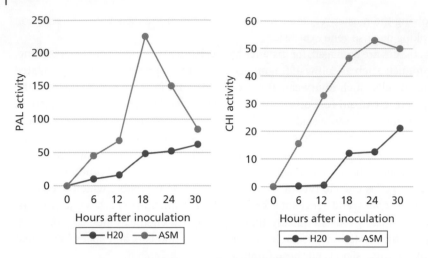

Figure 9.22 Effect of seed treatment with the defense activator acibenzolar-S-methyl (ASM) on extractable activities of phenylalanine ammonia lyase (PAL) and chalcone isomerase (CHI) in cowpea hypocotyls inoculated with the anthracnose fungus *Colletotrichum destructivum*. *Source:* Latunde-Dada and Lucas (2001).

Inevitably, the longevity and broad-spectrum efficacy of induced resistance have attracted interest as a natural and potentially valuable means of crop protection. In fact, there are a number of chemicals that can induce systemic resistance, some of them analogs of the endogenous signals involved in plant immunity. As these so-called **defense activators** can be applied to crops in a similar way to conventional pesticides, further discussion of their use is deferred until Chapter 11 on chemical approaches to disease management.

Conclusion

By comparison with animals, plants were originally regarded as passive organisms at the mercy of the pests and pathogens that feed on them or cause disease. This view has been profoundly changed by recent revelations of plant physiology, biochemistry, genetics, and molecular biology. It is now known that plants have a sophisticated suite of defenses, including an innate immune system capable of detecting and countering diverse threats. Many of the molecular components and properties of this system bear similarities to animal immunity, including the receptors recognizing nonself, parts of the signaling cascades activating immune response genes, and the capacity to "memorize" biotic threats and react more forcefully to subsequent encounters. Hence, there is some conservation of the structure and function of active defense between plants and animals. In turn, pathogens have evolved strategies to evade or subvert host immunity, and it is the interplay between pathogen offense and host defense that determines the final outcome of host–pathogen interactions. This topic is further explored in the following chapter.

Further Reading

Reviews and Papers

Ahuja, I., Kissen, R., and Bones, A.M. (2012). Phytoalexins in defense against pathogens. *Trends in Plant Science* 17 (2): 73–90. https://doi.org/10.1016/j.tplants.2011.11.002.

Bakker, P.A.H.M., Pieterse, C.M.J., and van Loon, L.C. (2007). Induced systemic resistance by fluorescent *Pseudomonas* spp. *Phytopathology* 97: 239–243.

Bowyer, P., Clarke, B.R., Lunness, P. et al. (1995). Host range of a plant pathogenic fungus determined by a saponin detoxifying enzyme. *Science* 267: 371–374.

Bradley, D.J., Kjellbom, P., and Lamb, C.J. (1992). Elicitor and wound-induced oxidative cross-linking of a proline-rich plant cell wall protein: a novel, rapid defense response. *Cell* 70: 21–30.

Camejo, D., Guzmán-Cedeño, A., and Moreno, A. (2016). Reactive oxygen species, essential molecules, during plant–pathogen interactions. *Plant Physiology and Biochemistry* 103: 10–23. https://doi.org/10.1016/j.plaphy.2016.02.035.

Hilleary, R. and Gilroy, S. (2018). Systemic signaling in response to wounding and pathogens. *Current Opinion in Plant Biology* 43: 57–62. https://doi.org/10.1016/j.pbi.2017.12.009.

Huysmans, M., Lema, A.S., Coll, N.S. et al. (2017). Dying two deaths – programmed cell death regulation in development and disease. *Current Opinion in Plant Biology* 35: 37–44. https://doi.org/10.1016/j.pbi.2016.11.005.

Karasov, T.L., Chae, E., Herman, J.J. et al. (2017). Mechanisms to mitigate the trade-off between growth and defense. *Plant Cell* 29: 666–680. https://doi.org/10.1105/tpc.16.00931.

Kusch, S. and Panstruga, R. (2017). mlo-based resistance: an apparently universal "weapon" to defeat powdery mildew disease. *Molecular Plant-Microbe Interactions* 30 (3): 179–189. https://doi.org/10.1094/MPMI-12-16-0255-CR.

Lacerda, A.F., Vasconcelos, E.A.R., Pelegrini, P.B. et al. (2014). Antifungal defensins and their role in plant defense. *Frontiers in Microbiology* 5: 116. https://doi.org/10.3389/fmicb.2014.00116.

Levine, A., Tenhaken, R., Dixon, R. et al. (1994). H_2O_2 from the oxidative burst orchestrates the plant hypersensitive disease resistance response. *Cell* 79 (4): 583–593. https://doi.org/10.1016/0092-8674(94)90544-4.

Morant, A.V., Jorgensen, K., Jorgensen, C. et al. (2008). β-glucosidases as detonators of plant chemical defense. *Phytochemistry* 69: 1795–1813.

Pastor, V., Luna, E., Mauch-Mani, B. et al. (2013). Primed plants do not forget. *Environmental and Experimental Botany* 94: 46–56.

Piasecka, A., Jedrzejczak-Rey, N., and Bednarek, P. (2015). Secondary metabolites in plant innate immunity: conserved function of divergent chemicals. *New Phytologist* 206: 948–964. https://doi.org/10.1111/nph.13325.

Ramirez-Prado, J.S., Abulfaraj, A.A., Rayapuram, N. et al. (2018). Plant immunity: from signaling to epigenetic control of defense. *Trends in Plant Science* 23 (9): 833–844.

Salguero-Linares, J. and Coll, N.S. (2019). Plant proteases in the control of the hypersensitive response. *Journal of Experimental Botany* 70 (7): 2087–2095. https://doi.org/10.1093/jxb/erz030.

Shewry, P.R. and Lucas, J.A. (1997). Plant proteins that confer resistance to pests and pathogens. *Advances in Botanical Research* 26: 136–192.

Sels, J., Mathys, J., de Coninck, B.M.A. et al. (2008). Plant pathogenesis-related proteins: a focus on PR peptides. *Plant Physiology and Biochemistry* 46: 941–950.

Thomma, B.P.H.J., Nelissen, I., Eggermont, K. et al. (1999). Deficiency in phytoalexin production causes enhanced susceptibility of *Arabidopsis thaliana* to the fungus *Alternaria brassicicola*. *Plant Journal* 19 (2): 163–171.

Ton, J., d'Allessandro, M., Jourdie, V. et al. (2006). Priming by airborne signals boosts direct and indirect resistance in maize. *Plant Journal* 49: 16–26.

Trapet, P., Kulik, A., Lamotte, O. et al. (2015). NO signaling in plant immunity: a tale of messengers. *Phytochemistry* 112: 72–79.

Underwood, W. (2012). The plant cell wall: a dynamic barrier against pathogen invasion. *Frontiers in Plant Science* 3: 85. https://doi.org/10.3389/fpls.2012.00085.

Van Loon, L.C. (2016). The intelligent behavior of plants. *Trends in Plant Science* 21 (4): 286–294. https://doi.org/10.1016/j.tplants.2015.11.009.

Voight, C.A. (2014). Callose-mediated resistance to pathogenic intruders in plant defense-related papillae. *Frontiers in Plant Science* 5: 168. https://doi.org/10.3389/fpls.2014.00168.

10

Host–Pathogen Specificity

The nature of specific or major gene resistance ... is one of the most challenging yet elusive problems in plant pathology. What are the mechanisms whereby the genes in the host and the parasite interact to control the development of the disease?

(H. Flor, 1900–1991)

Why certain microorganisms are able to cause disease, whilst the majority cannot, is a crucial question in both medical microbiology and plant pathology. Some of the properties distinguishing pathogens from nonpathogens have already been described in Chapter 8. Even where a microorganism has the ability to cause disease, it will not attack all plants equally. For instance, the bean rust fungus *Uromyces appendiculatus* infects bean plants but not coffee trees and the coffee rust fungus *Hemiliea vastatrix* vice versa. Hence each pathogen is restricted to particular plant types and has a characteristic **host range**.

Host–pathogen specificity involves not only those factors determining virulence in the pathogen, but also those conferring resistance on the host. Experimental analysis is therefore complex. Understanding specificity is, however, important for both practical and conceptual reasons; a complete analysis should suggest more precise and reliable ways of intervening in the disease process, either through chemicals designed to shift the balance of the interaction toward host resistance or by genetic procedures to produce novel types of resistant crops. Furthermore, insights into the nature of host–pathogen specificity are likely to prove of generic relevance to other biological processes involving compatibility, such as plant fertilization, or recognition between mutualistic microorganisms and their hosts.

Types of Specificity

Host–pathogen specificity involves several separate and probably different phenomena (Table 10.1). First, there is the distinction between pathogenic organisms and those which are unable to attack living hosts under any circumstances. In the example shown, *Cladosporium herbarum* is a saprophytic fungus which lives on leaf surfaces without causing disease, while the related species *Cladosporium fulvum* can invade leaves and grow in intercellular spaces, eventually causing necrotic lesions. Next, with a particular pathogen, there are host species, which are infected, and **nonhosts**, which are never infected.

Plant Pathology and Plant Pathogens, Fourth Edition. John A. Lucas.
© 2020 John Wiley & Sons Ltd. Published 2020 by John Wiley & Sons Ltd.
Companion website: www.wiley.com/go/Lucas_PlantPathology4

The tomato leaf mold pathogen *C. fulvum* attacks tomatoes but not other plants such as beans. Some pathogens have very wide host ranges, attacking many different host species, while others are highly specific, attacking only a few closely related plants. An example of such specialization is seen with the rust and powdery mildew fungi, where distinct form species occur on particular cereal hosts. For instance, the rust fungus *Puccinia graminis* occurs as forms attacking wheat (*Triticum* = form species *tritici*), barley (*Hordeum* = form species *hordei*) and other cereals. Even greater specificity may be detected in the interaction of particular pathogen isolates (often described as races) and specific host lines (cultivars) (Table 10.1b). Hence, a single genotype of a pathogen may infect only certain susceptible genotypes of the host. This race–cultivar specificity is described in genetic terms by the gene-for-gene model (see p.42).

Table 10.1 Some examples of host–pathogen specificity

(a) Species specificity

	Pathogen *Cladosporium fulvum*	Nonpathogen *Cladosporium herbarum*
Nonhost (bean)	−	−
Host (tomato)	+	−

(b) Race × cultivar specificity

	Cladosporium fulvum genotype					
Tomato genotype	**Race 0** *A1, A2, A5, A9*	**Race 1** *a1, A2, A5,* *A9*	**Race 2** *A1, a2, A5,* *A9*	**Race 1,2** *a1, a2, A5,* *A9*	**Race 5** *A1, A2, a5,* *A9*	**Race 9** *A1, A2, A5, a9*
No *R* genes	+	+	+	+	+	+
Cf1	−	+	−	+	−	−
Cf2	−	−	+	+	−	−
Cf5	−	−	−	−	+	−
Cf9	−	−	−	−	−	+

Cf, host resistance gene; A, pathogen avirulence gene (dominant allele); a, pathogen avirulence gene (recessive allele); + susceptible interaction, − resistant interaction.

(c) Tissue specificity

	Host tissue		
Pathogen	**Leaf**	**Root**	**Vascular system**
Cladosporium fulvum	+	−	−
Pyrenochaeta lycopsersici	−	+	−
Fusarium oxysporum f.sp. *lycopersici*	−	−	+
Tomato mosaic virus (ToMV)	+	+	+

A further form of specificity, often overlooked, is where a parasitic microorganism is restricted to particular host tissues or organs. Such **tissue specificity** is suggested in many of the common names for plant diseases, such as root rot, flower blight, and vascular wilt. Table 10.1c shows three fungal pathogens which colonize different types of tomato tissue, and a virus which can spread systemically throughout the plant. The factors underlying these patterns of colonization are poorly understood, although the entry route exploited by the pathogen is obviously important.

No single model of host–pathogen specificity is likely to explain all these phenomena. Nonetheless, concerted efforts have been made to identify the mechanisms involved, especially in cases where single genes of major effect appear to be present in the host, pathogen, or both.

Specificity Based on Gene-for-Gene Interactions

In a gene-for-gene system (Figure 10.1a), specific recognition occurs when a resistance (*R*) gene in the host plant is matched by a corresponding avirulence (*Avr*) gene in the pathogen. Almost always, plant resistance genes and pathogen avirulence genes segregate as single dominant characters. The simplest interpretation is that the product of the *Avr* gene interacts directly with the product of the *R* gene; models based on this predicted that the *R* gene product is likely to be a receptor located on the host cell membrane, while the *Avr* gene

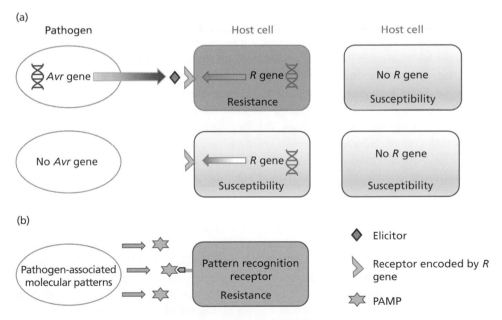

Figure 10.1 Genetic models of host–pathogen specificity. (a) Gene-for-gene system based on the interaction of a pathogen avirulence (*Avr*) gene with a host resistance (*R*) gene determines specific recognition. The product of the *Avr* gene (an elicitor) binds to the product of the *R* gene (a receptor), activating host resistance mechanisms. Absence of either the elicitor or receptor leads to susceptibility. (b) Broad-based immunity activated by interaction of a pathogen-associated molecular pattern (PAMP) with a host pattern recognition receptor (PRR).

encodes a secreted or surface-located molecule, initially described as an **elicitor**, which binds to the receptor. This molecular interaction then triggers the complex cascade of responses involved in active plant defense. Absence of the elicitor allows the pathogen to avoid recognition and leads to susceptibility.

Two experimental approaches were used to try to identify the genes and molecules involved in this specific interaction. The first was to attempt to clone the piece of host or pathogen DNA carrying the host *R* or pathogen *Avr* gene. Once located, the gene could be transformed into a susceptible host lacking the *R* gene, or pathogen genotype lacking the *Avr* gene, and any changes in host reaction type or pathogen host range noted. Conversion to resistance or avirulence, respectively, would confirm function. Alternatively, one can try to isolate the pathogen elicitor. This should be active only on host genotypes possessing the corresponding *R* gene, and also bind specifically to the host receptor (Figure 10.1a). The elicitor might therefore be used as a molecular probe able to locate the receptor itself.

Microbial and Plant Elicitors of Defense

The search for pathogen elicitors that met the necessary criteria of activity on specific host genotypes proved to be problematic. Microbial culture filtrates or extracts from microbial cells often act as potent inducers of plant defense responses, but in most cases such elicitors are nonspecific in action, inducing responses in nonhosts as well as resistant and susceptible host genotypes (Figure 10.1b). Attempts to purify such elicitors showed that they are chemically diverse, including polysaccharides, glycoproteins, and proteins (Table 10.2). Partial hydrolysis of fungal cell walls releases a mixture of breakdown products, some of which are highly active in switching on host defense. During infection, enzymes produced by the plant such as glucanase and chitinase may release similar fragments from invading hyphae. Analysis of the glucans extracted from *Phytophthora megasperma* f.sp. *glycinea*, an oomycete pathogen of soybean, identified the most active fraction as a branched oligosaccharide containing seven sugar residues. This heptaglucoside is the most potent elicitor of phytoalexins known, inducing glyceollin synthesis in soybean at concentrations as low as 10^{-8} M. Chitosan, a breakdown product from chitin, a component of fungal cell walls, is also an effective elicitor. In addition, potent elicitors have been isolated from nonpathogens, such as the yeast *Saccharomyces cerevisiae*.

With hindsight, these findings are predictable, as many of these extracts contain chemicals that would now be defined as pathogen-associated molecular patterns (PAMPs) or, in the case of nonpathogens, the related category microbe-associated molecular patterns (MAMPs). They therefore represent part of the nonspecific surveillance system enabling cells to recognize foreign organisms, and are not the products of *Avr* genes (Figure 10.1b).

A further class of oligosaccharide elicitors originates from the host cell wall, rather than from the pathogen (Table 10.2). It is well known that many plant-pathogenic bacteria and fungi produce cell wall-degrading enzymes during tissue colonization (see p.188), especially pectolytic enzymes which digest the middle lamella. Fragments of pectic polymers, ranging from 10 to 13 sugar residues in size, turn out to be very effective elicitors of plant defense. The products of a key process in pathogenesis might therefore alert the plant to the presence of an invading pathogen. An additional intriguing observation is that plant

Table 10.2 Some elicitors of plant immune responses

Source	Common name	Chemical type	Biological activity
Fungi			
Monilinia	Monilicolin	Peptide	Induces phytoalexins in bean
Saccharomyces	—	Glucan	Induces phytoalexins
Fungal cell walls	—	Chitosan	Induces plant defense and SAR
Oomycetes			
Phytophthora spp.	—	Glucan	Induces phytoalexins
P. parasitica	Elicitin	Peptide	Induces necrosis
Bacteria			
Erwinia amylovora and *Pseudomonas* spp.	Harpin	Protein	Induces hypersensitive response and SAR
Pseudomonas spp.	—	Lipopolysaccharide	Induces plant defense
Viruses			
Tobacco mosaic virus (TMV)	Coat protein	Protein	Induces hypersensitive response
Endogenous plant elicitors			
Oligosaccharins	—	Oligosaccharides	Induce plant defense
Elicitor peptides	—	Protein	Induce plant defense

SAR, systemic acquired resistance.

proteins which inhibit the activity of pectic enzymes such as polygalacturonase may also play a role by preventing complete hydrolysis of the polymer, thereby prolonging the half-life of elicitor-active fragments. These oligosaccharins, originally described as endogenous elicitors, are examples of damage-associated molecular patterns (DAMPs) that play a role in defense signaling between adjacent cells.

Bacterial *avr* Genes

The first *avr* genes to be successfully isolated were all from bacterial pathogens, which have relatively small genomes and are genetically tractable (Note: As bacteria are haploid and the recessive form of the gene confers virulence, the notation *avr* is often used). These genes were identified by mutations which altered host range, followed by testing of random clones from genomic libraries to find those which restored the original host specificity. Alternatively, one can change the host range of a particular pathogen race by introducing DNA clones from another, different race. The basic idea behind this procedure is shown in Figure 10.2 in which a pathogen race (i) initially able to infect cultivar B is converted to an incompatible pathogen by introduction of DNA from an incompatible race (ii). According to the gene-for-gene model, the piece of DNA responsible for this alteration in host specificity should contain an *avr* gene recognized by cultivar B. These types of experiments

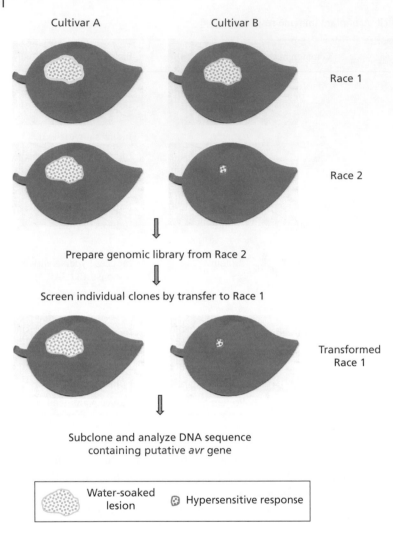

Cultivar A   Cultivar B

Race 1

Race 2

Prepare genomic library from Race 2

Screen individual clones by transfer to Race 1

Transformed
Race 1

Subclone and analyze DNA sequence
containing putative *avr* gene

Water-soaked
lesion   Hypersensitive response

Figure 10.2 Cloning bacterial *avr* genes by function. Race 2 is incompatible with cultivar B, inducing a hypersensitive response. Conversion of the normally compatible Race 1 into an incompatible phenotype is due to introduction of a specific *avr* gene.

quickly lead to the cloning and characterization of a whole series of *avr* genes from different pathovars of the major bacterial pathogens *Pseudomonas syringae* and *Xanthomonas campestris*.

One expectation of this work was that comparison of the sequences of these genes, and their predicted products, would reveal conserved features suggesting some common mode of action in eliciting host defense. In fact, most bacterial *avr* gene products do not show significant homology, and also lack the characteristics one would expect of a signal molecule exported out of the cell to act as an elicitor. Instead, they appeared to be cytoplasmic proteins located inside the bacterium. Critically, the gene products themselves did not prove to be active in inducing a hypersensitive response (HR) in hosts containing the corresponding *R* gene. One possibility was that these genes were indirectly involved in the production of a specific elicitor. Support for this idea came from analysis of the *avrD* gene derived from *Ps. syringae* pathovars infecting tomato or soybean. The AvrD protein itself lacks elicitor activity but instead directs the production of a family of low molecular weight nonpeptide elicitors known as syringolides, which induce the HR in soybeans containing the corresponding *R* gene. However, this type of interaction is now known to be exceptional. Instead, further experiments lead to the observation that close and persistent contact between the bacteria and plant cells was an essential requirement for the transfer of elicitors. This suggested that internalization of the elicitor was necessary and additional pathogen genes might be involved (see *hrp* genes, p.250).

Fungal *Avr* Genes

The search for elicitors exhibiting gene-for-gene specificity was also pursued in fungal pathogens, initially with inconclusive results. The breakthrough came with the leaf mold disease of tomatoes, *C. fulvum*. This fungus enters via stomata and grows in the intercellular spaces of the leaf, eventually sporulating on the surface. The pathogen occurs as a series of races which interact with different tomato cultivars according to a well-defined gene-for-gene system (Table 10.1b). When a race containing a specific *Avr* gene is inoculated into a tomato leaf containing the corresponding *R* gene (designated Cf), resistance is expressed as a HR.

Early work on this system suggested that culture filtrates from avirulent races might contain factors which induced HR when infiltrated into leaves with the appropriate Cf gene. However, attempts to repeat these experiments proved unsuccessful. One reason might be that *Avr* genes are only expressed in the internal environment of the host plant, rather than in culture media. If this hypothesis was correct, the gene products should, in theory, be secreted into the intercellular spaces. To test this, a procedure was developed for extracting fluids from leaves inoculated with different races of the fungus (Figure 10.3). Initial experiments of this type concentrated on the interaction between the tomato gene *Cf9* and races of the fungus possessing the putative avirulence gene *Avr9* (Table 10.1b). First, the pathogen was inoculated onto a susceptible cultivar lacking the *Cf9* gene (Figure 10.3). Infected leaves were then infiltrated with water under vacuum and the fluids recovered by centrifugation. Purified fractions from these fluids were then injected into leaves containing *Cf9*. As predicted, a necrotic reaction typical of a positive HR was observed. Fluids recovered from compatible races, that is, those able to infect *Cf9* cultivars, did not induce necrosis. Furthermore, the fluids inducing a positive HR on *Cf9* tomatoes were inactive on cultivars lacking this gene, or containing alternative *Cf* genes. Altogether, this was strong evidence that races of the fungus incompatible with *Cf9* cultivars produced a specific elicitor recognized by this host gene.

The next step was to purify the active molecule from the leaf extracts. The specific elicitor turned out to be a peptide containing 28 amino acids. Once the sequence of this peptide was known, it could then be used to design an oligonucleotide probe complementary to the DNA sequence coding for the peptide, that is, the putative *Avr9* gene. In fact, this probe was initially used to identify a cDNA clone prepared from mRNA isolated from leaves infected with a pathogen race containing *Avr9*. The cDNA could then be used to hybridize with samples of genomic DNA extracted from the pathogen to identify the *Avr9* sequence.

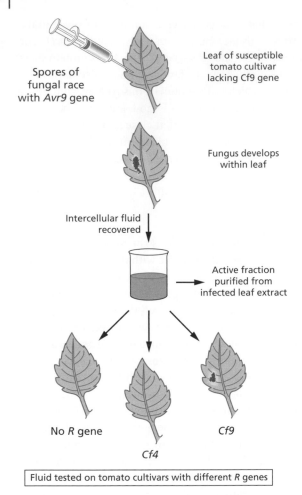

Spores of fungal race with *Avr9* gene

Leaf of susceptible tomato cultivar lacking Cf9 gene

Fungus develops within leaf

Intercellular fluid recovered

Active fraction purified from infected leaf extract

No *R* gene

Cf4

Cf9

Fluid tested on tomato cultivars with different *R* genes

Figure 10.3   Procedure used to isolate a specific elicitor of host resistance produced by the tomato leaf mold fungus *Cladosporium fulvum*.

This elegant experiment led to the first successful cloning of an avirulence gene from a fungal pathogen.

To confirm that *Avr9* functioned according to the predictions of the gene-for-gene model, several further experiments were necessary. Tests showed that races of *C. fulvum* virulent on *Cf9* tomato genotypes lacked the *Avr9* gene; introducing the gene into such races by transformation converted them into avirulent forms. Conversely, when the *Avr9* gene was inactivated by a gene disruption procedure, a race previously avirulent on *Cf9* genotypes was converted into a virulent form. Hence, *C. fulvum* can evade recognition on *Cf9* tomatoes either if it lacks the specific avirulence gene or if the gene is unable to direct synthesis of the peptide elicitor. Subsequent work on another avirulence gene, *Avr4*, suggested that nonfunctional versions of this gene may be present in pathogen races virulent on tomato cultivars containing the resistance gene *Cf4*. Further studies showed that a single base-pair change is sufficient to abolish recognition on tomato cultivars containing the corresponding resistance gene.

Some further features of the *Avr9* gene merit comment. The primary product is a 63 amino acid precursor peptide, which is cleaved to yield the smaller 28 amino acid elicitor found in the intercellular fluids of leaves. Experiments on regulation of the *Avr9* gene show that it is expressed in the host plant, but not in nutrient-rich culture media. This appears to be typical of pathogen genes which play a role in specificity; they are only switched on in nutrient-poor environments such as those prevailing in plant tissues. Only a single copy of the avirulence gene is present which provides an explanation for the frequent changes in the virulence of fungal pathogens encountered in the field. A single mutation, or deletion, of the avirulence gene will result in a gain of virulence on a previously resistant host containing the matching *R* gene.

Further avirulence genes from fungi were subsequently cloned, including from *Magnaporthe oryzae*, causal agent of the destructive blast disease of rice. Mapping of one of these rice blast genes, *Avr2*, showed it to be located very close to the end of a chromosome, within 50 base-pairs of the telomere. This region of the genome is known to be relatively unstable, and deletion or inactivation of the gene can therefore take place. Such changes seem likely to account for the variability in host range often observed in this fungus. Interestingly, a further gene identified in *M. oryzae* (known as *PWL2*) determines pathogenicity or nonpathogenicity to a wild relative of rice, weeping love grass. This is an example of a gene determining host species specificity, rather than host range on different genotypes of the same species, which is the usual pattern with *Avr* genes. The gene product is a secreted protein composed of 145 amino acids. Expression of this gene is again unstable, with frequent mutations converting forms of the fungus nonpathogenic to love grass to pathogenic forms. Investigation of a collection of different isolates of *M. oryzae* showed that variants of the *PWL2* gene are widely distributed in the pathogen population, but only some are actually functional and influence host range.

Extracellular Versus Intracellular Infection Routes

Fungi like *C. fulvum* that grow between host cells without penetrating them have been described as **apoplastic** pathogens, in which molecular communication between the pathogen and the host plant usually involves secreted and diffusible proteins and cell surface-located receptors. A number of other important crop diseases such as Septoria tritici blotch of wheat *Zymoseptoria tritici*, and stem canker of oilseed rape, *Leptosphaeria maculans*, also show this type of relationship. Hence, the products of *Avr* genes in these pathogens are predicted to be small secreted apoplastic proteins.

A similar scenario occurs with vascular wilt pathogens which colonize xylem vessels. An example is *Fusarium oxysporum* infecting tomatoes in which a series of proteins, designated as Six (secreted into xylem), have been detected in xylem fluid. Some of these are now known to act as Avr factors. Race-specific resistance to another tomato wilt pathogen, *Verticillum dahliae*, has been shown to involve a secreted fungal protein, Ave1, interacting with a cell surface-located immune receptor, Ve1. These pathosystems contrast with biotrophic fungi such as the rusts and powdery mildews, in which host cells are penetrated by haustoria, and hemibiotrophs like *Colletotrichum* and *M. oryzae,* where the initial host–pathogen interface is an intracellular vesicle formed within epidermal cells (see Chapter 6, Figure 6.9). These fungi possess alternative pathways for the delivery of *Avr* gene products into the host.

Other Functions for *Avr* Genes?

The identification and further characterization of *Avr* genes and their products initially posed an evolutionary question. Why, if *Avr* genes encode products directly or indirectly recognized by the plant, and hence restrict the host range of the pathogen, are these genes not strongly selected against and eliminated from the pathogen population? Surely it would be an advantage for the pathogen to infect as many different host types as possible? The most likely explanation was that these genes must be playing some other, perhaps vital role in the life of the pathogen. Loss of the gene might therefore compromise the fitness of the organism.

According to this view, the products of *Avr* genes are proteins or other molecules which the host plant has "learned" to recognize. Their role in specific recognition is not, therefore, their primary function. Unlike their plant counterparts, the *R* genes, different *Avr* genes showed little homology and appeared to be randomly distributed in the pathogen genome. This indicated that they encoded diverse products with different functions. The widespread occurrence of nonfunctional versions of *Avr* genes in pathogen races virulent on plants with corresponding *R* genes also suggested that some important property may be conserved, but the products themselves may have been modified under selection to evade recognition. by the host plant. Altogether, it was likely that *Avr* genes had an alternative and positive role during invasion of the host.

Bacterial *hrp* Genes

Following chemical or insertional mutagenesis of bacteria with transposons (see Chapter 8, Figure 8.1), a class of mutants was recovered in which pathogenicity to the host plant and the ability to induce a HR in resistant hosts, or also in nonhosts, were both simultaneously lost. Isolates with this phenotype, described as **hrp** (hypersensitive reaction and pathogenicity) mutants, were identified in several of the most important genera of plant-pathogenic bacteria including *Erwinia*, *Pseudomonas*, *Xanthomonas*, and *Ralstonia*. Genetic analysis of such mutants added a further piece to the jigsaw of molecular plant–pathogen interactions.

The bacterium *Xanthomonas axonopodis* pv. *vesicatoria* (Xav, formerly *X. campestris* pv. *vesicatoria*), which causes spot disease of pepper and tomato, serves as an example. Complementation analysis, in which DNA clones from a genomic library of the wild type were tested for their ability to restore function in *hrp* mutants, showed that the genes responsible are clustered in the bacterial chromosome spanning a region of about 25 kb (Figure 10.4a). Transposon mutagenesis confirmed that there are at least six *hrp* genes, coded A–F, in this region. Inactivation of each *hrp* gene by insertion of a transposon led to reduced multiplication of the mutant in the host plant, while insertions outside the *hrp* cluster did not affect pathogenicity (Figure 10.4b). A similar clustered arrangement of *hrp* genes was found in several other plant–pathogenic bacteria.

Like *avr* genes, *hrp* genes are not expressed when the bacteria are grown in nutrient-rich culture media, but are switched on when the pathogen grows under low nutrient conditions typical of the environment encountered within the host plant. When *Ps. syringae* is inoculated into tobacco leaves, enhanced expression of *hrp* genes can be detected within one hour. Filtrates from plant cells are effective inducers of the *hrp* cluster, indicating that

(a)

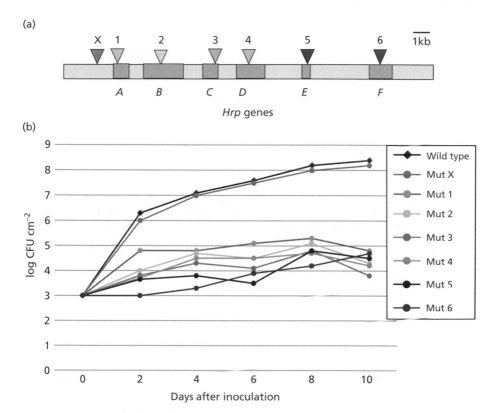

Hrp genes

(b)

Figure 10.4 (a) A cluster of *hrp* genes (*A–F*) from *Xanthomonas axonopodis* (formerly *X. campestris*). Insertion of a transposon in any of these genes, to give mutants 1–6, abolishes the ability to induce a hypersensitive response, and reduces multiplication of the bacterium in host leaves. Insertions outside the cluster (e.g., mutant X) have no effect on HR activity or pathogenicity. (b) Growth curves in leaves for mutants 1–6 compared with wild type and mutant X. (CFU = colony-forming units). *Source:* After Bonas et al. (1991).

these genes are regulated by plant signals. Synthesis of *hrp* gene products therefore represents an early event in the molecular "cross-talk" between host and pathogen.

But what are the actual products of these genes? Bioinformatic analysis and comparison of *hrp* gene sequences between different bacteria gave some surprising results. Several of the predicted plant pathogen gene products had close homology with proteins implicated in the pathogenicity of bacteria to animal hosts, including well-known human pathogens such as *Salmonella*, *Shigella*, and *Yersinia pestis*, the causal agent of bubonic plague, the notorious Black Death. These highly conserved genes encode components of a bacterial secretory pathway required for pathogenicity, known as the Type III secretion system (T3SS). The T3SS is a specialized apparatus for delivering pathogen products into host cells (see Figure 10.8). Hence, there are striking similarities between the mechanisms responsible for bacterial pathogenesis in both plants and animals. This discovery also finally resolved the question as to why many *avr* gene products, when tested directly on host cells, failed to induce an HR. The elicitor molecules need to be introduced into the cytoplasm of the cell to trigger a response.

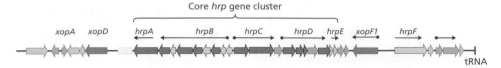

Figure 10.5 Diagrammatic representation of the hypersensitive response and pathogenicity (*hrp*) gene cluster from a fully sequenced isolate of *Xanthomonas axonopodis* pv. *vesicatoria*. The red genes in the core cluster are conserved between plant and animal pathogenic bacteria (*hrc*), blue genes are further *hrp* genes, while green genes are *hrp* associated (*hpa*), not all of which are essential for pathogenesis. *Hrp* F lies outside the core cluster. *Xop* (**X**anthomonas **o**uter **p**rotein) genes encode further secreted products. Black arrows above the sequence indicate the operon structure and direction of transcription, and black dots the presence of promoters regulated by plant signals. The yellow box indicates a transposable element (insertion sequence, IS) and the tRNA genes at the right border are features characteristic of a pathogenicity island that can be acquired from other bacteria by lateral gene transfer. *Source:* Hammond-Kosack and Jones (2015).

The sequencing of genomes of bacterial pathogens, including several species and pathovars of *Xanthomonas*, has now provided further insights into the occurrence and organization of genes involved in host–pathogen interactions. The *hrp* gene cluster from strains of *Xav* comprises around 24 genes with a core group of highly conserved *hrp* genes, and a more variable region centered on *hrpF* (Figure 10.5). The cluster also contains several *hrp*-associated (*hpa*) genes, as well as an insertion (IS) element and transfer RNA gene typical of a pathogenicity island that may have been laterally transferred from another plant-pathogenic bacterium. The genome also has further cassettes of genes implicated in pathogenesis, including the biosynthesis of lipopolysaccharide (LPS) and the *gum* cluster producing xanthan, an exopolysaccharide involved in biofilm formation and host colonization. The clustering of pathogenicity genes in potentially mobile regions of the genome, such as pathogenicity islands, may enable these pathogens to adapt quickly to a changing environment, including new host plants.

A Unifying Concept: Pathogen Effectors and Plant Immunity

The studies described above helped to lay the foundation for a more integrated view of host–pathogen interactions and supported a new paradigm described by the so-called zig-zag model (Figure 10.6). This reinforced the idea that there are two branches of the plant immune system and, importantly, recognized that the products of *Avr* genes play significant roles as **effectors** targeting key components of plant recognition and immune response pathways. The model also provides an evolutionary timeline suggesting how such molecular interactions might have developed.

The first layer of plant immunity involves detection of conserved pathogen signatures (**PAMPS**), by pattern recognition receptors (PRRs). This triggers a cascade of responses that increase the level of host resistance (**PAMP-triggered immunity, PTI**). In turn, pathogens evolved a diverse molecular armory to counter the initial immune response. This includes many secreted products, mainly small proteins which either act as effectors in the apoplast or enter the host cell to suppress immune responses, restoring susceptibility to infection (**effector-triggered susceptibility**, ETS). A second layer of plant immunity

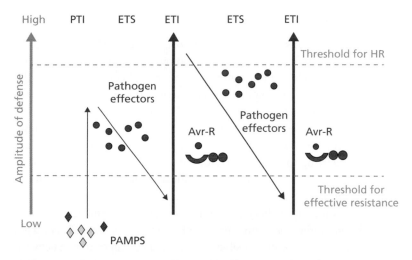

High   PTI   ETS   ETI   ETS   ETI

Figure 10.6   A zig-zag model of plant immunity. In the first phase, the presence of typical microbial molecules (MAMPs/PAMPs) is detected by host pattern recognition receptors (PRRs) activating PAMP-triggered immunity (PTI). In the second phase, pathogen effector molecules target components of the plant immune system suppressing PTI (effector-triggered susceptibility, ETS). In phase 3, a plant receptor detecting a specific effector activates a strong immune response restoring resistance (effector-triggered immunity, ETI). Novel effectors may then evolve, countering ETI, and the evolutionary process continues. *Source:* Adapted from Jones and Dangl (2006).

then emerged with development of plant molecular sensors able to detect pathogen effectors, triggering a strong defense response usually characterized by hypersensitive cell death (**effector-triggered immunity, ETI**). ETI is activated by the specific interaction of an effector with a host resistance (R) protein and restores a high level of resistance. The gene-for-gene system, first proposed on the basis of resistant and susceptible disease phenotypes (Figure 10.1), is therefore determined by molecular interactions between plant R proteins and specific pathogen effectors encoded by *Avr* genes. However, if the cognate effector undergoes changes or the *Avr* gene is lost from the pathogen genome, ETI breaks down. In addition, new effectors may evolve, restoring susceptibility. Hence the zig-zag model portrays a molecular arms race in which the balance between resistance and susceptibility fluctuates over time.

The zig-zag model has been widely adopted as a framework for further study but has some limitations. These include the assumption that PAMPs are less variable than other microbial products recognized by plants, and that the two branches of the immune system, based on PRRs versus R genes, are somehow distinct. The existence of particular R genes that function against different pathogens and even different types of biotic agents such as insects and nematodes also confounds the idea that R–Avr interactions are highly specific. It has been argued that a more comprehensive model of plant immunity that avoids the PTI and ETI dichotomy is required. There is also the need to accommodate interactions with necrotrophic pathogens in which recognition of a pathogen toxin effector by a plant immune receptor leads to susceptibility rather than resistance (see inverse gene-for-gene relationships, p.259). Nonetheless, the zig-zag model has helped to integrate some key aspects of host–pathogen interactions and stimulated debate about how they may have evolved.

Pathogen Effector Repertoires

It is now realized that many organisms colonizing plant hosts produce effector molecules aiding exploitation of the host plant. They include fungi, oomycetes, bacteria, and invertebrates such as nematodes and aphids. As most effectors are small, secreted proteins, the availability of full genome sequences has enabled searches to identify genes encoding candidate effectors. These have shown that pathogens produce multiple effectors, numbering from 30 to 80 in bacterial pathogens, to more than 200 in some oomycetes. Assigning functions to all these proteins is problematical as knocking out individual genes often produces no change in phenotype, suggesting that other effectors may substitute for the loss. Such redundancy is most likely due to the expansion and diversification of effector gene families during host–pathogen evolution.

Due to the large number and diversity of effectors produced by pathogens, and in many cases uncertainty over their molecular targets, a comprehensive classification of the different types based on their mode of action is not currently possible. One distinction can be drawn between protein and nonprotein effectors, the latter exemplified by the bacterial toxin coronatine which reverses stomatal closure induced by PAMP detection. Another is between apoplastic effectors that act outside host cells and cytoplasmic effectors that enter cells and interfere with defense pathways. Some effectors are even targeted to the nucleus where they reprogram host gene expression. So, there are many ways in which effectors can subvert plant immunity.

Table 10.3 lists some examples of effectors produced by bacterial, fungal, and oomycete pathogens, along with information on their modes of action where known.

Apoplastic Effectors

Apoplastic effectors are typically produced by pathogens that grow through intercellular spaces or in other locations such as just beneath the cuticle, without penetrating host cells (see Chapter 6, Table 6.1). Examples listed in Table 10.3 include the Avr proteins and Ecp6 from *C. fulvum*, and Zt3LysM from *Z. tritici*. Avr2 inhibits plant protease enzymes secreted into the apoplast, while Ecp6 and Zt3LysM both interfere with immunity based on detection of the fungal PAMP chitin (Figure 10.7). Fragments of this conserved component of fungal cell walls are released into the apoplast during pathogen invasion, amplified by the action of host chitinase enzymes. These breakdown products are bound by pattern recognition receptors in the host plasma membrane, activating a PTI defense response. Secretion of the LysM effector into the apoplast sequesters chitin breakdown products, preventing them from reaching the host receptors. This is an example of host invasion by stealth, whereby a pathogen PAMP is removed from the host–pathogen interface, disguising the enemy. The Lys3M protein can also serve a defensive role by protecting fungal hyphae against plant-derived hydrolytic enzymes.

Delivery of Cytoplasmic Effectors

Many pathogen effectors enter host cells and target proteins in a range of locations, including membranes, organelles such as chloroplasts and mitochondria, and in the cytoplasm. Some, such as the transcription activator-like (TAL) effectors produced by *Ralstonia* and

Table 10.3 Some examples of pathogen effectors and their interactions with plant hosts

Pathogen species	Effector	Host interactions
Bacteria		
Pseudomonas syringae	Hop1	Disrupts chloroplast function and SA synthesis
	HopM1	Inhibits vesicle trafficking and promotes water soaking of tissues
	HopF2	Suppresses ROS production and stomatal closure
	Avrpto	Targets flagellin receptor complex and PTI
	AvrB	Affects JA signaling and induces stomatal opening
Ralstonia solanacearum	RipG1–8	Targets ubiquitin protein degradation pathway
	RipTAL1	TAL effector – enters nucleus, mimics eukaryotic transcription factors, and alters gene expression
Xanthomonas campestris	AvrBs3	TAL effector recognized by Bs3 R gene
Fungi		
Cladosporium fulvum	Avr2	Inhibits cysteine protease enzymes in host apoplast
	Ecp6	Interferes with chitin-triggered immunity
Zymoseptoria tritici	Zt3LysM	Interferes with chitin-triggered immunity and protects against chitinases
Magnaporthe oryzae	AVR-Piz-t	Targets ubiquitin protein degradation pathway
	AVR-Pita	Protease enzyme exported from infection hypha into host cell cytoplasm
Fusarium oxysporum	Avr7	**S**ecreted **i**nto **X**ylem (SiX) protein suppresses host cell death
Pyrenophora tritici-repentis	PtrToxA	Induces host cell death in toxin-sensitive wheat genotypes
Cochliobolus victoriae	Victorin	Induces host cell death in plants with corresponding toxin receptor
Ustilago maydis	Cmu1	Inhibits SA synthesis
Puccinia striiformis	Pst-8713	Suppresses cell death and PTI
Oomycetes		
Phytophthora infestans	Avrblb2	Suppresses secretion of protease enzymes
	Avr3a	RXLR effector translocated into host cells, inhibits PTI?
Phytophthora parasitica	PpRxLR2	Suppresses host cell death in tobacco
Phytophthora sojae	GIP1	Inhibits glucanase enzymes
Hyaloperonospora arabidopsidis	HaRxL44	Attenuates SA defense signaling in *Arabidopsis*

Hop, hrp outer protein; JA, jasmonic acid; PTI, PAMP-triggered immunity; Rip, *Ralstonia* injected protein; ROS, reactive oxygen species; RXLR, oomycete effector with conserved amino acid motif; SA, salicylic acid; TAL, transcription activator-like.

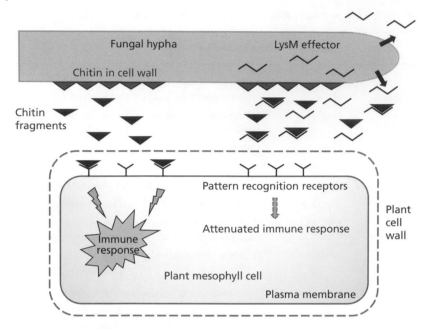

Figure 10.7 Diagrammatic representation of suppression of PAMP-triggered immunity by a chitin-binding pathogen effector (Zt3Lysm from *Zymoseptoria tritici*). Fragments of chitin from the fungal cell wall released into the apoplast (*left*) are detected by receptors in the plant cell plasma membrane, triggering an immune response. The pathogen effector protein secreted into the apoplast (*right*) sequesters chitin fragments, preventing detection of the fungus and allowing further development in host leaves by stealth.

Xanthomonas, are targeted to the host nucleus where they mimic transcription factors (TF) and alter host gene expression. This unique activity is based on a DNA-binding domain in the proteins that binds to the promoter region of the induced host genes.

But how do effector proteins actually get into host cells? Several specialized devices and delivery pathways are involved (Figure 10.8). Many pathogenic bacteria form a kind of molecular hypodermic syringe, known as the T3SS, which injects effectors into cells. Several of the genes found in bacterial *hrp* clusters encode structural components of the T3SS. For instance, the *hrpA* protein of *Ps. syringae* self-assembles into a pilus, forming the needle part of the syringe. Many, but not all bacterial effectors are exported via the T3SS. Alternative secretion systems are involved in the delivery of cell wall-degrading enzymes, the toxin coronatine and *Agrobacterium* effectors involved in the transfer and integration of the T-DNA into the host genome. Biotrophic fungi and oomycetes produce haustoria that penetrate the host cell wall and invaginate the plasma membrane (Figure 10.8b), providing an intimate interface for the uptake of nutrients and export of effectors (as described in Chapter 6). Hemibiotrophic fungi such as *Colletotrichum* and *Magnaporthe* penetrate the epidermal cells of their respective hosts and form swollen primary hyphae (also described as infection vesicles) during the initial biotrophic phase of colonization. Effector proteins of *Magnaporthe* have been shown to accumulate in a biotrophic interfacial complex prior to delivery into the host cytoplasm (Figure 10.8c). Subsequently, secondary hyphae invade adjacent cells and initiate the more destructive, necrotrophic phase of infection.

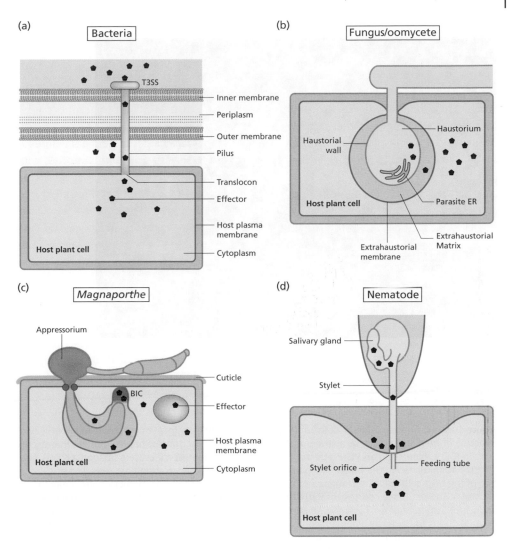

Figure 10.8 Delivery systems for cytoplasmic effectors. (a) Bacterial type III secretion system (T3SS) with pilus translocating effectors across plant cell wall and membrane. (b) Biotrophic fungus or oomycete forming haustoria penetrating the plant cell wall and invaginating membrane with secretion of effectors into the cytoplasm. (c) Primary hypha of the rice blast fungus *Magnaporthe oryzae* invading a rice epidermal cell and secreting effectors via a biotrophic interfacial complex (BIC) formed at the hyphal tip adjacent to the host membrane. (d) Nematode stylet penetrating host cell wall and membrane to deliver effectors from the salivary glands. *Source:* (a,b,d) adapted from Torto-Alalibo et al. (2010), (c) from Giraldo and Valent (2013).

Several lines of evidence have confirmed that haustoria are sites of effector delivery. Gene libraries prepared from isolated haustoria are enriched for effector proteins, for instance the *Avr* gene products produced by the flax rust fungus *Melampsora lini*. Experiments in which specific effectors are labeled with fluorescent proteins have also shown that they are localized to haustoria. An example is shown in Figure 10.9a. where an effector produced by *Phytophthora infestans* was tagged with a fluorescent protein

(a)

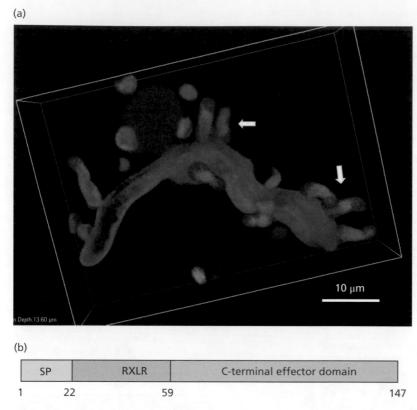

10 μm

n Depth: 13.60 μm

(b)

SP	RXLR	C-terminal effector domain

1   22   59   147

Figure 10.9 Oomycete effector delivery. (a) 3D projection of a confocal microscope image of infection of host leaf tissues by a strain of *Phytophthora infestans* transformed with green fluorescent protein (GFP), expressing an RLXR effector fused to another fluorescent protein (magenta color). The effector accumulates in the finger-like haustoria (*white arrows*) penetrating host cells. The nucleus of one of the host cells (*blue*), surrounded by chloroplasts (*yellow*) can be seen in close proximity to haustoria. *Source:* Image courtesy of Petra Boevink, Hutton Institute; further details in Wang et al. (2018). (b) Structure of Avr3a oomycete effector from *P. infestans*. An N-terminal secretory signal peptide (SP) is followed by a domain containing the RXLR motif believed to be involved in entry into host cells. The C-terminal effector domain interacts with an intracellular plant target. Numbers indicate amino acids. *Source:* Adapted from Petre and Kamoun (2014).

and imaged by confocal fluorescence microscopy. The magenta signal is associated with the finger-like haustoria extending into host cells. The precise mechanism by which fungal and oomycete effectors cross the host–pathogen interface is not yet clear. Many oomycete effectors (Figure 10.9b) have a conserved amino acid domain consisting of the sequence arginine-any aa-leucine, arginine (the RXLR motif) that is believed to play a role in effector delivery, although recent evidence suggests that this is cleaved from the molecule prior to entry into the host cell. Large families of effectors possessing the RXLR motif are widely distributed among oomycete pathogens such as *Phytophthora* species and downy mildews.

There are interesting parallels with some invertebrate pests that colonize plants, which form modified mouthparts to deliver secretions into host tissues. Plant-parasitic nematodes produce a structure known as a stylet (Figure 10.8d), analogous to the bacterial delivery system. Proteins secreted by nematode glands pass through the stylet and can be detected in plant cells around nematode feeding sites. Some of these proteins have been shown to subvert plant immunity. Another type of stylet is used by aphids to probe the plant host and deliver salivary fluids into phloem tissues (see Chapter 3, Figure 3.13). These also contain effector proteins that modify host physiology to aid feeding. Such structural adaptations are examples of convergent evolution to perform a particular biological function during exploitation of the host plant.

Necrotrophic Effectors and the Inverse Gene-for-Gene Relationship

Necrotrophic pathogens often produce toxic molecules that kill host tissues. The host-specific toxins first described in Chapter 8 are now considered to be a type of pathogen effector as they contribute to virulence by inducing host cell death. A well-characterized example is the peptide toxin victorin produced by the necrotrophic fungus *Cochliobolus victoriae* (Table 10.3). Sensitivity to the toxin in *Arabidopsis* has been shown to depend on a host gene *LOV1*, which encodes an NLR protein, a type usually associated with resistance to biotrophic pathogens. Another necrotrophic effector is the protein toxin PtrToxA produced by the wheat-infecting fungus *Pyrenophora tritici-repentis*. Sensitivity to this toxin is also host specific; wheat genotypes containing the corresponding gene *Tsn1* are toxin sensitive and susceptible to tan-spot disease. Genotypes lacking the *Tsn1* gene are insensitive to the toxin and resistant to tan-spot. The interaction between the fungus and different wheat genotypes is therefore the opposite of the conventional gene-for-gene relationship in which resistance is determined by recognition of a pathogen effector by a host *R* gene protein (Table 10.1). Instead, sensitivity to a pathogen effector determines disease susceptibility, dependent on the presence of a host sensitivity gene (Figure 10.10). This relationship has been described as an inverse gene-for-gene interaction.

Genome sequencing has shown that *Tox A* genes are also present in other cereal-infecting fungi such as *Parastaganospora nodorum* and *Bipolaris sorokiniana*. In fact, sequence data suggest that the *ToxA* genes may have been acquired by lateral gene transfer from *P. nodorum* to the other species (see Chapter 8, Figure 8.9). The wheat sensitivity gene *Tsn1* has recently been cloned and turns out to encode an NLR protein,

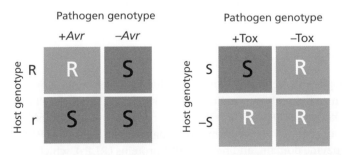

Figure 10.10 The classic gene-for-gene relationship (*left*), defined by the interaction of a pathogen *Avr* gene product (effector) with a plant *R* gene, compared with the inverse gene-for-gene relationship (*right*) based on the interaction of a necrotrophic toxin effector with the product of a host sensitivity gene. R, resistant combination; S, susceptible combination. *Source:* Adapted from Faris et al. (2013).

similar to *LOV1*. It therefore seems that these toxin effectors exploit a key component of the plant immune surveillance system to trigger cell death and hence favor the necro-trophic lifestyle of the pathogen.

Genes for Pathogen Recognition and Resistance

While the discovery of pathogen effectors and identification of some host targets was a major breakthrough in understanding the molecular basis of host–pathogen specificity, an even greater prize was to find the plant genes determining resistance. This is because *R* genes are the raw material selected for by plant breeders, and isolation of such genes might create novel possibilities for their use in practice. Deciphering the characteristics of such genes and how they work should accelerate the process of screening germplasm and suggest alternative ways of engineering crop resistance. Furthermore, it might resolve the vital question as to why some genes provide durable resistance while others do not. Hence there was intense interest in cloning *R* genes and discovering their functions.

The existence of dominant genes conferring resistance to pathogens has been recognized by plant breeders for more than a century; indeed, many such genes have been utilized in conventional breeding programs aimed at crop improvement (see p.310). But techniques for locating and physically isolating these genes only became available with advances in molecular genetics and, more recently, genome sequencing. The main challenge is the large size of plant genomes. Even the relatively small genetic model *Arabidopsis*, the first plant genome sequence completed in 2000, has around 100 million bases and an estimated 26 000 protein-encoding genes. The first crops sequenced, maize and rice, have around 30 000 and 40 000 genes respectively while wheat, a hexaploid species, has more than 90 000. So, finding a particular gene is a demanding task. Nonetheless, several *R* genes were successfully isolated prior to the availability of any complete plant genome sequence, either by the painstaking approach of map-based cloning using genetic markers in segregating populations or transposon tagging whereby insertion of a transposon into an *R* gene con-verts the phenotype from resistance to susceptibility. As this is a relatively rare event, large numbers of plants need to be screened, and further analysis is required to isolate the gene and confirm function by complementation.

Pathogen Recognition Genes

In the Z model of plant-pathogen interactions (Figure 10.6), the first layer of plant immu-nity involves detection of conserved pathogen molecules (known as PAMPs) by cell sur-face-located plant receptors (known as PRRs). Plant PRRs occur as two types: receptor-like kinase (RLK) proteins, with an extracellular recognition domain which binds the PAMP and an intracellular kinase enzyme domain, and receptor-like proteins (RLPs), which lack the cytoplasmic kinase domain.

Some examples of PAMPS and their corresponding plant receptors are listed in Table 10.4. In *Arabidopsis*, a transmembrane receptor kinase known as FLAGELLIN SENSITIVE 2 (FLS2) specifically recognizes the protein component of flagella, the filamentous structures

Table 10.4 Some examples of microbial polymers recognized by plants, and their corresponding receptors

Pathogen type	Pathogen-associated molecular pattern	Plant recognition receptor and type[a]
Bacteria	Flagellin	FLS2 (RLK)
	EF-Tu (elongation factor)	EFR (RLK)
	Peptidoglycan	LYM1/3 (RLP)
	Lipopolysaccharide	LORE (RLK)
Fungi	Chitin	CERK1 (RLK)
		LYK4/5 (RLK)
	β-glucan	CERK1? (RLK)
	Polygalacturonase	RLP42 (RLP)
Oomycetes (and other microbes)	NLPs (necrosis and ethylene-inducing peptide-like proteins)	RLP23 (RLP)

[a] RLK, receptor-like kinase; RLP, receptor-like protein.
Source: Adapted from Yun et al. (2018).

responsible for bacterial motility. The leucine-rich extracellular domain of the receptor binds a 22-amino acid sequence in the flagellin molecule. Other "molecular signatures" of bacteria detected by plants include an elongation factor involved in selection and binding of amino acids to the ribosome (EF-Tu, one of the most abundant prokaryotic proteins), and the peptidoglycan component of bacterial cell walls, as well as secreted LPSs. A similar surveillance system operates against fungi and oomycetes. For instance, CERK1, a further RLK from *Arabidopsis*, detects the fungal cell wall polymer chitin. The NLP family of cytotoxic proteins, which causes necrosis in plants, have been found in a wide range of microbes, including oomycetes such as *Phytophthora* species, and are also sensed by plant receptors. A consistent theme, exemplified by Table 10.4, is that plants have evolved the ability to detect molecules that are highly abundant and widely distributed among the major groups of microorganisms.

The molecular binding of a PAMP by a PRR is usually insufficient to trigger plant immunity and often requires a second protein, a co-receptor, to initiate defense signaling (Figure 10.11). For instance, in *Arabidopsis* the activation of FLS2 by flagellin involves formation of a receptor complex with another receptor kinase known as BAK1. Once activated, the receptor complex triggers a downstream signaling cascade with phosphorylation of multiple kinase enzymes, leading to defense gene expression and associated immune responses. It has been shown that BAK1 is also required for the immune responses triggered by multiple PAMPs, including elongation factor EF-Tu, and hence acts as a positive regulator of PAMP signaling. Further studies on other plants such as rice have suggested that there may be overlapping functions of PRRs in perception of bacterial and fungal PAMPS including chitin and peptidoglycan. It is also becoming clear that the participation of protein complexes in immune signaling ensures that the process is tightly regulated to avoid any potentially wasteful deployment of plant defenses unless they are really needed.

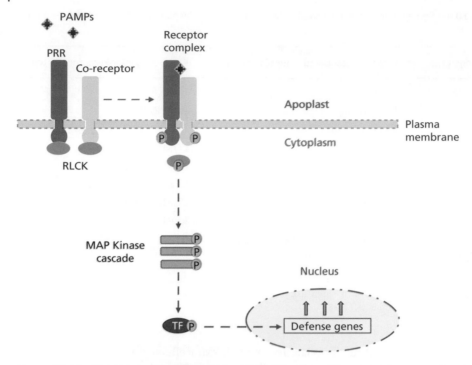

Figure 10.11 Simplified scheme of processes in PAMP-triggered immunity. A conserved pathogen molecule binds to a pattern recognition receptor (PRR) in the plant cell plasma membrane, initiating a dynamic interaction with a co-receptor and receptor-like cytoplasmic kinases (RLCKs), and phosphorylation (P) of the kinase domain. Downstream signaling occurs via kinase enzyme cascades, activating several targets including transcription factors (TF) leading to defense gene expression in the nucleus. *Source:* Adapted from Saijo et al. (2018).

Plant Resistance (*R*) Genes

The cloning and characterization of many of the *R* genes first identified by breeders in plant populations segregating for resistance to different pathogens provided further insights into the molecular basis of plant immunity. The gene-for-gene model predicted that *R* genes most likely encoded plant receptors that interacted with products of pathogen *Avr* genes to confer race-specific resistance. Initial results suggested that the story was not that simple, as the proteins encoded by the first sequenced genes varied in their structure and potential function. Table 10.5 illustrates some of the diversity of *R* genes cloned to date, including examples acting against fungal, bacterial, and viral pathogens.

The first *R* gene to be cloned was *Hm1* from maize, conferring resistance to *Cochliobolus (Helminthosporium) carbonum*. The gene product turned out to be a reductase enzyme degrading a cyclic peptide toxin produced by the fungus. As in this case the pathogenicity of the fungus is correlated with production of the host-specific HC toxin, then plants able to inactivate the toxin are unaffected by the fungus. Isolation and analysis of *Hm1* added convincing genetic and biochemical proof to the argument that host-specific toxins are the sole determinants of disease with these types of fungal pathogens (see p.194).

The next *R* gene to be isolated was *Pto* from tomato, which confers resistance to strains of the bacterial pathogen *Ps. syringae* possessing the avirulence gene *avrPto*. This was therefore

Table 10.5 Some cloned plant disease resistance genes

Gene	Host	Pathogen	Location	Function
Hm1	Maize	*Cochliobolus*	Cytoplasmic	Toxin reductase
Pto	Tomato	*Pseudomonas syringae*	Cytoplasmic	Serine/threonine kinase
N	Tobacco	Tobacco mosaic virus	Cytoplasmic	NLR receptor
Cf9	Tomato	*Cladosporium fulvum*	Cell membrane	Receptor-like protein
RPS2	Arabidopsis	*Pseudomonas syringae*	Cell membrane	NLR receptor
RPP5	Arabidopsis	*Hyaloperonospora*	Cytoplasmic	NLR receptor
L6	Flax	*Melampsora lini*	Cytoplasmic	NLR receptor
Xa21	Rice	*Xanthomonas*	Cell membrane	Receptor-like protein
Rpg1	Barley	*Puccinia graminis*	Cytoplasmic	Receptor kinase
Stb6	Wheat	*Zymoseptoria tritici*	Cell membrane	Receptor kinase
Lr34	Wheat	Multiple pathogens	Cell membrane	ABC transporter
Mlo	Barley and many other plants	Powdery mildew fungi	Cell membrane	Transmembrane receptor-like protein

the first example of a characterized *R* gene conforming to the gene-for-gene model. *Pto* was located by map-based cloning procedures, using molecular landmarks in the tomato genome to eventually isolate DNA clones containing the gene. Analysis of the sequence revealed that *Pto* encodes a protein with a serine–threonine kinase domain, but without any typical features of a receptor. Nonetheless, experiments confirmed that *Pto* interacts directly with the type III secreted effector AvrPto, leading to induction of defense responses including the HR. It was subsequently shown that signaling by *Pto* requires the presence of a partner protein known as Prf.

Several other *R* genes have since been found to encode RLKs, including *Rpg1*, which has provided durable protection against black stem rust *P. graminis* in barley cultivars in North America, and more recently *Stb6*, a widely used source of resistance in wheat to Septoria tritici blotch, *Z. tritici*. The gene product of *Stb6* is a receptor kinase-like protein associated with the plasma membrane and cell wall which detects an effector protein AvrStb6 secreted into the apoplast by the fungus. Interestingly, resistance due to *Stb6* does not result in an HR, and may instead prevent the pathogen from transitioning to necrotrophic growth dependent on the death of host cells (see Chapter 7, Figure 7.12).

Other atypical *R* genes include *Lr34* which confers adult plant resistance to several rust and powdery mildew pathogens in wheat, and *Mlo* controlling powdery mildew, first discovered in barley and subsequently many other plant species. *Lr34* encodes a putative ABC transporter protein in the cell membrane which may export metabolites inhibiting fungal growth, thereby accounting for its efficacy against several pathogens. *Mlo* encodes an integral membrane protein with seven transmembrane domains which bears some resemblance to a G-protein coupled receptor. However, resistance is in this case recessive and acts by a loss of function associated with the *mlo* allele. The Mlo protein may be a general suppressor of cell death and hence deregulation of a cell death pathway leads to resistance.

The remainder of the *R* genes listed in Table 10.5 all have some conserved features, notably a leucine-rich repeat (LRR) domain determining pathogen recognition specificity. A distinction can be drawn, however, between those with extracellular LRRs detecting pathogen products in the apoplast, or intracellular LRRs, detecting effectors delivered into the host cell (Figure 10.8). For instance, the *Cf* genes in tomato recognizing *C. fulvum* effectors secreted into intercellular spaces encode proteins with an extracellular LRR domain anchored to the plasma membrane. The Xa21 protein from rice recognizing the bacterium *Xanthomonas oryzae* is similar but also has a cytoplasmic kinase domain involved in downstream signaling. The tobacco resistance gene *N* and *L6* from flax encode proteins with cytoplasmic LRRs recognizing the virus replicase enzyme from tobacco mosaic virus (TMV), and an effector introduced into host cells via flax rust haustoria respectively.

Figure 10.12 illustrates some of this diversity in the structure of plant *R* gene proteins. As further plant species are sequenced and new *R* genes cloned, it is likely that additional types of proteins involved in pathogen recognition will be discovered.

Structure and Function of NLR Immune Receptors

While, as described above, various classes of *R* genes are now known to be involved in plant defense, the majority (to date c.60%) belong to a particular family of immune receptors with a modular organization, characterized by nucleotide binding site and leucine-rich domains,

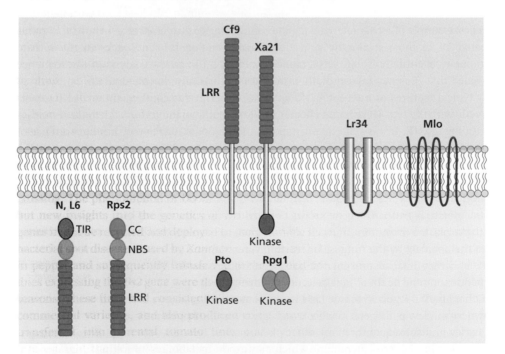

Figure 10.12 Structure and cellular location of some proteins encoded by plant R genes. N, L6 and Rps2 are cytoplasmic NLRs. Cf9 and Xa21 are membrane anchored with extracellular LRR domains. Pto and Rpg1 are cytoplasmic kinases. Lr34 is a membrane transporter and Mlo a protein with seven transmembrane domains. Abbreviations: CC coiled coil domain; TIR toll-interleukin receptor; NBS nucleotide binding site; LRR leucine rich repeat.

originally described as NBS-LRR but now known as NLR proteins (Figure 10.13a). These receptors recognize specific pathogen effectors and act, either directly or indirectly, as molecular switches turning on the second layer of plant immunity (ETI), as proposed in the Z scheme (Figure 10.6). Recent phylogenetic studies suggest that NLRs are an ancient protein class that can be traced all the way back to the early land plants and their progenitors, the green algae. The conserved features of NLRs have facilitated detailed searches of plant genomes to identify *R* gene candidates with potential roles in resistance. Such comparisons have shown that plant genomes harbor hundreds of NLRs, but the total complement varies widely between plant species. *Arabidopsis*, for instance, has almost 200, while rice has over 500 and bread wheat more than 1000. The number is not, however, directly linked to genome size and presumably reflects expansion of the gene family in some plants and not others.

How exactly NLRs function to recognize effectors and activate ETI is not yet fully understood. What is known is that detection of the effector switches the NLR from an inactive to an active state but this can take place in more than one way. In many cases, the receptor directly binds the effector via the LRR domain and undergoes a conformational change, opening the molecule. This is followed by formation of a dimer or oligomer of the NLR, along with a change in the nucleotide bound by the NBS domain from adenosine diphosphate (ADP) to adenosine triphosphate (ATP) (Figure 10.13b). Downstream defense signaling then takes place from the N-terminal domain. In other cases, rather than directly recognizing the pathogen effector, NLRs detect molecular changes in the host protein targeted by the effector

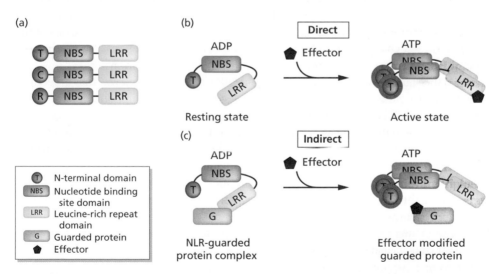

Figure 10.13 Structure and activation of NLR immune receptors. (a) NLRs are modular proteins with N-terminal, nucleotide-binding site, and leucine-rich domains. (b) Direct activation of the NLR. In the inactive resting state, the protein is folded and the bound nucleotide is adenosine diphosphate (ADP). Recognition of the pathogen effector by the LRR domain activates the protein, which opens, and the bound nucleotide is now adenosine triphosphate (ATP). In the example shown, the NLR forms a dimer with another NLR; in some cases multimers are formed. Defense signaling is initiated from the N-terminal domain (*light red glow*). (c) Indirect activation of the NLR through modification of a guarded protein targeted by the effector. In this case, the effector interacts with the guarded protein and a conformational change is detected by the NLR, which switches to the active state. *Source:* Adapted from Monteiro and Nishimura (2018).

(Figure 10.13c). In essence, the NLR acts as a "guard" monitoring key components of the plant immune system for signs of any interference by a pathogen molecule.

One advantage of this indirect form of recognition is that a single NLR can detect different pathogen effectors acting on the same host target, rather than the plant requiring specific NLRs for all the various effectors it might encounter. It is also becoming clear that NLRs often act in pairs, with one partner binding the effector and the other performing the downstream signaling. Remarkably, there is some evidence that the receptor partner in such pairs may have evolved to resemble the original plant target of the effector, thereby acting as a molecular decoy.

Genomic Organization of *R* Genes

A further revelation emerging from work on the mapping and cloning of *R* genes is that they often occur in clusters grouped at particular points in the plant genome. In *Arabidopsis*, specific genes for downy mildew resistance (RPP – **R**ecognition of **P**eronospora **p**arasit-ica – genes) are present on all five chromosomes, but map to loci close to other *RPP* genes. A similar clustering together of *R* genes with different specificities for the same pathogen is seen in the interaction of lettuce with downy mildew, *Bremia lactucae*, and flax with rust, *M. lini*. Furthermore, such clusters can contain not only several *R* genes recognizing different races of the same pathogen, but also genes specific for quite different pathogens, such as fungi, bacteria, and nematodes. In maize, for example, there is a cluster of tightly linked genes on chromosome three which confer resistance to rust and two different viruses, as well as other genes which influence reaction type to fungal stalk rot and an insect pest, the European corn borer. This suggests that there are particular regions of the plant genome which encode multiple recognition functions. Such regions also appear to be rich in genes for kinase proteins which might function in signal pathways.

What is the significance of this genetic organization? An initial idea was that different *R* genes belong to multigene families which have evolved from a common ancestral gene. Duplication of such genes, coupled with mutation or recombination between genes, could generate new variants capable of recognizing different pathogen signal molecules. According to this theory, a prototype receptor could diversify in time to bind a range of different pathogen effectors. More recent bioinformatic studies comparing *NLR* genes from different plant families concluded that the thousands of currently known *NLR*s probably evolved from three distinct ancestral lineages. Furthermore, genetically clustered *NLR* genes frequently swap sequences, thereby creating novel specificities. Plants are therefore able to shuffle their pack of immune receptors to counter the constant changes taking place in the pathogen arsenal of virulence factors.

Specificity of Virus Pathogens

Plant viruses also exhibit specificity in the range of host plants which they can successfully infect. The host–pathogen relationship in this case is unique, as successful reproduction of these subcellular parasites depends upon a precise sequence of molecular events (Figure 10.14). Thus, the virus must gain entry to the host cell, the virus genome must be released from its protective protein coat, then be translated and replicated, and finally new particles must be assembled. In addition, continuous spread of the virus to other cells and tissues is necessary if systemic colonization of the host is to occur. Specificity might be

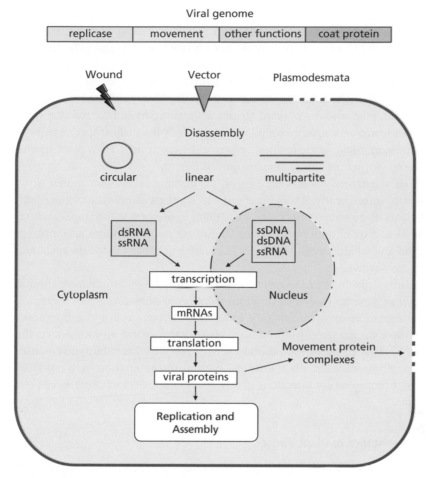

Figure 10.14 Overview of the main steps in a plant virus life cycle. The virus genome (*top*) has a small number of genes encoding proteins involved in replication, movement, the coat protein and other functions such as vector transmission. Entry into the host cell via wounds, vector feeding, and natural channels between cells (plasmodesmata) is followed by disassembly to release the genome, which can be circular, linear, or multipartite (composed of more than one segment). Depending on genome composition, the replication cycle takes place in the nucleus or cytoplasm (*yellow shaded boxes*). Regulation of virus gene expression can occur at the transcriptional or translational level, with posttranslational modification of proteins in some cases. Cell-to-cell movement via plasmodesmata may involve whole virus particles or complexes of the virus genome and movement protein. *Source:* After Scholthof et al. (1993).

determined at any of these stages in the infection cycle. Virus genomes encode only a small number of proteins which has two important consequences. First, completion of the virus infection cycle relies heavily on the participation of host components. Resistance might therefore be due to the lack of a host factor required for reproduction of the virus. Second, virus proteins are often multifunctional and play several roles during replication and colonization of the host, including suppression of plant defense.

The need for an appropriate vector to transmit infection adds a further dimension to the phenomenon of virus specificity. The relationship of the virus with the vector, and the vector with the host plant population, may limit the host range of a virus to a greater extent than the ability of the pathogen to replicate in host cells. Mechanical inoculation of viruses that are normally transmitted by a vector has been shown to extend their natural host range. Similarly, inoculation of isolated protoplasts with virus can give rise to infection in cells of host species not known to be susceptible in nature. These findings suggest that the limiting factor with many plant viruses is their transmission and introduction into plant cells, rather than their capacity to exploit the biosynthetic machinery of the host. A further difficulty in evaluating the specifity of plant viruses is the diversity of host reaction types that may occur. For instance, a virus may replicate successfully in one host species without causing any visual symptoms, while in another severe disease results. Both plant species are hosts, but the consequences of infection are quite different.

Some viruses have extremely wide host ranges; for instance, cucumber mosaic virus (CMV) is known to infect nearly 800 plant species in 85 plant families, including both monocots and dicots. Even more remarkably, some plant viruses are able to replicate both in their host plant and within the insect vector. Their host range therefore encompasses both the plant and animal kingdoms. This poses fascinating questions as to how such relationships might have evolved.

New insights into virus–host interactions have been gained by manipulating specific regions of the virus genome through mutation, deletion or substitution, or swapping parts between different viruses, for instance to replace the gene coding for the movement protein with a similar gene from another virus. Such hybrid viruses gained the ability to spread in other hosts, but sometimes were reduced in their ability to colonize the original host. These experiments mimic the natural recombination between virus strains which can occur in mixed infections of different plant hosts including weeds and other wild species.

Mechanisms of Resistance to Plant Viruses

Plants possess three main kinds of defense against infection by viruses. The first is recognition of viral products in the same way as fungal and bacterial effectors are detected by plant immune receptors. The NLRs involved are encoded by dominant plant genes and interact with a specific viral protein to give a hypersensitive-like response, thereby restricting further multiplication of the virus, which depends upon living host cells. Well-characterized examples include the *N* gene in tobacco which recognizes the replicase protein of TMV (Table 10.5), the *Rx* gene from potato and tomato which detects the coat protein of potato virus X and some other viruses, and the *Tm-2* gene in tomato that detects the movement protein of TMV. Hence plants have evolved the ability to recognize and respond to different molecular components of the virus.

There has been some debate as to whether such viral proteins should be classed as effectors, as suppression of host defense is not their primary role during infection. However, it is now becoming apparent that some multifunctional proteins involved in viral replication or other steps in the infection cycle may also interfere with plant defense pathways, and hence act like effectors.

The second type of resistance to viruses is encoded by recessive genes and is due to the absence of appropriate host factors required for the virus to complete its infection cycle. The third type is based on gene silencing by RNA interference (RNAi), a widely conserved process in eukaryotes which may have evolved as a defense mechanism against viruses but now also plays a wider role in the regulation of endogenous gene expression.

Recessive Resistance to Plant Viruses

Around half of the known resistance genes acting against plant viruses are recessively inherited. Where these genes have been cloned, they are often found to encode host components involved in the translation of messenger RNA by ribosomes. Such eukaryotic initiation factors (eIFs) recruit mRNA to the ribosome via interaction with a mRNA cap structure so that translation can occur. Some RNA viruses hijack this process by producing a small protein attached to the end of their genome which mimics the mRNA cap, binds to the appropriate eIF, and allows translation to proceed (Figure 10.15). The eIF therefore acts as a host susceptibility factor. However, an alteration in the eIF can prevent the viral genome from associating with the translation initiation complex (TIC), and hence stops the virus from producing the proteins required for replication and spread. The recessive genes responsible for resistance to the virus are generally mutant alleles encoding structural variants of the eIF. In turn, such resistance may itself break down due to the emergence of novel forms of the viral mRNA cap protein in the virus population. A practical example of

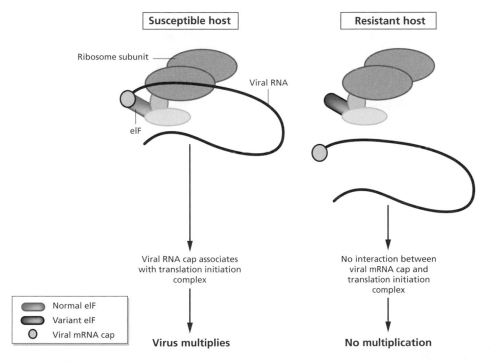

Figure 10.15   Resistance to a virus due to lack of a compatible host susceptibility factor. (*Left*) The viral mRNA cap protein associates with the eIF in the translation initiation complex (TIC) and translation of the viral RNA takes place. (*Right*) A variant eIF is present and there is no interaction with the viral mRNA cap protein and no translation of viral genes. Note: The TIC is shown in simplified form. In reality, there are additional proteins in the complex. *Source:* Adapted from Hammond-Kosack and Jones (2015).

the use of variant alleles of eIFs by plant breeders to control a virus disease is described in Chapter 12 (see Figure 12.2).

Resistance Based on RNAi

Both animal and plant cells have evolved sophisticated mechanisms to detect and destroy foreign nucleic acids, such as those present in invading viruses. RNAi plays a critical role in defense responses against virus infection by targeting the double-stranded RNA (dsRNA) intermediates formed during replication of viruses. A detailed description of the many facets of RNAi is beyond the scope of this book, but some key points are given here.

The majority of plant viruses have RNA genomes that often contain double-stranded secondary structures or produce them through the action of viral RNA-dependent RNA polymerases during replication. DNA viruses also produce dsRNAs at certain points in the infection cycle. These dsRNA molecules are cut by host ribonuclease enzymes such as DICER into small RNAs (21–24 nucleotides in length, termed short interfering RNAs – siRNAs), which are incorporated into an RNA-induced cytoplasmic silencing complex (RISC). The siRNAs then pair with their complementary mRNA sequences and induce their degradation. Another key feature of this RNAi pathway is generation of a mobile silencing signal that spreads from cell to cell via plasmodesmata, and over longer distances through the phloem. The signal activates RNA silencing in noninfected cells, and this systemic process can lead to reduced virus titers in infected tissues and eventual recovery from symptoms.

As RNAi is such a potent antiviral mechanism, perhaps unsurprisingly viruses have developed various ways of countering it. Some viruses may replicate in subcellular compartments partly protected from silencing, but the most notable strategy is production of proteins, called RNA silencing suppressors, which can block or interfere with antiviral silencing. In fact, it has been shown that some viral proteins initially defined by other roles in the infection cycle, such as virus movement or plant-to-plant transmission, can also function as silencing suppressors. One example is the helper component protein (HCPro) from potyviruses, which not only facilitates virus transmission by aphid vectors but can also interfere with several steps in the silencing pathway, including preventing loading of siRNAs into the RISC. Another is the P6 protein from cauliflower mosaic virus (CaMV), which was first shown to function in translation of the virus genomic RNA, but later was discovered to suppress several aspects of plant innate immunity, including RNA silencing by interfering with the processing of siRNAs.

In most cases, the exact mechanisms involved in silencing suppression by viral proteins are not fully understood. It has also been suggested that the multiple small RNA molecules produced during the virus replication cycle might themselves act as decoys for silencing while the large messenger RNAs survive and propagate.

Conclusion

A recurrent theme in this analysis of host–pathogen specificity is the dynamic molecular interplay between pathogen factors promoting infection and host factors determining plant immunity. On the host side of the interaction is the need to maintain constant surveillance of the environment to detect biotic threats, but also tight regulation of active defense to ensure that resources are only deployed when necessary. Loss of control can trigger spontaneous cell death and related deleterious responses with potentially serious consequences. On the pathogen side is the need to evade or counter host defense, but also to maintain sufficient genetic flexibility to adapt to an immune system that can evolve new recognition specificities. Dependence on a living host without such flexibility could lead to extinction. One of the revelations of genome sequencing is that genes involved in host–pathogen interactions are under diversifying selection and often show greater polymorphism than those encoding other functions. This diversity reflects the ongoing co-evolution of plants and their natural enemies.

The practical challenge for agriculture now is to devise effective strategies to slow down or halt this evolutionary process, thereby ensuring sustainable control of plant disease. The following section of the book discusses the main methods available for crop protection and how our increased understanding of host–pathogen interactions may help to meet this challenge.

Further Reading

Books

Hammond-Kosack, K.E. and Jones, J.D.G. (2015). Responses to plant pathogens. In: *Biochemistry and Molecular Biology of Plants* (eds. B.B. Buchanan, W. Gruissem and R.L. Jones), 984–1050. Rockville, Maryland: American Society of Plant Physiology.

Reviews and Papers

Boch, J., Scholze, H., Schornack, S. et al. (2009). Breaking the code of DNA binding specificity of TAL-type III effectors. *Science* 326: 1509–1512. https://doi.org/10.1126/science.1178811.

Carella, P., Evangelisti, E., and Schornack, S. (2018). Sticking to it; phytopathogen effector molecules may converge on evolutionarily conserved host targets in green plants. *Current Opinion in Plant Biology* 44: 175–180.

Chen, J., Upadhyaya, N.M., Ortiz, D. et al. (2017). Loss of *AvrSr50* by somatic exchange in stem rust leads to virulence for *Sr50* resistance in wheat. *Science* 358: 1607–1610.

Giraldo, M.C. and Valent, B. (2013). Filamentous plant pathogen effectors in action. *Nature Reviews Microbiology* 11 (11): 800–814.

Hein, I., Gilroy, E.M., Armstrong, M.R. et al. (2009). The zig-zag-zig in Oomycete-plant interactions. *Molecular Plant Pathology* 10 (4): 547–562.

Jones, J.D.G. and Dangl, J.L. (2006). The plant immune system. *Nature* 444 (7117): 323–329.

Jones, J.D.G., Vance, R.E., and Dangl, J.I. (2016). Intracellular innate immune surveillance devices in plants and animals. *Science* 354 (6316): aaf6395.

Leisner, S.M. and Schoelz, J.E. (2018). Joining the crowd: integrating plant virus proteins into the larger world of pathogen effectors. *Annual Review of Phytopathology* 56: 89–110. https://doi.org/10.1146/annurev-phyto-080417-050151.

Mansfield, J.W. (2009). From bacterial avirulence genes to effector functions via the *hrp* delivery system: an overview of 25 years of progress in our understanding of innate plant immunity. *Molecular Plant Pathology* 10 (6): 721–734.

Monteiro, F. and Nishimura, M.T. (2018). Structural, functional, and genomic diversity of plant NLR proteins: an evolved resource for rational engineering of plant immunity. *Annual Review of Phytopathology* 56: 243–267.

Nicaise, V. (2014). Crop immunity against viruses; outcomes and future challenges. *Frontiers in Plant Science* 5: 660. https://doi.org/10.3389/fpls.2014.00660.

Ryan, R.P., Vorhölter, F.-J., Potnis, N. et al. (2011). Pathogenomics of *Xanthomonas*: understanding bacterium-plant interactions. *Nature Reviews/Microbiology* 9: 344–355. https://doi.org/10.1038/nrmicro2558.

Sanfaçon, H. (2015). Plant translation factors and virus resistance. *Viruses* 7: 3392–3419. https://doi.org/10.3390/v7072778.

Santenac, C., Lee, W.-S., Cambon, F. et al. (2018). Wheat receptor-kinase-like protein Stb6 controls gene-for-gene resistance to fungal pathogen *Zymoseptoria tritici*. *Nature Genetics* 50: 368–374. https://doi.org/10.1038/s41588-018-0051-x.

Torto-Alalibo, T., Collmer, C.W., Gwinn-Giglio, M. et al. (2010). Unifying themes in microbial associations with animal and plant hosts described using the gene ontology. *Microbiology and Molecular Biology Reviews* 74 (4): 479–503.

Wang, S., Welsh, L., Thorpe, P. et al. (2018). The *Phytophthora infestans* haustorium is a site for secretion of diverse classes of infection-associated proteins. *mBio* 9 (4): e01216-18. https://doi.org/10.1128/mBio.01216-18.

Whisson, S.C., Boevink, P.C., Moleleki, L. et al. (2007). A translocation signal for delivery of oomycete effector proteins into host plant cells. *Nature* 450: 115–117. https://doi.org/10.1038/nature06203.

Wu, C.-H., Abd-El-Haliemb, A., Bozkurt, T.O. et al. (2017). NLR network mediates immunity to diverse plant pathogens. *Proceedings of the National Academy of Sciences, United States of America* 114 (30): 8113–8118.

Part III

Disease Management

Plant pathology has many successes to its credit during its brief history, and methods have been developed to prevent the great losses some diseases used to cause. However, it would be vain to maintain that all is well.

(F.C. Bawden, 1908–1972)

Agricultural systems are in many ways the opposite of natural biological communities. Modern crop husbandry entails growing huge numbers of genetically similar plants crowded together over large areas. Nutrients, water or both are applied in quantities sufficient to ensure vigorous crop growth. Any pest, pathogen or weed able to thrive under these ideal conditions therefore has the potential for explosive multiplication and spread, free from the usual constraints encountered in a diverse natural community. One important aspect of crop production is to prevent such disease outbreaks or, if they occur, to restrict populations of pathogens or pests to levels which do not adversely affect the performance of the crop. Strategies for disease management are therefore an essential part of any crop production system.

Options for Disease Control

A number of different approaches are available to the grower to prevent or limit the damage caused by plant pathogens. These are grouped in Table 1 under three main headings. First, it may be possible to avoid infection altogether by ensuring that the pathogen is excluded from the crop. One obvious strategy here is to grow the crop in a region in which the pathogen does not occur; alternatively, vectors required for transmission of a pathogen may be absent. In the UK, for example, seed potatoes are mainly produced in Scotland where the chances of aphids introducing viruses are much less than in warmer southern regions (see p.99). Exclusion of the pathogen is an important strategy in crops grown in controlled environments such as glasshouses. Maintaining a pathogen-free area usually requires constant vigilance to avoid introduction of disease agents via seed or propagation material. Hence there may be legislative measures to ensure clean seed or to quarantine any plants being moved between regions or countries. These aspects have already been discussed in Chapter 5.

Plant Pathology and Plant Pathogens, Fourth Edition. John A. Lucas.
© 2020 John Wiley & Sons Ltd. Published 2020 by John Wiley & Sons Ltd.
Companion website: www.wiley.com/go/Lucas_PlantPathology4

Table 1   Options for control of plant disease

1) Exclusion of the pathogen
   - Pathogen-free area
   - Clean seed or propagation material (certification)
   - Quarantine
2) Reduction of inoculum
   - Soil treatment
   - Treated seed or propagation material
   - Cultural practices, e.g., crop rotation
   - Biological control
3) Reduce rate of pathogen multiplication
   - Rate-reducing plant resistance
   - Cultivar diversification/mixtures
   - Chemical control
   - Biological control
4) Integrated control programs
   - Combination of several approaches

The second group of control options concerns reducing the amount of pathogen inoculum available to cause disease. Measures such as soil sterilization or disinfection of seeds are aimed at eradicating or reducing the number of pathogen propagules present before the crop is planted. Many cultural practices, such as crop rotation, also reduce the carry-over of spores or other pathogen survival structures between seasons. This decline in pathogen populations occurring between crops depends to a large extent upon natural processes of biological control, such as predation or parasitism of propagules by other organisms. There is considerable interest therefore in devising methods for enhancing this natural antagonism in soil or on crop residues. Such methods of biological control are discussed in Chapter 13.

Once a crop is exposed to a pathogen, a third set of control options must be considered (Table 1). These are all aimed, directly or indirectly, at reducing the rate of disease development in the crop. Genetic resistance to the pathogen is one of the most important and effective ways of achieving this (see Chapter 12). In cases where resistance is not fully effective against all genotypes of the pathogen population, benefits may be obtained by growing mixtures of different cultivars or diversifying crops between fields, farms or regions (see p.322). Another highly effective strategy to restrict epidemic development is to treat the crop with a chemical active against the pathogen or its vector. Traditionally, such treatments were best applied to protect the crop before significant exposure to a pathogen, but the newer generation fungicides have curative properties and thus may be worth applying even after infection is established in a crop (see Chapter 11). Concerns over the possible environmental effects of pesticides have recently placed increased emphasis on biological approaches to disease control, although developing biological agents with an efficacy comparable to the best available chemicals remains a difficult task.

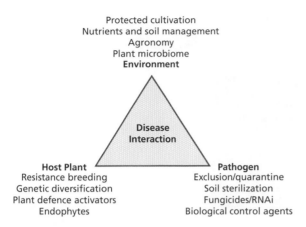

Protected cultivation
Nutrients and soil management
Agronomy
Plant microbiome
Environment

**Disease
Interaction**

Host Plant
Resistance breeding
Genetic diversification
Plant defence activators
Endophytes

Pathogen
Exclusion/quarantine
Soil sterilization
Fungicides/RNAi
Biological control agents

Figure 1 The disease triangle showing options for intervention targeting the host plant, pathogen or environment.

An alternative way of thinking about the control of plant disease is to return to the disease triangle, introduced in Part I (see p.3), and to ask what interventions might be available to alter the three components of the interaction: the host, the pathogen, and the environment (Figure 1). Options for altering the environment may be limited for field crops, but management of soil and nutrients can affect the incidence and severity of some diseases. The agronomic system used, for instance crop rotation, sowing date, or irrigation, will also influence disease. With protected cultivation in greenhouses or polyethylene tunnels, more control over environmental conditions can be achieved, for instance to alter temperature or humidity, and this in turn will affect the risk of disease.

The host plant part of the triangle can be altered in several ways, including breeding disease-resistant cultivars, and deploying them in various ways to maximize the evolutionary hurdles the pathogen has to surmount to break down resistance (see Chapter 12). Host plant defense can also be boosted by using chemical or biological agents that induce resistance. Direct impacts on the pathogen can be achieved by exclusion or removal of inoculum, or by using chemicals such as fungicides that inhibit or delay infection or reproduction, hence reducing the rate of epidemic development.

Finally, the most cost-effective, durable, and environmentally acceptable strategies for disease management usually entail the use of several, complementary approaches. Such integrated control programs require a sound understanding of the diverse factors underlying disease development, including pathogen ecology, crop growth and physiology, soil status, and climatic conditions. For most plant diseases, the ideal scenario of a fully integrated control program remains elusive, but progress toward this goal is discussed in Chapter 14.

The Economics of Control

The options available to the grower to control plant disease depend not only upon the type of crop and the nature of the pathogen concerned, but also upon the economic context. Almost all control measures require significant expenditure in terms of manpower and

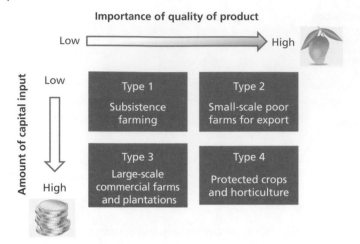

Figure 2 Farming systems compared. *Source:* Adapted from Bunders (1988).

materials, and these costs need to be set against the benefits gained by applying such measures. There is no point recommending a control strategy, however effective, if the resources required cost more than any likely return in crop yield or quality. The only exception to this rule is in cases where some wider strategic significance applies, for instance where eradication of a pathogen removes a long-term threat to crop production in a particular region.

It is also important to appreciate that agricultural systems differ widely, and disease control options appropriate for one system may not be economic in another. Figure 2 shows a simple classification of agricultural systems based upon the level of inputs into the system and the market value of the product. In subsistence farming, which is still practiced over large areas of the developing world, the value of the crop is insufficient to sustain expensive inputs of fertilizers or pesticides. Control measures must be cheap and easy to apply. In contrast, intensive production of high-value horticultural crops in glasshouses involves sophisticated systems for regulation of the crop environment, including the use of biological and chemical disease and pest control agents. Most agricultural systems fall between these two extremes, and the level of inputs varies depending upon market conditions.

In some regions, such as within the European Union, certain crops attract subsidies which maintain the market value of the product above its normal baseline. In such instances, the grower may be able to justify the use of extra inputs to control disease. But market conditions change from year to year, and options for disease control have to be adjusted accordingly. For instance, some commodity prices have fallen in both Europe and North America, and hence there is increasing pressure to reduce the costs of inputs by, for instance, improving the efficiency of use of fungicides and fertilizers.

A generalized relationship between control costs and crop income is plotted in Figure 3. This is a simplified model but shows that increasing expenditure is required to improve the degree of control achieved. While it may be possible to reduce the amount of disease, or populations of pests or weeds, to insignificant levels, this may entail repeat treatments or much larger doses of pesticides. The **economic threshold**, the point at which the cost of control is balanced by the income from the extra crop produced, is usually at an intermediate point, where some disease is still present. Practical disease management in most crops therefore involves an element of compromise, in which some damage by pests or pathogens is tolerated.

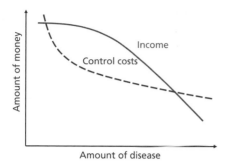

Figure 3 Crop income versus control costs.

The model above is based solely on direct costs, such as the manpower and materials used to apply a treatment. Increasingly, however, these analyses are being required to include calculations of indirect costs, for instance of chemical residues in food or ground water, or effects on nontarget species in agricultural ecosystems. Obtaining such impact estimates is, of course, a complex and often controversial business, but it does highlight the need for a sophisticated approach to disease control, taking account of environmental as well as socioeconomic considerations.

Further Reading

Reviews and Articles

Bunders, J. (1988). Appropriate biotechnology for sustainable agriculture in developing countries. *Trends in Biotechnology* 6: 173–180.

Lucas, J.A. (2011). Advances in plant disease and pest management. *Journal of Agricultural Science* 149 (S1): 91–114. https://doi.org/10.1017/S0021859610000997.

Lucas, J.A. (2017b). Fungi, food crops, and biosecurity: advances and challenges. In: *Advances in Food Security and Sustainability*, vol. 2 (ed. D. Barling), 1–40. Burlington, Virginia: Academic Press.

Strange, R.N. and Scott, P.R. (2005). Plant disease: a threat to global food security. *Annual Review of Phytopathology* 43: 83–116.

Sundin, G.W., Castiblanco, L.F., Yuan, X. et al. (2016). Bacterial disease management: challenges, experience, innovation and future prospects. *Molecular Plant Pathology* 17 (9): 1506–1518. https://doi.org/10.1111/mpp.12436.

11

Disease Management by Chemicals

It was, finally, a matter of trying a large number of remedies, assuring myself of their merits, making them easy of execution, of taking into special account their economy of application, and of winning the confidence of the farmer.

(M. Tillet, 1714–1791)

The use of chemicals to control pests, pathogens, and weeds is in many respects a success story. The discovery and commercial development of a succession of bioactive compounds over many years has provided growers with an arsenal of highly effective products that have helped to boost crop yields and ensure stability of production in developed agriculture. The availability of chemical treatments that can be quickly deployed over large areas has helped to reduce the impact of disease outbreaks that previously caused widespread losses and hardship.

But this scenario is now changing. Some commentators have argued that the convenience and efficacy of crop protection chemicals have resulted in overreliance on them rather than alternative disease control strategies. Concerns over the potential effects of chemicals in agriculture on human health and the environment have led to tighter regulation and restrictions on their use. This in turn has led to a reduction in the diversity of chemistry available, and limited the pipeline of new products. To add to these difficulties, many of the target pests and pathogens that were previously controlled by pesticides have now developed resistance to them, analogous to the situation with antibiotics in medical use. Hence there is now an increased emphasis on integrated, more sustainable approaches to disease control, in which targeted use of chemicals will continue to play a part.

The following account will focus on fungicides, as these are the chemicals most frequently used to control plant diseases. Many of the basic principles discussed, however, apply equally well to other important types of crop protection chemicals, such as insecticides and herbicides.

The Evolution of Fungicides

The fungicidal properties of certain chemicals have been known for many years. The first fungicides, based on sulfur and copper, were discovered in 1846 and 1882 respectively. The discovery of Bordeaux mixture, based on copper sulfate and lime, by Millardet in France, is

Plant Pathology and Plant Pathogens, Fourth Edition. John A. Lucas.
© 2020 John Wiley & Sons Ltd. Published 2020 by John Wiley & Sons Ltd.
Companion website: www.wiley.com/go/Lucas_PlantPathology4

one of the most familiar stories in plant pathology, starting with the chance observation that copper salts applied to grapevines to deter thieves also controlled infection by the downy mildew pathogen *Plasmopara viticola*. Millardet's achievement was to translate this observation into practical use by developing formulations of copper for effective commercial application on crops.

Over the years since this discovery, fungicides have diversified and changed dramatically (Figure 11.1 and Tables 11.1 and 11.2). The early inorganic compounds have now been superseded by organic chemicals which are active at very low doses, effective against a wide range of fungal pathogens, and can be applied with precision by machinery appropriate to a small plot or a 1000 ha plantation. However, if pioneers such as Millardet were still alive today, two features of the current fungicide market (Figure 11.1) would surprise them. First, some of the older compounds are still widely used. Second, many of the modern-generation fungicides were discovered by a process not dissimilar to Millardet's initial observation, with the most effective compounds being selected by screening for activity against a few pathogens chosen to represent the most important disease targets. Only in recent years, with advances in molecular biology, genomics, and cheminformatics (the use of computational techniques to solve problems in structural chemistry), has the prospect of designing molecules to perform specific tasks become a reality.

The Perfect Fungicide

It is relatively easy to compile a list of the properties one would like any ideal fungicide to possess (Table 11.3). In practice, it is difficult to satisfy all these requirements. The biological aspects listed may be attainable for fungicides applied to aerial tissues or seeds, but targeting pathogens found in the soil is more problematic. Soil sterilants such as chloropicrin and methyl bromide are broad-spectrum biocides which impact the wider soil microflora and fauna, and have now been withdrawn in many countries due to concerns about their

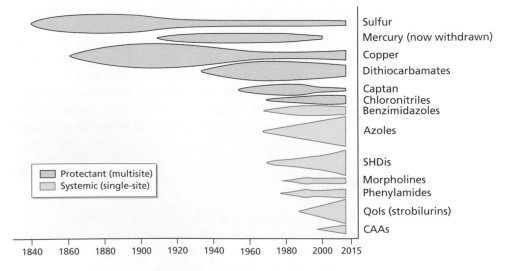

Figure 11.1 The evolution of fungicides. Main types available, origins, and relative importance.

Table 11.1 Examples of protectant, multisite fungicides

Type	Example	Mode of action (where known)
Metal-based fungicides		
Copper fungicides	Bordeaux mixture $CuSO_4 + Ca(OH)_2$	Non-specific
Tin fungicides	Fentin acetate $Ph_3SnOCOCH_3$	Non-specific?
Mercury fungicides	Phenyl mercury acetate $PhHgOCOCH_3$	Non-specific
Sulphur fungicides	Elemental sulphur	Respiration
Dithiocarbamates		Thiol proteins
	Zineb (Zn), Maneb (Mn)	
Others		
Pthalimides	Captan	Proteins
Dicarboximides	Iprodione	?
Chloronitriles	Chlorothalonil	Multisite contact action

safety. Similarly, the use of certain systemic compounds which are taken up by plants raises questions over the persistence of residues in crop products. Such residues may be beneficial in terms of reducing postharvest diseases, such as fruit rots, but persistence into the food chain is regarded as less acceptable. All new agrochemicals are subjected to rigorous toxicology testing (see later in this chapter), and the most common reason for a compound

Table 11.2 Some examples of systemic, single-site fungicides

Type	Example	Mode of action (where known)
Methyl benzimidazoles carbamates (MBCs)	Carbendazim	β-Tubulin
Sterol biosynthesis inhibitors		
Azoles	Prothioconazole	Sterol 14α-demethylase
Imidazoles	Prochloraz	Sterol 14α-demethylase
Morpholines	Fenpropimorph	Sterol isomerase and reductase
Others		
Phenylamides	Metalaxyl	RNA polymerase
Phosphonates	Fosetyl-Al	?
Melanin biosynthesis	Tricyclazole	Inhibits polyketide pathway

Table 11.2 (Continued)

Type	Example	Mode of action (where known)
Anilinopyrimidines	Pyrimethanil	Methionine biosynthesis
Qols (strobilurins)	Azoxystrobin	Mitochondrial electron transport
Succinate dehydrogenase inhibitors (SDHIs)		
Carboxin	Bixafen	Enzyme in citric acid cycle
Carboxylic acid amides (CAAs)	Mandipropamid	Cellulose biosynthesis

failing to make it to the market is some question mark, however small, about safety. Alternatively, the cost of production, or the economics of use by comparison with existing products, may lead to the demise of otherwise promising novel compounds.

Finally, given the time, effort, and expense of developing a new fungicide, one hopes that it will remain effective in the longer term and not be compromised by the evolution of resistance in the pathogen population. Whether this aspiration is realistic given the adaptability of fungi is discussed further in this chapter.

Table 11.3 Specifications of the perfect fungicide

Biological properties

It must provide effective and consistent disease control

It should not be phytotoxic at the recommended dose

It should not adversely affect nontarget organisms

Toxicology

It must not be hazardous to apply

Residues in the crop should not pose a problem for consumers

Residues in the wider environment should not pose a threat to human health or biodiversity

Formulation

It should be stable and safe to store and transport

It should be simple to apply at a precise dose rate

The formulation should enhance its efficiency as a fungicide

Economic considerations

It should be affordable and give a financial return on use

Evolutionary considerations

Continuous use should not lead to the emergence of resistance to the fungicide

---

The Discovery Process

The starting point for the development of a new fungicide is chemistry. There are several routes by which novel compounds can be obtained (Figure 11.2). Traditionally, teams of chemists synthesized a diversity of compounds for screening, using methods such as combinatorial chemistry which can generate huge numbers of new molecules. Another rich area of chemistry to explore is natural products from plants, animals, and microorganisms. Already, some of the most effective fungicides, such as the strobilurins (QoIs), have been derived from natural compounds, in this case formed by some species of basidiomycete fungi. The biochemical diversity of living organisms is now seen as a potentially limitless source of novel activity. The chemical libraries obtained by these various routes are tested for activity against a range of target organisms, including plant pathogens, pests, and weeds. Typically, high-throughput screening for fungicides is done on inoculated plants in small modules in the glasshouse as the behavior of the compound might differ from tests done in culture media. The most promising candidate molecules are then taken forward for further evaluation.

This apparently random selection process, known as **empirical screening**, has been refined over time to increase the chances of finding potentially useful products. Once a new chemical with promising properties has been identified, analog synthesis to produce a series of related compounds is done to see if such properties can be further improved. There are many examples in drug discovery or agrochemical development where the potency, spectrum of activity, and other characteristics of a molecule, including safety, have been enhanced by chemical modification. In addition, once a new chemical lead has been found, computer-based methods such as predictive modeling can be applied to optimize potential activity. Increasingly, such *in silico* approaches are now being used to aid decision making and choice

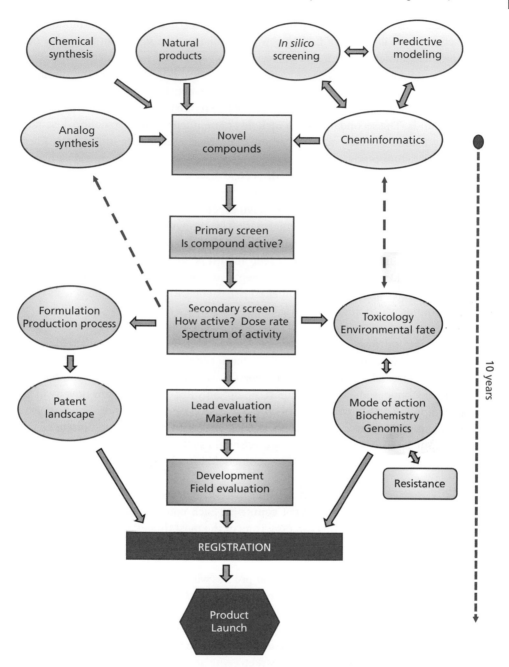

Figure 11.2 Activities involved in the discovery and development of new agrochemicals.

of candidate compounds (Figure 11.2). Large amounts of relevant data on millions of compounds can be mined and integrated using procedures described as cheminformatics.

The aim of all these approaches is to increase the likelihood of finding compounds with the right properties to make it onto the market. Only a tiny proportion of chemicals screened eventually become commercial products. Original estimates of one in 10 000 chemicals tested have been revised downwards to 1 or 2 per 100 000, at an overall cost of more than £200 million. This is a reflection of the difficulty of finding new compounds with a significant advantage over existing products, along with the increased stringency of registration requirements, particularly concerning product safety. As a consequence, agrochemical development is a multimillion dollar exercise which can only be tackled by large, multinational companies.

Following the primary screen, a series of further steps are taken to progress the selected lead compounds. The secondary screen includes more intensive tests to define dose rates and activity toward a wider range of targets than those used in the initial tests. These bioassays will include representative diseases caused by different types of fungal pathogens such as rusts and powdery mildews, *Botrytis*, cereal diseases like Septoria tritici leaf blotch (*Zymoseptoria tritici*) and rice blast, *Magnaporthe oryzae*, as well as oomycetes including a *Phytophthora* sp. and downy mildew of grapevines, *Plasmopara viticola*. Choice of targets is largely guided by the size of any potential market for the new product, and varies between companies. The potency and spectrum of activity of the compound will be compared with existing products and if no advantage is seen then it is unlikely the lead will be further progressed. Work also starts on any possible negative properties of the compound, such as toxicity to nontarget species or environmental persistence. Rigorous toxicology tests reject those compounds which might fail on safety grounds.

Once the efficacy and safety of the chemical have been confirmed, work on appropriate formulations is done to optimize activity and other properties, such as delivery to the target and uptake into the plant. The logistics and potential costs of production will also be considered. Initial research on mode of action begins, supported by biochemistry, molecular genetic and genomic studies. This is important, not only to demonstrate any novelty in the way the compound acts but also to assess the risk of resistance developing. There is also a legal requirement to protect the discovery as soon as possible with an appropriate patent, taking account of the existing patent literature. Should all these hurdles in the development pathway be cleared, testing of the product in the field is undertaken, at different sites and over more than one season.

Finally, a dossier of all the relevant information and data is submitted to support registration of the product in different countries. Each country has its own regulatory requirements and in practice this can prove one of the most time-consuming and difficult steps in launching a new agrochemical. Altogether, the pathway from initial screening to product launch takes around 10 years.

The scheme in Figure 11.2 portrays a linear process with a series of discrete stages, but in reality there are many points at which feedback and interactions can occur, such as in analog synthesis, and between the more fundamental studies on mode of action, toxicology, and product development. The sooner any problems with a potential product are identified the better, to avoid investing further time and expenditure in a fungicide that is unlikely to secure registration or to compete in the market.

Biorational Design

An alternative approach to empirical screening is a mainly computer-based exercise in which chemicals are designed to optimize activity against a specific molecular target. The latter might be an enzyme essential for the growth and development of the fungus, or production of a pathogenicity factor such as a toxin. To date, *in silico* screening has mainly been used by the pharmaceutical industry to optimize drugs targeting protein kinases and hormone receptors involved in a range of human diseases. The availability of full-genome sequences of many plant-pathogenic fungi has, however, provided an opportunity to explore molecular variation in known fungicide targets as well as identifying novel targets in currently hard-to-control species. There is also the possibility to revisit and modify chemistry that might have previously been discarded. Any novel compounds identified or refined by such methods will still have to pass the acid tests of activity *in planta* and the field, hence computational and genomic approaches are now an integral part of the discovery pathway (Figure 11.2) rather than a separate undertaking.

In spite of the complexity and expense of this long process, the agrochemical industry has been remarkably successful in finding and developing a succession of new types of molecules to aid the fight against plant disease. The notion that agrochemical companies are all involved in some competitive discovery race is, however, misleading. In reality, they cooperate as well as compete, through licensing agreements to permit manufacture or marketing of each other's compounds, often in mixtures. The virtue of such collaboration is not only commercial, as it can encourage concerted action to combat problems such as fungicide resistance.

But a significant number of problems remain. To date, there are no commercially effective compounds active against plant viruses, and few effective bacteriocides. Several fungal targets remain elusive, including some ear blights, vascular wilts, root pathogens, and other soil-borne diseases. Only one class of compounds, the phosphonates, has significant shoot-to-root mobility, providing control of root infection following aerial application. And added to the commercial and legal pressures of patent life and product registration are the increasing problems of environmental acceptability, tighter regulation, and pathogen resistance.

While predictions of the rapid demise of chemical control have proved premature, there has been a contraction of the industry, with fewer players and a diminishing pipeline of new products. This is a concern, as plant pathogens have not gone away, and many of the substitutes for conventional fungicides are not yet as effective in preventing crop loss.

Types of Fungicides

There are several different ways of classifying the diverse range of chemical compounds currently used as commercial fungicides. One is by chemical class, for instance inorganic versus organic compounds. Another is by mode of action, such as compounds which have toxic effects on a variety of cellular processes versus those which interfere with a specific process (see later in this chapter). A further classification is based on the behavior of the compound on or in the plant.

Prior to 1960, nearly all the fungicides discovered came under the general description of **protectant** compounds (Table 11.1). These materials supplement the defenses of the plant by forming a superficial chemical barrier to prevent, or protect against, infection. While protectant compounds are effective against a wide range of fungi, they have limitations in practical use. By the very nature of their mode of action, they must be applied before the

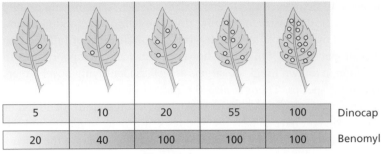

| 5 | 10 | 20 | 55 | 100 | Dinocap |
| 20 | 40 | 100 | 100 | 100 | Benomyl |

% control

Figure 11.3 Effect of distribution on the efficiency of control of powdery mildew (*Sphaerotheca fuliginea*) on cucumber leaves using droplets of the protectant fungicide dinocap and the systemic fungicide benomyl. For each fungicide, the total amount applied per leaf was the same in either 1 or up to 16 drops. The effects of the fungicides were measured in terms of the area of leaf infected, and the results are expressed as percentage reduction compared with untreated control leaves. *Source:* Adapted from Evans (1977).

pathogen attempts to penetrate the host. There is a need, therefore, for reliable, early warning of an infection risk (see p.104) if protectant compounds are to be used effectively and economically. Because they form surface coatings, such fungicides are subject to degradation and erosion by light, rain, and other environmental factors. Last but not least, applications to growing plants rapidly become ineffective as new leaves, flowers, and fruits continue to develop. For this reason, protectant fungicides may need to be applied to a crop at regular intervals throughout the season.

Due to these limitations, there was a sustained hunt for compounds with a different type of activity, which are actually taken up by the plant and inhibit the pathogen from within. The advantages of such **systemic** chemicals should be obvious (Table 11.2). These compounds offer opportunities for therapy, that is, they may eradicate an established infection. In this way, the use of chemicals in plant pathology would resemble human medicine, where the emphasis has always been on developing cures as well as preventing disease.

Some of the first compounds to be used as systemic pesticides were in fact antibiotics such as streptomycin, but the real breakthrough came in the 1960s with the advent of the benzimidazoles such as benomyl and carbendazim. These compounds are active against many different plant-pathogenic fungi, and also move through tissues so that even distribution of the chemical over the host is not necessary (Figure 11.3). Better still, it is possible to apply such fungicides as seed dressings, where continual uptake by the growing seedling can protect the plant for many weeks after germination. The large majority of systemic compounds, however, move only in the apoplast, and hence tend to travel from the base to the top of the plant, accumulating in leaves and shoot apices. For this reason, they are usually ineffective against soil-borne pathogens infecting roots or other subterranean organs. The one notable exception is the phosphonates, such as fosetyl-Al (Table 11.2), which are phloem mobile and can therefore travel from the shoot to the root, providing effective treatment for diseases such as root rot caused by *Phytophthora* species (see p.364).

Mode of Action of Fungicides

Most first-generation, protectant fungicides (Table 11.1) are known to be **multisite** inhibitors, which interfere with the central metabolic processes of the target fungus. Metal-based fungicides such as copper or mercury inhibit a wide range of enzymes involved in various metabolic pathways. Similarly, dithiocarbamates complex with thiol groups on proteins, thereby inactivating enzymes and ultimately causing cell death. This fatal disruption of core processes probably explains why few fungi have evolved mechanisms able to circumvent the toxic effects of these fungicides. Thus, for decades the copper, sulfur, and dithiocarbamate fungicides have remained as effective as when they were first discovered. One remarkable exception to this rule is the case of *Pyrenophora avenae* which managed to overcome the toxic effects of mercury applied as a seed dressing to oats.

By contrast, most of the systemic compounds discovered to date act at a single site in the cell, inhibiting a specific enzyme or process, and hence are known as **single-site** fungicides. For example, early work on the mode of action of benzimidazole fungicides showed that these compounds inhibited cell division. It was later shown that the site of action is β-tubulin, a polymeric protein found in microtubules, an essential component of the cytoskeleton. Binding of the fungicide to the tubulin molecule prevents polymerization and hence disrupts the normal activities of the cytoskeleton, including spindle formation during cell division. The widely used azole fungicides interfere with the biosynthesis of sterols, which are molecules found in fungal cell membranes. These "sterol biosynthesis inhibitors" (SBIs) affect membrane structure and function, with widespread consequences for the cell. The specific interaction is with a cytochrome P450 enzyme (lanosterol 14α-demethylase, usually shortened to CYP51) catalyzing a single demethylation step in the sterol biosynthesis pathway. Such azole fungicides are therefore referred to more precisely as "demethylation inhibitors" or DMIs.

A second class of SBIs are the morpholines (Table 11.2), which act on the same pathway but at different steps affecting sterol isomerization and reduction. This property is useful as fungi which have become insensitive to DMIs are often still sensitive to morpholine fungicides. Other examples of single-site inhibitors listed in Table 11.2 include the acylalanines, which affect nucleic acid synthesis, and the more recently introduced QoI (strobilurin) and succinate dehydrogenase inhibitor (SDHI) fungicides which both block mitochondrial electron transport, but at different sites (Complex II and Complex III respectively).

The mode of action of single-site inhibitors means, perhaps inevitably, that small changes in the fungus, for instance in the target protein, may alter the efficacy of the compound. In many cases, only a single mutation in the fungus is sufficient to abolish activity, and hence lead to resistance. The implications of this are discussed in more detail later in this chapter.

The contrasting properties of multisite and single-site fungicides are summarized in Table 11.4.

Alternative Modes of Action

Conventional fungicides either kill fungi directly or inhibit their growth, a distinction described as fungicidal versus fungistatic activity. Some other agrochemicals show no direct effects on pathogens but nonetheless reduce the severity of disease when applied to

Table 11.4 Multisite and single-site fungicides compared

	Multisite	Single-site
Activity	Prophylactic	Therapeutic
Mode of action	Many metabolic processes affected	Single metabolic process affected
Phytotoxicity	Common, especially if applied to wrong host or tissue	Less common
Pathogens affected	Numerous	Variable, some highly specific, others broad spectrum
Pathogen resistance	Rare	Common
Movement	Confined to redistribution on surface	Translocated, usually in apoplast (xylem, cell walls)
Alternative approximate description[a]	Protectant fungicides	Systemic fungicides

[a] *Note:* This is not an exact synonym as the extent of systemicity can vary between the site-specific types.

Table 11.5 Some examples of plant defense activators

Common name	Type	Commercial name	Comments
Probenazole (PBZ)	Benzisothiazole	Oryzemate®	Rice blast control
Acibenzolar-S-methyl (ASM)	Benzothiadiazole (BTH)	Bion®, Actigard®	Active on many diseases
Menadione sodium bisulfite (MSB)	Vitamin K analog	Act-2®	*Fusarium* wilt
β-aminobutyric acid (BABA)	Nonprotein amino acid	Experimental use only	Oomycetes and others
Methyl jasmonate	Jasmonic acid analog	Experimental use only	Necrotrophic fungi
Harpin	Gram-negative bacterial protein	Messenger®	Active against many diseases

plants. These compounds work by stimulating the plant immune system and are therefore known as **plant defense activators**. Some examples are shown in Table 11.5.

Probenazole has been used in the control of the damaging rice blast disease for more than 40 years. Its mode of action is not completely understood, but some enzymes involved in the synthesis of antifungal metabolites are stimulated by treatment, and studies in *Arabidopsis* have suggested it may increase salicylic acid levels and induce systemic acquired resistance (SAR). In the field, it appears to be only effective on rice. Acibenzolar-S-methyl (ASM) was developed via research on salicylic acid analogs which act as signal molecules in SAR. It was commercialized as Bion® in Europe and Actigard® in the USA. The product is active on a wide range of diseases and can be used as a spray or seed treatment. Other defense activators that have been commercialized include a vitamin analog menadione sodium bisulfite (MSB) and the bacterial protein harpin, while further chemicals are at

Table 11.6 Some advantages and disadvantages of plant defense activators

---

Advantages

- Effective against a wide range of pathogens including some bacteria and viruses
- Long-lasting protection for several weeks or months
- Development of resistance is less likely than with single-site fungicides

Disadvantages

- Variable performance due to physiological state of plant
- Yield penalty in absence of disease pressure

---

an experimental stage (Table 11.5). It should also be noted that some microbial inoculants act by stimulating plant defense pathways.

The development of this class of crop protection products was seen to provide some potential advantages over conventional pesticides (Table 11.6). In principle, activation of plant defense is nonspecific and gives protection against a broad range of diseases and different pathogen types. The protection is also long-lasting, sometimes for the duration of the key growth stages of the crop, and hence only a single treatment needs to be applied. Crucially, as the chemicals act via the endogenous defense systems of the host plant, the development of resistance to this class of agrochemicals is unlikely. Hence, defense activators should provide a sustainable alternative mode of action for use in crop protection programs. In practice, there have been some problems with their deployment in the field. The level of disease control can be variable, most likely as it is affected by the physiological state of the plant, including growth stage and other factors. When applied to young plants, there can be a check in growth.

Overall, treated crops may incur a yield penalty, especially if disease pressure is low. So to date, defense activators have not been as widely used as their discovery and development promised. There is, however, ongoing interest in their potential, especially if the variability in performance can be resolved and the beneficial effects can be uncoupled from any impact on yield. For example, the defense priming activity of the nonprotein amino acid β-aminobutyric acid (BABA) (Table 11.5) has been known for many years, and demonstrated to be effective on many crops against numerous pathogens, but development as a commercial product has been constrained by phytotoxicity and other deleterious effects on plants. Recent research has, however, identified structural analogs of BABA that prime different defense pathways without stunting plant growth, and may therefore be a source of novel crop protection products.

Control by Gene Silencing

Recent research has explored the possibility that other natural plant defense pathways such as RNA interference (RNAi) might be manipulated to control infection by pathogens. The demonstration that RNA molecules can move between organisms laid the foundation for experiments in which expression of double-stranded RNA (dsRNA) in transgenic plants was shown to silence complementary target genes in fungi invading such plants, thereby reducing disease. However, this host-induced gene silencing (HIGS), while effective, involves genetic manipulation of the host plant and is likely to encounter the same regulatory constraints as other genetically modified (GM) crops. An alternative non-GM approach, exploiting the same RNAi machinery, is to directly spray long dsRNA and small-interfering RNA (siRNA) molecules onto infected plants. The potential mechanisms involved in such spray-induced gene silencing (SIGS) are shown in Figure 11.4. The RNA may be taken up by both the plant and the invading fungus, and siRNA can also move from plant cells to the fungus, initiating silencing of the complementary fungal gene.

A proof-of-concept experiment was done with *Fusarium graminearum*, causal agent of the damaging and hard to control *Fusarium* head blight disease of cereals (see Chapter 4, Figure 4.16b). First, a fungal gene known to affect growth was selected as the target: CYP51 encoding an enzyme involved in sterol biosynthesis, and the site of action of azole fungicides. dsRNA complementary to CYP51 was sprayed onto barley leaves inoculated with *F. graminearum*, with a subsequent reduction in lesion size in treated leaves. There was also an effect on lesions on untreated areas of treated leaves, indicating that the silencing signal had moved. This encouraging result shows that application of exogenous RNA has the potential to inhibit disease development. Further work is required to test this approach with infections of cereal ears rather than leaves under field conditions. There are also questions about the longevity of control, and the costs and formulation of RNA sprays for commercial use. Nonetheless, RNAi technology has emerged as a promising alternative to conventional pesticides, and is being actively explored as a potential method to control weeds, insect pests, and fungal pathogens.

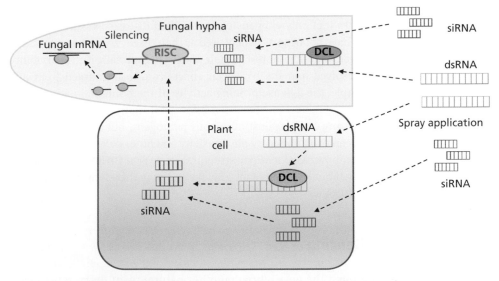

Figure 11.4 Proposed mechanism of spray-induced gene silencing (SIGS). Exogenous long dsRNAs and siRNAs are sprayed and can be taken up by both plant cells and fungal hyphae. The long dsRNAs are processed by plant or fungal Dicer-like (DCL) enzymes to produce small-interfering RNAs (siRNAs). The siRNAs are incorporated into a RNA-induced silencing complex (RISC) that binds to complementary sequences of messenger RNA in the target fungus, resulting in degradation of the mRNA and gene silencing. *Source:* Adapted from Machado et al. (2017).

Formulation and Application

Discovering chemicals with useful biological activity is only one part of fungicide development. It is also necessary to produce the compound in a form suitable for storage and subsequent application to the crop. Therefore, alongside the constant hunt for better compounds, efforts are continually made to optimize activity in the field by improving formulation and devising better ways of delivering the chemical to the target fungus. The goal is not only to ensure effective control of the disease but also to reduce the amount of fungicide applied to the crop. This lowers costs and also minimizes any risk to nontarget species in the crop ecosystem.

Much of the mystique of the agrochemical industry concerns formulation. The tricks of how to get insoluble compounds into a form suitable for application, and then distribute them over a crop in such a way that they stick to water-repellent plant surfaces and remain active for days or even weeks, are closely guarded secrets. With protectant compounds, the main problem is to ensure an even coverage of the plant, and to prevent loss of the active ingredient through weathering or degradation. The biologically active chemical is therefore mixed with carrier compounds which aid dispersion and adhesion to the crop. Such ingredients are often described as "stickers" and "spreaders." They include surface-active detergents and polymers such as carboxymethylcellulose and alginates. With systemic compounds, uptake by the plant is important, and ingredients may therefore be added to aid penetration across the external cuticle of the plant.

The problems of formulation are closely allied to the method of application. Clearly, this depends in large part on the type of pathogen to be controlled. Formulations designed to deliver an active ingredient to the leaf surface are unlikely to be fully effective as seed dressings or treatments for soil. There is also the question of scale. Applying fungicides in a controlled environment such as a glasshouse is a very different proposition to treating a whole field or plantation. In the former case, it may be possible to add a chemical to the irrigation system or to fumigate the crop atmosphere, while the latter usually requires spray application from a tractor or the air. Figure 11.5 illustrates some of the different methods used to apply agrochemicals. The most common approach is to spray a diluted solution or suspension of the active ingredient through a hydraulic nozzle (Figure 11.5a). Such conventional, high-volume sprays are relatively inefficient as a wide range of droplet sizes is generated and much of the active ingredient misses the target. The larger droplets tend to run off the plant, while the smaller droplets may be carried away in turbulent air, a phenomenon known as "drift." Only a small proportion of the chemical may reach the right place.

There is also the difficulty of reaching disease agents which may be present on the underside of leaves or, as in the case of the eyespot pathogens of wheat, *Oculimacula* spp., infect the stem base. Interestingly, control of this disease has been shown to be more effective in instances where rainfall follows application of a fungicide spray, presumably due to the compound being washed down the leaf sheaths to the infection site.

Due to the inefficiency of conventional hydraulic sprays, much effort has been expended on developing techniques for controlled droplet application (CDA), using improved nozzle designs or air-assisted rotating discs that generate sprays of defined droplet size (Figure 11.5b). Such improvements permit a lower volume of pesticide to be applied to the crop, and are particularly useful for applying compounds in protected cultivation, such as glasshouses.

Figure 11.5 Some ways of applying pesticides to crops. (a) Tractor-mounted boom sprayer delivering crop protection products via hydraulic nozzles. (b) Controlled droplet application using a spinning disc rotary atomizer (Electrafan) delivering low and ultralow spray volumes. *Source:* Image courtesy of Micron Group. (c) Electrostatic spray applicator creating charged droplets. *Source:* Courtesy of G. Matthews. (d) Trunk injection of avocado tree using a plastic syringe containing an aqueous solution of the fungicide fosetyl-Al to control *Phytophthora* root rot. The compound is phloem mobile and hence can reach the roots. *Source:* Coffey (1987).

Other ultra-low volume methods have been developed, such as electrodynamic sprayers (Figure 11.5c). This technology generates electrically charged droplets which are attracted to plant surfaces, giving improved coverage from a smaller volume of liquid. The active chemical can be formulated in oil rather than water, which prevents evaporation and is also an advantage in regions where rainfall and hence the water supply is limiting. Electrostatic spraying has been successfully used to apply pesticides in tropical crops such as cotton and cowpea.

While spraying is the most widely used way of applying pesticides, other methods may be more suitable depending on the compound or the target pathogen. Sulfur, for instance, is easily applied as a dust rather than in suspension, while fumigation can be highly effective in enclosed environments including glasshouses and grain stores. Seed dressing is a very efficient way of applying compounds to prevent seed-borne diseases as well as soil-borne pathogens which infect emerging roots or cause damping-off diseases of seedlings. Example fungicides used as seed treatments include carboxin, effective against cereal smuts (see later), and the more recently introduced SDHI compounds fluopyram, to control sudden death syndrome of soybeans caused by *Fusarium virgiforme*, and sedexane to reduce infection by *Rhizoctonia* on a broad range of crops. Where the fungicide is sufficiently systemic to move into the germinating plant, seed treatment may also provide some protection against early-season foliar pathogens. Some azole fungicides, for instance, can reduce infection by rusts, powdery mildews, and Septoria tritici leaf blotch on the first emerging leaves of wheat plants, hence delaying epidemic development later in the season.

Treating pathogens in soil presents particular challenges. Drenching soil with a pesticide or treating by fumigation is costly and often affects nontarget species. In this situation, granular formulations of the chemical may be more appropriate, designed to gradually release the active ingredient into soil over a period of time. Granular fungicide treatments are often used to control diseases of amenity turf, such as dollar spot (*Sclerotinia (Clarireedia) homeocarpa*) on golf courses. Techniques have also been developed to deliver pesticides or microbial inoculants into soil at the same time as seeds or tubers are planted. Such in-furrow treatments can protect against seedling diseases or early infection by soil-borne potato pathogens such as *Rhizoctonia*.

Pathogens which grow in inaccessible parts of the plant, such as deep within seeds or in vascular tissues, may also present problems. Seed-borne fungi that simply contaminate the seed coat can be eradicated by soaking in a fungicide solution, while those that infect internal tissues are more difficult to control. In the early years of the twentieth century, smut fungi such as species of *Ustilago* and *Tilletia* were among the most damaging disease agents of cereal crops, due to transmission from season to season through seed. The introduction of organomercury treatments helped to control bunt, *Tilletia tritici*, where spores are carried on the seed surface, but control of loose smut, *Ustilago nuda*, which infects the embryo, was not achieved until the advent of truly systemic compounds such as carboxin which could penetrate and move within host tissues. Ironically, some smut diseases are now staging a comeback due to the withdrawal of organomercury fungicides on environmental grounds. Effective alternatives are available but are more expensive, and hence the use of seed treatments has declined in some crops.

Vascular wilt fungi are also difficult to control with chemicals, especially those infecting woody perennial hosts. A good example is Dutch elm disease, caused by *Ophiostoma novo-ulmi* (see Chapter 5, Figure 5.9), which has decimated elm populations across much of the northern hemisphere. Attempts were made to protect specimen elms, for instance those in city parks, by injecting fungicides into the sapwood, where the chemicals are carried upwards in the transpiration stream. Some success was achieved using benzimidazole fungicides such as benomyl, but the treatment proved ineffective on large trees, mainly due to the relative insolubility of these compounds, which prevented them moving in sufficient concentrations to prevent infection of the upper branches. However, in some instances trunk injection can be a very effective way of applying a fungicide to a tree. The most destructive disease of avocado groves is root rot, caused by *Phytophthora cinnamomi*. This soil-borne pathogen attacks the small, feeder roots of the tree, with often fatal consequences (see p.364). *Phytophthora* root rot was not amenable to chemical control until the introduction of systemic fungicides active against oomycetes such as metalaxyl. This compound, applied as a root drench, controlled disease in young trees but the prognosis for more mature trees remained gloomy.

The advent of phosphonates such as fosetyl-Al, which can move from shoot to root, radically changed the situation, with foliar sprays at monthly intervals giving good control. Even better results were achieved by applying the compound by injection with a syringe which is left inserted in the trunk of the tree (Figure 11.5d). The fungicide solution is gradually taken up by the tree, and only two injections a year may be sufficient to ensure control of this previously lethal disease. It should be obvious that labor-intensive applications, such as injection, are only feasible with high-value crops where protection of individual plants can be justified on economic or amenity grounds.

Fungicide Programs

Field-grown crops are exposed to a range of biotic threats including insect pests and weeds as well as plant pathogens, so fungicides are often applied as part of an integrated crop protection program that might include insecticides and herbicides, as well as defense activators and biological control agents (see Chapter 14). Scheduling of treatments usually takes account of the crop growth stage and the range of diseases that are likely to pose a threat. The cultivation of wheat in the UK and other countries in north-west Europe serves as an example (Figure 11.6). Most of the wheat grown in these areas now consists of winter varieties that are sown in the autumn and mature for harvesting the following summer. Seed may be treated with pesticides, including fungicides to control root diseases such as take-all (*Gaeumannomyces graminis*), and to provide some protection against early foliar infection. The main treatments are, however, applied as sprays to the developing crop during the spring and summer. Typically, two or three main spray applications

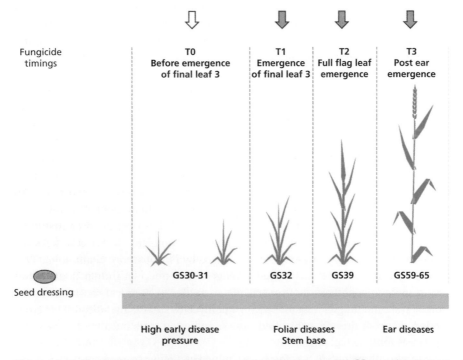

Figure 11.6 Fungicide treatment timings on winter wheat in the UK. GS + number indicates the growth stage of the crop based on a numerical scale. *Source:* Adapted from AHDB wheat disease management guide.

are made (T1–T3) (Figure 11.6). These are timed to coincide with the emergence of different leaf layers as the stem extends and the ear emerges. The T1 and T2 sprays provide protection for the vital top three leaves that generate most of the photosynthate for grain development, while the T3 treatment is mainly targeted at ear diseases such as *Fusarium* ear blight (see Chapter 4, Figure 4.16).

The principal foliar pathogens requiring control in the UK are yellow and brown rusts (*Puccinia striiformis* and *P. triticina*), Septoria tritici leaf blotch (*Zymoseptoria tritici*), and powdery mildew (*Blumeria graminis*). The relative severity of these diseases varies from season to season, so the fungicide program may be adjusted to take account of the degree of perceived risk. In years where conditions lead to early epidemic development, an additional early spray (T0) may be applied.

Prior to the 1970s, few wheat crops in the UK were treated with fungicides. However, once the benefits of good crop protection were understood, mainly in terms of higher yields and improved grain quality, more crops received a fungicide program. This trend coincided with progressive increases in wheat yields due to the sequential introduction of higher yielding varieties, combined with better control of pests and diseases. There were also intermittent changes in the types of chemicals used, starting with early systemic fungicides such as the methyl benzimidazoles (MBCs) and then introduction of DMIs such as the azoles, the strobilurins (otherwise known as QoIs), and most recently the SDHIs. These innovations represented advances in the relative efficacy of the different fungicide classes against the main target pathogens, with the newer chemistry usually providing superior levels of disease control. Such genetic and chemical improvements have led to gains in overall wheat yields from around 4 t per hectare in the 1970s to more than 7 t per hectare by the year 2000, but at a cost in terms of the degree of selection pressure applied to pathogen populations. Hence there have been corresponding changes in the virulence of key pathogens such as the rusts, along with the emergence of pathogen strains with resistance to one or more of the main fungicide classes.

Fungicide Resistance

Prior to the discovery of the first systemic fungicides, there were very few instances when application of a protectant compound, at the correct time and dose rate, failed to control a pathogen. Hence for decades the copper, sulfur, and dithiocarbamate fungicides remained as effective as when they were first discovered. The few exceptions to this include the development of resistance to mercury-based seed dressings in *Pyrenophora*, dodine in apple scab, *Venturia inaequalis*, and problems with the use of aromatic diphenyl compounds used to control postharvest rots of citrus fruits caused by *Penicillium* species (Table 11.7). But in general, the protectant, multisite inhibitors have given long-term, durable control of many crop diseases.

Practical experience with the newer, single-site fungicides has been different. There have been numerous cases in which an initially highly effective compound has subsequently failed to control a pathogen in a crop (Table 11.7). Sometimes such failures have occurred within a short time of first use of the compound. For example, the pyrimidine fungicide dimethirimol, introduced in 1968, showed outstanding activity against powdery mildews and was recommended to control the cucumber powdery mildew pathogen *Sphaerotheca*

Table 11.7 Timeline of fungicide resistance

	Fungicide type	Date	Pathogen
Multisite fungicides	Bordeaux mixture	1882	
	Organomercury	1964	*Pyrenophora*
	Dodine (guanidine)	1969	*Venturis inaequalis*
Single-site fungicides	Methyl benzamidazoles	1971	*Botrytis cinerea*
		1973	*Blumeria graminis* (pm)
	Pyrimidines	1971	*Sphaerotheca fuliginea* (pm)
	Dicarboximides	1977	*Botrytis cinerea*
	Phenylamides	1979	*Phytophthora infestans*
	Demethylation inhibitors	1982	*Blumeria graminis* (pm)
		2006	*Venturia nashicola*
	Morpholines	1994	*Blumeria graminis* (pm)
	Strobilurins (QoIs)	1999	*Blumeria graminis* (pm)
		2002	*Mycosphaerella graminicola*
	Succinate dehydrogenase inhibitors	2008	*Alternaria alternata*
		2012	*Pyrenophora teres*

Prior to the introduction of single-site fungicides there were few reported instances of resistance.
pm, powdery mildew.

fuliginea in glasshouses. Intensive use quickly led to the emergence of highly resistant strains of the pathogen, and by 1971 the compound was withdrawn in the Netherlands. Similarly, the phenylamide fungicide metalaxyl (Table 11.2) was launched in 1977 and recommended for control of many important oomycete pathogens such as the downy mildews, *Phytophthora*, and *Pythium*. By 1979, isolates of cucumber downy mildew, *Pseudoperonospora cubensis*, able to tolerate 20 times the initially effective dose of the fungicide had been recorded and, more dramatically, in the following season failures to control potato blight occurred in Ireland and the Netherlands. This was shown to be due to the incidence of metalaxyl-resistant strains of *Phytophthora infestans* in the field and shortly after, formulations of fungicide containing metalaxyl alone were withdrawn. Similar experiences have occurred with more recently introduced fungicide classes such as the QoIs and SDHIs.

Why did this happen and what lessons can be learned from these setbacks? More importantly, can anything be done to prevent such "boom and bust" episodes in the future?

Some Definitions

Before attempting to answer these questions, some terms used to describe the problem should be defined. **Intrinsic resistance** refers to the natural insensitivity of many fungi or oomycetes to different classes of fungicides. For instance, the phenylamide fungicides are active against oomycetes but ineffective against true fungi, while the pyrimidines such as dimethirimol are only highly active against powdery mildews. Each class of fungicide therefore has a particular **spectrum** of activity with a range of efficacy against different pathogens. **Acquired resistance** refers to the phenomenon whereby an initially sensitive species of fungus becomes insensitive after repeated exposure to the fungicide. It is also important to distinguish between instances where the sensitivity of a target fungus to a particular chemical has changed and the actual loss of efficacy of a fungicide in practical use. There are many examples of resistance being detected in some strains of fungi, yet the fungicide still gives effective control of the disease in the field. The phrase **resistance in practice** has therefore been recommended to describe situations where reduced sensitivity of a fungal pathogen to a fungicide results in poor disease control in the crop.

Cross-resistance is where development of resistance to one chemical in a particular class also confers resistance to other related chemicals. For example, strains of *Botrytis cinerea* (see Figure 11.11) resistant to benomyl are also resistant to other benzimidazole fungicides such as carbendazim and thiabendazole. This has important practical implications, as strains altered in sensitivity to one fungicide in a particular group often, at the same time, become resistant to all other compounds in that group.

The Evolution of Resistance

The development of resistance to fungicides is an example of an evolutionary change in a fungal population caused by a human activity – the application of a chemical to a crop. The raw material for this evolution is genetic variation in the fungus; mutations conferring resistance may already exist in the pathogen population (so-called standing variation) or may arise after introduction of a new fungicide. Use of the fungicide exerts selection pressure for insensitive individuals. Factors influencing selection will include the dose rate and frequency of use of the fungicide, and whether it is used alone or with other compounds with a different mode of action. The rate of emergence of resistance in a pathogen population will also depend on the relative fitness of resistant strains by comparison with sensitive wild-type strains. If a mutation to resistance carries little or no fitness cost, resistant individuals will have a competitive advantage in the presence of the fungicide and will eventually predominate in the population (as seen in Figure 11.7c, azoxystrobin).

The dynamics of resistance development vary depending on the mechanisms involved. In cases where a single mutation of major effect takes place, for instance affecting the target of a single-site fungicide, the pathogen population separates into sensitive and resistant subgroups with a bimodal distribution (Figure 11.7a). Where more than one gene or mechanism contributes to resistance, a unimodal distribution can occur (Figure 11.7b). In both cases, there is directional selection for resistance, but in the first case it is acting on discrete variation, as opposed to the continuous distribution seen in the second, characterized by gradual shifts toward resistance over time.

Figure 11.7(c,d) show actual examples of discrete versus continuous selection for resistance in the wheat leaf blotch fungus *Zymoseptoria tritici*. Resistance to the QoI fungicide azoxystrobin is conferred by a single amino acid change in the mitochondrial cytochrome b target protein (G143A, denoting a glycine to alanine substitution at position 143 in the protein). Isolates of the fungus possessing this mutation are highly resistant to the fungicide. At the start of the sampling period (2003), the population was separated into R and S groups,

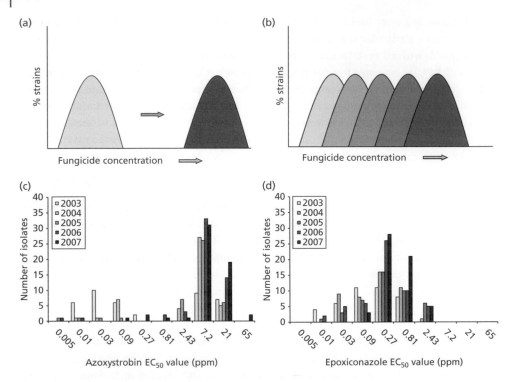

Figure 11.7 The dynamics of resistance development to fungicides. Discrete versus continuous sensitivity shifts. (a) Separation into sensitive and resistant subpopulations, typical of resistance due to a single genetic change of major effect. (b) Unimodal sensitivity distribution shifting toward resistance over time, due to multiple genetic changes of smaller effect. *Source:* Based on Georgopoulus and Skylakakis (1986). (c,d) Sensitivity distribution of isolates of *Zymoseptoria tritici* obtained from the field at Rothamsted over a five-year period (2003–2007). (c) Response to the QoI fungicide azoxystrobin, showing increase in proportion of highly resistant isolates over time. (d) Response to the azole fungicide epoxiconazole showing progressive shift toward resistance over the sampling period. EC_{50} values are averages for a range within each category. *Source:* Adapted from Lucas et al. (2015).

with or without the mutation. Azoxystrobin use over subsequent seasons selected a progressively greater proportion of R isolates to the point where relatively few S isolates remained in the population. In contrast, the development of resistance to the azole fungicide epoxiconazole was a gradual process, with the population shifting slightly toward insensitivity from season to season (Figure 11.7d). Subsequently, this process has continued so that the efficacy of azole fungicides in controlling *Z. tritici* has been gradually eroded. This pattern of resistance development is due to multiple mutations of minor effect in the CYP51 protein target of azole fungicides, combined with some additional genetic changes affecting sensitivity.

Mechanisms of Fungicide Resistance

Genetic and molecular studies on fungi acquiring resistance to different fungicides have shown that several mechanisms are involved (Figure 11.8). With plant pathogens, the most common type is target site resistance, such as the G143A change in the cytochrome

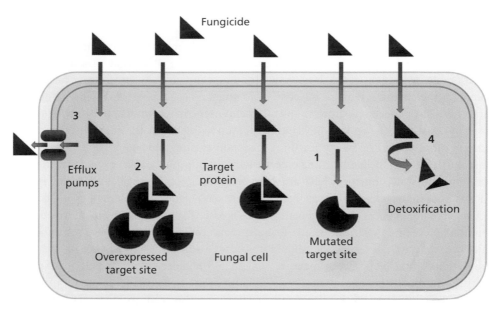

Figure 11.8 Mechanisms of resistance to single-site fungicides. 1. Alteration of the target protein prevents fungicide binding (target site resistance). 2. Overexpression of target protein increases amount of fungicide necessary for inhibition. 3. Efflux pumps expel fungicide from cell. 4. Breakdown of fungicide by metabolic enzymes. *Source:* Lucas et al. (2015).

b protein targeted by QoI fungicides described earlier. Such mutations alter the conformation of the protein concerned and thereby reduce or abolish binding of the fungicide to the target. Typically, this confers a high level of resistance to the fungicide. For instance, with the benzimidazole fungicides (Table 11.2), mutations in the β-tubulin target led to the emergence of strains that could withstand fungicide doses 1000-fold higher than sensitive strains. A second mechanism is overexpression of the gene encoding the target protein, so there is a higher concentration in the cell and consequently more fungicide is required to inhibit it. This increases resistance but to a lesser extent than target site changes. Fungicide can also be removed from the cell through the action of efflux pumps in the plasma membrane. This mechanism is common in fungi causing mycoses of humans, such as the yeast *Candida*, but is less often reported in plant-pathogenic fungi. Finally, there is the possibility that fungicide might be broken down by metabolic enzymes to detoxify it. This resistance mechanism occurs frequently in weeds developing resistance to herbicides, and insects and insecticides, but is rarely found in fungi, which is surprising given the propensity of many fungi to degrade a diverse range of recalcitrant chemicals in the environment.

The different resistance mechanisms shown in Figure 11.8 can also occur together in some plant pathogens, leading to additive increases in resistance. For example, the progressive shifts in the sensitivity of *Zymoseptoria tritici* to azole fungicides are due mainly to an accumulation of genetic changes in the CYP51 target protein, but some isolates with higher levels of resistance also have overexpression of CYP51, as well as increased activity of efflux pumps. In this case, fungicide selection over time has led to complex, more resistant phenotypes through mutation and recombination in the pathogen population.

Management of Fungicide Resistance

Resistance to pesticides is now a fact of life for the agrochemical industry, and measures to combat it are a key part of product stewardship and use. It is vital to safeguard the considerable investment in the development of a new fungicide by ensuring, as far as possible, that the chemical remains effective in commercial use. The sustainability of disease control is also of concern to growers and other end-users. In response to the growing threat posed by resistance, the agrochemical industry set up a joint forum, the Fungicide Resistance Action Committee (FRAC; www.frac.info), to identify existing and potential resistance problems and develop guidelines for fungicide use which minimize the risk of resistance in practice. Similar action committees have also been established for insecticides and herbicides.

Resistance management involves three main aspects (Figure 11.9). Initially, can one predict the likelihood of resistance occurring? This is essentially a risk assessment to preempt potential problems. Next, one needs to have measures in place to detect resistance as quickly and accurately as possible should it occur. This involves surveillance of pathogen populations using bioassays to measure sensitivity or molecular diagnostic methods to identify the genetic changes underlying resistance. Finally, as the development of resistance is an evolutionary process driven by selection, one also needs to adopt tactics that can reduce or ideally remove the selection pressure. This objective is the conundrum at the heart of resistance management. How can one reduce selection while still maintaining adequate levels of disease control?

Estimating Resistance Risk

Three factors determine the risk of resistance arising and the extent to which it will spread and persist in the pathogen population. These are the nature of the fungicide, the biology of the pathogen concerned, and the way in which the fungicide is used. These criteria have been combined in a matrix to derive an estimate of overall risk (Figure 11.10). The different types of fungicides and pathogens are awarded individual risk scores, together with agronomic factors influencing the intensity of fungicide use. For example, multisite fungicides represent a lower

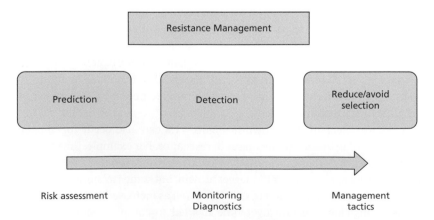

Figure 11.9 Components of resistance management. *Source:* Lucas (2017a).

Fungicide risk	Risk	Combined risk			Agronomic risk
Benzimidazoles Dicarboximides Phenylamides Qols	High = 3	3 1.5 0.75	6 3 1.5	9 4.5 2.25	High = 1 Medium = 0.5 Low = 0.25
Carboxanilides SDHI fungicides SBI fungicides Anilinopyrimidines Phenylpyrroles	Medium = 2	2 1 0.5	4 2 1	6 3 1.5	High = 1 Medium = 0.5 Low = 0.25
Multisite fungicides Melanin biosynthesis inhibitors Plant defense activators	Low = 1	0.5 0.25 0.125	1 0.5 0.25	1.5 0.75 0.3	High = 1 Medium = 0.5 Low = 0.25
		Low = 1	Medium = 2	High = 3	**Pathogen risk**
		Seed-borne and soil-borne fungi Rusts	*Oculimacula Rhynchosporium Leptosphaeria Mycosphaerella*	*Blumeria Botrytis Plasmopara Magnaporthe Zymoseptoria*	

Figure 11.10 A risk matrix for estimating the likelihood of fungicide resistance developing for different plant pathogens. Combined scores are derived from individual estimates for fungicide, pathogen, and agronomic factors. Note: Only a few example pathogens are included, and ratings can change as more evidence for particular diseases and/or fungicide groups is collected. *Source:* Adapted from Kuck and Russell (2006), Grimmer et al. (2014).

risk than single-site compounds, fungi with rapid life cycles producing large numbers of spores are higher risk than slowly reproducing soil-borne species, and open-field cultivation is considered a lower risk than production in enclosed environments such as glasshouses.

Practical experience also plays a part in estimating risk. Some fungi, such as powdery mildews and the gray mold pathogen *Botrytis cinerea* (Figure 11.11), have proved to be particularly prone to resistance development to several classes of fungicides (see Table 11.7). Recently, *B. cinerea* populations from commercial strawberry fields in Spain and Germany have been found to possess resistance to as many as six different fungicide classes.

The risk matrix has proved useful in obtaining generalized estimates but lacks precision when comparing particular cases, such as different types of single-site fungicides. Refined models are now being developed which incorporate a larger number of fungicide, pathogen, and agronomic system factors associated with the rate of evolution of resistance, to produce estimates of the likely time from introduction of a fungicide to the first detection of resistance.

Such conceptual models can be supplemented by experimental studies in the laboratory. These include culturing fungi in the presence of a fungicide, with or without a mutagenic treatment such as exposure to UV light or a chemical mutagen, to select resistant individuals. If the site of action of the fungicide is known, selected mutants can be quickly screened by sequencing to detect genetic changes in the target gene or, if unknown, whole-genome sequencing may provide useful clues to mode of action. There is also increasing interest in

Figure 11.11 Courgette infected with gray mold *Botrytis cinerea.* This versatile pathogen infects a wide range of hosts, including grapevine and many horticultural crops, and has also developed resistance to multiple fungicides. *Source:* Photo by Brian Case.

"experimental evolution" in laboratory culture, such as passaging pathogens on increasing concentrations of fungicide over many generations.

Needless to say, these *in vitro* approaches are only possible with microorganisms that can be cultured, and have other limitations in terms of predicting what might happen in the field. Nonetheless, they have proved useful in identifying the range of potential mutations that might affect fungicide efficacy, and also designing molecular assays to aid early detection of any changes. But as yet, there is no foolproof way to predict when or where resistance might occur in practice.

Monitoring Resistance

Monitoring changes in the response of fungal populations to fungicides, season by season, has become an important activity, both in defining potential problems and predicting future events. This relies on surveys of fungal isolates taken from the field to determine their dose response to the chemical concerned. The usual way to measure fungicide sensitivity is to estimate the dose which inhibits 50% of a particular physiological parameter. For example, if growth or spore germination is reduced by half, then this is described as the ED_{50} (effective dose) or EC_{50} (effective concentration) which gives 50% inhibition. Alternatively, the minimum inhibitory concentration (MIC), the lowest concentration completely preventing growth of the pathogen, may be determined. Such assays are used to define the baseline sensitivity of fungi, and field isolates can then be compared to detect any changes in their dose response.

Laboratory bioassays are, however, labor intensive and time consuming and are now being supplanted by molecular diagnostic methods detecting the genetic changes responsible for the resistant phenotype. The majority of these rapid molecular assays are based on polymerase chain reaction (PCR) or sequencing technologies, and can be directly applied to samples taken from infected plants in the field without the need to culture the pathogen.

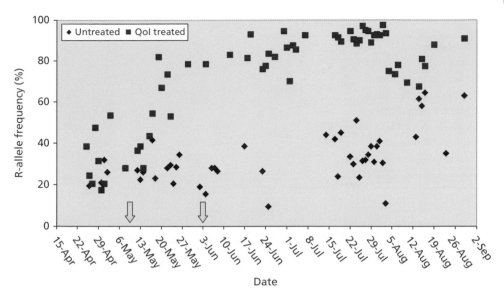

Figure 11.12 Quantification of fungicide resistance in ascospores of *Mycosphaerella graminicola* (*Zymoseptoria tritici*) sampled by spore traps located in wheat plots either treated or untreated with a QoI fungicide. Spore populations trapped at each sampling date were analyzed using a real-time PCR assay estimating the frequency of the mutation (G143A) of the cytochrome b QoI target gene. Two sprays were applied to treated plots on the dates indicated (*yellow arrows*). Note the large increase in the proportion of the resistant allele in the population following the first spray, and continuation after the second. Proportions in untreated plots remained lower until harvest when mixing of spores between plots may have occurred. *Source:* Fraaije et al. (2005).

Another advantage of such assays is that the amount of resistance present in a sample can also be measured. Where there is a mixture of resistant and sensitive strains, the proportion carrying the mutant form (R-allele) of the target gene as opposed to the sensitive form (S-allele) can be determined. An example is shown in Figure 11.12. In this case, air-borne spores of *Mycosphaerella graminicola* (the sexual stage of *Zymoseptoria tritici*) were sampled in field plots of wheat using spore traps (see Chapter 4, Figure 4.11) during the growing season. One plot was sprayed twice with a QoI fungicide and the other left untreated. Samples of trapped spores were analyzed using a real-time PCR assay discriminating the R- and S-alleles of the G143A mutation conferring resistance to QoIs. There was a rapid increase in the proportion of air-borne spores carrying the R-allele in the treated plot. These techniques can therefore be used not only to detect resistant genotypes but also to map their incidence and spatial spread within fields, on farms, and across regions. Such information is of practical value to inform fungicide treatment regimes or alternative control measures in affected regions.

Strategies to Reduce Selection

The practical challenge in resistance management is to find ways to reduce selection for resistant strains while maintaining an acceptable level of disease control. A number of approaches have been suggested (Table 11.8). The most obvious tactic is to limit exposure of the pathogen to the fungicide, by reducing either the dose rate or the number of times the

Table 11.8 Resistance management strategies

- Reduce dose rate
- Reduce number of sprays
- Use fungicide mixtures
- Use fungicide alternations
- Use multisite inhibitors
- Use plant defense activators or RNAi
- Use biocontrol agents
- Improve cultivar resistance
- Adopt agronomic measures

chemical is applied. The issue of dose rate has been debated for some time, as while in evolutionary terms it is clear that higher doses will exert greater selection pressure when resistance is already present, it could be argued that higher doses might prevent resistance emerging in the first place. In most instances, however, reducing the dose is likely to reduce selection and delay the development of resistance. Limiting the number of applications of the fungicide is also recommended, and especially the number of times in a season chemicals with the same mode of action are used. Most fungicide programs on cereals (see Figure 11.6) therefore advise limiting a particular mode of action to one or two treatments. Another tactic is to only use the fungicide when a disease risk is present. This approach depends on having a reliable disease forecasting system in place. The selection pressure can also be reduced by mixing or alternating fungicides with different modes of action. This is strongly recommended when using a single-site fungicide that is considered at high risk of resistance.

Many manufacturers now co-formulate such fungicides with partner compounds and avoid supplying formulations with only one active ingredient. The choice of partners for mixed formulations is also important. Ideally, mixtures should contain a low-risk fungicide to protect the high-risk chemical. Hence some of the older, multisite fungicides are now being used in mixtures with the more recent single-site compounds. The benefit of mixing with a multisite fungicide may depend on the dose rate used. Studies modeling the development of resistance suggest that best results are achieved when the low-risk partner is used at the full recommended rate, while the dose of the high-risk partner is adjusted to the level required to give effective disease control. Alternating fungicides with different modes of action is generally a less reliable tactic, especially when the decay rate of the low-risk fungicide might limit any overlap between treatments.

The other strategies listed in Table 11.8 depend on the availability of alternative methods to control disease. As described earlier, one of the advantages of using plant defense activators is that they should pose a much lower risk of resistance. The situation with novel approaches such as RNAi and gene silencing is not yet clear, although a few sequence mismatches caused by mutations are less likely to compromise efficacy than with changes affecting conventional fungicide protein targets. The number of biocontrol agents available for commercial use on broad-acre crops is currently limited (see Chapter 13). Improvements in cultivar resistance, which can delay epidemic development, should permit the use of fewer fungicide applications and/or lower doses, while agronomic measures lowering disease risk will also provide opportunities to reduce reliance on chemicals alone.

The Future of Fungicides

Pesticides, including fungicides, continue to play an important role in crop protection worldwide. Fungicide use has helped to maintain high yields in staple crops such as cereals in Europe, and elsewhere has provided valuable protection against emerging diseases. When Asian soybean rust invaded South America in the 2000s, severe losses were initially incurred but fungicide use restored production to economic levels to buy time for plant breeders to select more resistant soybean varieties.

But there are questions about the reliance of intensive production systems on the routine use of agrochemicals, mainly due to potential impacts on nontarget species, human health and the environment, but also the increasing problems posed by resistance. Recent surveys in Europe have shown that the incidence of resistance to QoI and azole fungicides is lower in countries with less intensive treatment regimes. Overall, there is now an increased interest in designing crop production systems which are less dependent on chemical inputs (see Chapter 14).

In the twenty-first century, agrochemical companies face challenges due to an increasingly stringent regulatory framework and the cost and difficulty of discovering novel modes of action. The change from a risk-based to hazard-based assessment of chemical safety has led to the withdrawal of many previously available compounds, especially in Europe, and at the same time there is a diminishing pipeline of new products reaching the market. In response to these challenges, there has been a consolidation of the agrochemical industry into fewer, larger companies, and also diversification into seeds and other genetic resources, including biotechnology. Hence there is a trend toward integration of crop protection chemicals with genetic traits, creating an opportunity to offer growers a package of products and processes to optimize production from seed to harvest. For the foreseeable future, conventional pesticides and newer crop protection agents such as plant defense activators and RNAi are likely to be part of this more integrated approach.

The emergence of fungicide resistance is just one facet of a worldwide problem of increasing resistance to antimicrobial chemicals used in medicine as well as agriculture. The selection pressure exerted by overuse of antibiotics and fungicides has led to many pathogens developing resistance, sometimes to multiple modes of action. The doomsday scenario is that we may lose the ability to control many infectious diseases that were until recently consigned to history. In this context, the decline in chemical diversity caused by the loss of existing products, combined with the reduced rate of registration of new chemistry, is a serious concern. There need to be renewed efforts to discover novel antimicrobial agents, including biopesticides, along with more awareness of best practice in their use. There is also a requirement for greater realism about the costs and benefits of chemical control in crop protection, and its ongoing contribution to global food security.

Further Reading

Books

Oliver, R.P. and Hewitt, H.G. (2014). *Fungicides in Crop Protection*, 2e. Wallingford: CABI.

Matthews, G.A., Bateman, R., and Miller, P. (2014). *Pesticide Application Methods*, 4e. Oxford: Wiley Blackwell.

Stevenson, K.L, McGrath, M.T., Wyenandt, C.A. (Eds) (2019). Fungicide resistance in North America. American Phytopathological Society Press, St Paul, Minnesota.

Thind, T.S. (ed.) (2012). *Fungicide Resistance in Crop Protection: Risk and Management.* Wallingford: CABI.

Reviews and Papers

Buswell, W., Schwarzenbacher, R.E., Luna, E. et al. (2018). Chemical priming of immunity without costs to plant growth. *New Phytologist* 218: 1205–1216. https://doi.org/10.1111/nph.15062.

Cools, H.J. and Hammond-Kosack, K.E. (2013). Exploitation of genomics in fungicide research: current status and future perspectives. *Molecular Plant Pathology* 14 (2): 197–210. https://doi.org/10.1111/mpp.12001.

Cooper, J. and Dobson, S. (2007). The benefits of pesticides to mankind and the environment. *Crop Protection* 26 (9): 1337–1348. https://doi.org/10.1016/j.cropro.2007.03.022.

Grimmer, M.K., van den Bosch, F., Powers, S.J., and Paveley, N.D. (2014). Fungicide resistance risk assessment based on traits associated with the rate of pathogen evolution. *Pest Management Science* 71: 207–215. https://doi.org/10.1002/ps.3781.

Hollomon, D.W. and Brent, K.J. (2009). Combating plant diseases – the Darwin connection. *Pest Management Science* 65 (11): 1156–1163. https://doi.org/10.1002/ps.1845.

Jorgensen, L.N., van den Bosch, F., Oliver, R.P. et al. (2017). Targeting fungicide inputs according to need. *Annual Review of Phytopathology* 55: 181–203. https://doi.org/10.1146/annurev-phyto-080516-035357.

Lamberth, C., Jeanmart, S., Luksch, T. et al. (2013). Current challenges and trends in the discovery of agrochemicals. *Science* 341: 742–746.

Lucas, J.A., Hawkins, N.J., and Fraaije, B.A. (2015). The evolution of fungicide resistance. *Advances in Applied Microbiology* 90: 29–92. https://doi.org/10.1016/bs.aambs.2014.09.001.

Ma, Z.H. and Michailides, T.J. (2005). Advances in understanding molecular mechanisms of fungicide resistance and molecular detection of resistant genotypes in phytopathogenic fungi. *Crop Protection* 24 (10): 853–863. https://doi.org/10.1016/j.cropro.2005.01.011.

Russell, P.E. (2005). A century of fungicide evolution. *Journal of Agricultural Science* 143: 11–25. https://doi.org/10.1017/s0021859605004971.

Sparks, T.C., Hahn, D.R., and Garizi, N.V. (2016). Natural products, their derivatives, mimics and synthetic equivalents: role in agrochemical discovery. *Pest Management Science* 73: 700–715. https://doi.org/10.1002/ps4458.

Stevenson, K.L., McGrath, M.T., and Wyenandt, C.A. (Eds) (2019). Fungicide resistance in North America. American Phytopathological Society Press, St Paul, Minnesota.

van den Bosch, F., Oliver, R., van den Berg, F., and Paveley, N. (2014). Governing principles can guide fungicide resistance management tactics. *Annual Review of Phytopathology* 55: 181–203. https://doi.org/10.1146/annurev-phyto-102313-050158.

Website

Fungicide Resistance Action Committee: www.frac.info/home

12

Disease Management by Host Resistance

Those who believed only in genes postulated the existence of an eternal quality, R, which they could take from wild plants, build into the genetical constitution of cultivated ones, as to make them disease-resistant forever. Those who thought ... of the green flux of ever-changing nature, saw little hope of such permanancy, and no end to man's labours in defending the crops.

(E.C. Large, 1902–1976)

The natural genetic resistance of plants to pests and pathogens has no doubt played a key role in crop protection since the dawn of agriculture. Indeed, it is doubtful if systematic cultivation of crops would have been possible were it not for the ability of most plants to withstand attack by pathogens. Up until the end of the nineteenth century, plant breeding was an empirical process. Nevertheless, it must have been important in restricting disease losses. In years when disease epidemics wiped out most of the crop, only the most resistant plants survived to provide seed for the next season. This sequence was merely an extension of the process of natural selection which maintains an equilibrium between a pathogen and its host in natural communities. At the same time, the gradual development of the intuitive art of plant breeding, in which higher-yielding and more resilient individuals were progressively selected, undoubtedly included an element of selection for resistance to disease (Figure 12.1).

The foundations of the scientific process of breeding for disease resistance were laid by Biffen early in the twentieth century. By 1912, he had shown that resistance to yellow stripe rust, *Puccinia striiformis*, in the wheat cultivar Rivet was determined by a single recessive gene which was inherited according to mendelian laws. The realization that disease resistance was a genetic trait which could be manipulated in a breeding program triggered a sustained effort to introduce similar resistance into other major crops. The introduction of crop cultivars containing new genes has since become a normal part of modern agriculture and, apart from concerns about the conservation of genetic diversity, and more recently recombinant DNA, the process does not have serious environmental implications. Indeed, at one time, breeding for resistance seemed to promise an ideal and permananent solution to the problem of plant disease. What had not been anticipated, however, was the ability of pathogens themselves to adapt and evolve in response to genetic changes in the crop. The

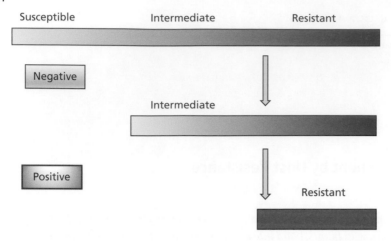

Figure 12.1 Selection for disease resistance in a population of plants. Negative selection removes the most susceptible individuals, while positive selection, as practiced by the breeder, chooses only the most resistant plants.

permanent solution proved to be elusive, and instead resistance breeding began to resemble a race between plant breeders and the pathogens.

Breeding for Disease Resistance

As described in Chapter 9, there are many different ways in which plants defend themselves against pathogens. For the plant breeder, resistance can be defined as any inherited characteristic of the host plant which lessens the effect of disease. From a practical viewpoint, any character which improves the performance of the plant in the presence of a pathogen may be useful, provided that it is heritable and stable. It is also worth pointing out that resistance to disease is only one objective of the breeder, and for many crops is of lower priority than other characters such as increased yield or improved quality.

The basic requirements to produce a novel, disease-resistant crop cultivar are:

- a source of genetic resistance
- a method for identifying and selecting this resistance
- a method for combining this resistance with other, agronomically desirable characters to produce a commercially acceptable crop genotype.

Sources of Resistance

The raw material for any breeding program is genetic variation. In the case of disease resistance, there are a number of potential sources of this variation that can be explored (Table 12.1). For a start, there may already be significant variation within the gene pool of the crop itself. For instance, *Verticillium* wilt is an important disease of the forage legume alfalfa (lucerne), but because this crop is outbreeding and genetically variable, individual plants differ in their susceptibility to infection. In this case, recurrent selection, season by

Table 12.1   Sources of disease resistance

1) Within crop cultivars
2) Within crop species
3) In closely related plant species that can be crossed with the crop
4) In more distantly related species where compatibility may be an issue
5) In unrelated plants or other organisms via genetic modification (GM)
6) Novel genes engineered by gene editing?

season, of the most resistant plants can lead to significant improvements in the overall wilt resistance of the crop.

But what if the crop itself lacks any useful variation in disease reaction type? In this case, it is often necessary to look for alternative sources, such as in wild relatives of the crop species. This approach has been widely used in many important crops, including cereals, potatoes and tomatoes, and tropical plantation trees such as coffee and avocado. When potato blight, *Phytophthora infestans*, swept Europe and North America in the nineteenth century, contemporary potato varieties were almost universally vulnerable to disease. Later attempts to introduce resistance therefore focused on wild relatives in Central and South America. One species in particular, *Solanum demissum*, found in Mexico, proved to be a useful source of resistance (R) genes which were introgressed into the cultivated potato, *Solanum tuberosum*, by hybridization and repeated back-crossing. The genes were dominant, gave a high level of resistance, and were therefore easily selected in progeny plants from a cross. Unfortunately, the resistance proved to be specific to particular races of the blight pathogen, and broke down once new genotypes of the pathogen appeared (see later). However, wild *Solanum* species continue to be a useful source of genes for late blight resistance, for instance in more recent work using a genetic modification (GM) approach (see p.329). They have also provided some resistance to insect pests. For example, *Solanum berthaultii* possesses surface hairs (trichomes) which impair the activity of aphids. This not only reduces damage due to the aphid itself, but also restricts transmission of important virus diseases such as potato leaf roll virus (PLRV).

The most fruitful source of novel resistance to a disease is usually assumed to be the geographic center of origin of the pathogen, where co-evolution with the host has taken place over a long period. Thus, considerable diversity for late blight resistance has been found in wild potatoes in Central America, while for rust and powdery mildew diseases of wheat and barley, useful sources of resistance have been discovered in wild grasses in the Middle East. Provided the wild species will hybridize with the crop, such genes can be introduced into commercial genotypes by conventional crossing. An important barrier to introduction of novel resistance genes from wild relatives is interspecific incompatibility. Often, in wide crosses, the embryos produced fail to survive. Methods have been developed, however, to excise the embryos and culture them *in vitro*. This technique, called **embryo rescue**, has played a significant role in modern plant breeding by producing hybrids that could not have been obtained by traditional plant breeding.

The timespan required to introgress new genes into a commercial crop cultivar by hybridization is, however, a drawback of this process, as several rounds of back-crossing and selection are usually necessary to dilute out other, unwanted genes from the wild relative, to end up with a disease-resistant cultivar suitable for commercial use. It may take more than 10 years to complete this process, depending on the life cycle of the crop concerned. Hence, there is continued interest in biotechnological methods as a means to introduce desirable traits with greater speed and precision (Table 12.1).

The important issue of the conservation of plant genetic resources is directly relevant to resistance breeding. Germplasm collections of both cultivated species and their wild relatives should, ideally, embrace all the natural variation for resistance occurring in such species. One concern is that with most of the current emphasis being placed on elite, high-yielding genotypes of a crop species, much of the useful variation for other characters, including resistance to pests and pathogens, may be discarded. Furthermore, the continuing destruction of natural plant communities, and consequent loss of biodiversity, might also limit future options for introducing novel variation into crops.

Some further sources of novel resistance should also be considered. The first is to try to actually induce or create a new resistance within the crop using techniques designed to cause genetic changes. For instance, **mutation breeding** is based on the use of mutagenic treatments, such as irradiation or chemical mutagens, which induce changes in the DNA. This approach was in vogue for a period following World War II, but overall has produced relatively little variation of potential value in breeding programs. One difficulty is that most mutations cause a loss of function rather than a gain of a new property such as increased resistance. It is also necessary to screen very large numbers of plants to detect potentially useful mutations.

More recently, techniques of plant tissue culture have been used to extend the range of variation found in crops. If one regenerates plants from protoplasts, or callus cultures maintained *in vitro*, the regenerants often show much greater variation than equivalent populations grown from seed. Such **somaclonal variation** includes alterations in the reaction of plants to pathogens. This approach is potentially of value in extending the range of variation available to the breeder within a commercially acceptable genetic background, and is of particular interest with vegetatively propagated crops such as bananas, sugarcane, and strawberries.

Biotechnological methods have since been developed to transfer genes into plants and animals from outside their natural gene pool via **genetic transformation**, and more recently techniques for **gene editing** have become available. These technologies open up a completely new set of possibilities for engineering novel disease resistance, and will be discussed in more detail later in this chapter.

Selection of Disease Resistance

Traditionally, resistance breeding has relied on selection of genotypes with improved resistance either in the field or in mass inoculation experiments where plants are exposed to the pathogen concerned. These methods have often succeeded in identifying the most promising individuals, but they are time-consuming and relatively inefficient. It would be an advantage if one could screen for resistance in some more efficient way.

One possibility which has been explored is to select cell cultures rather than intact plants. In theory, this should permit screening of much larger numbers of individuals, and accelerate the selection process. In practice, there are difficulties, mainly because the correlation between the resistance of individual cells and whole plants is often poor. The only convincing exceptions are diseases in which symptoms are mainly or exclusively due to production of a toxin (see p.194). In these cases, insensitivity to the toxin in culture is usually correlated with insensitivity in the plant, and hence the toxin can actually be used as a chemical selection agent to identify resistant cells which are then regenerated to obtain resistant plants. Host-specific toxins have also been used to screen germplasm to identify and eliminate susceptible individuals from breeding lines. An example is tan spot disease of wheat, caused by *Pyrenophora tritici-repentis*. This fungus produces ToxA, a secreted protein toxin that causes cell death in susceptible wheat genotypes (see p.197). Sensitivity to this toxin is governed by a plant gene designated *Tsn1*. Plant breeders in Australia have successfully used purified ToxA as a tool to select insensitive wheat genotypes that lack this susceptibility gene.

A more generally applicable procedure, increasingly used by breeders, is **marker-assisted selection**. This is based on the molecular techniques now available for detecting variation at the genomic level. Rather than looking for the phenotype, that is, individuals with less disease in an infection trial, one searches for a genetic marker which segregates with this phenotype. The basic idea is simple. Two parent plants, one resistant and the other susceptible to a particular pathogen, are crossed and the progeny derived segregate for resistance. If the resistant progeny are then pooled in one group and the susceptible progeny in another, and the DNA from these two pools is compared, any differences should be linked to the presence or absence of resistance. This method relies on identifying a DNA polymorphism (i.e., a sequence difference) which is correlated with resistance or lack of resistance (i.e., susceptibility). If such a molecular marker can be found, then selection for the marker can be used instead of selection for resistance in an infection trial. The main advantage here is the speed and convenience of selection, which can be applied to large numbers of seedling plants or even seeds.

The genetic maps initially constructed using phenotypic markers have now been super-seded by maps saturated with molecular markers such as single nucleotide polymorphisms (SNPs) and microsatellites (otherwise known as simple sequence repeats – SSRs), so that the location of genes of interest can be more precisely narrowed down. This also provides an opportunity to analyze more complex traits controlled by several genes. Genetic regions on chromosomes contributing a proportion of such a trait can be identified, described as quantitative trait loci (QTLs). It is also possible to identify genetic variants in populations where molecular markers such as SNPs are strongly associated with a particular trait, such as disease resistance, an approach known as a genome-wide association study (GWAS). These methods have greatly improved the resolution of molecular breeding by increasing the probability that markers are closely linked to the gene(s) of interest. The ideal scenario is that a polymorphism can be found within the gene itself. An example is shown in Figure 12.2.

Cereal crops such as wheat and barley are infected by several soil-borne viruses causing mosaic symptoms. These viruses are spread by a microbial vector, *Polymyxa graminis*, that produces resting spores which can survive in soil for many years. Chemical methods to eradicate the vector are expensive and largely ineffective, so breeding resistant crop varieties is the only feasible control measure for these diseases. Plant breeders initially identified barley lines with resistance to the viruses and mapped the gene responsible to a specific

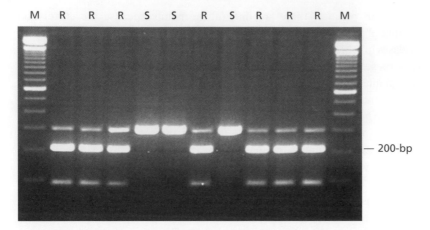

Figure 12.2   Marker-assisted selection for resistance to a soil-borne mosaic virus in barley. A DNA sequence polymorphism can be detected in cultivars possessing the rym4 (resistance to yellow mosaic) allele by PCR and digestion with an endonuclease enzyme to produce a diagnostic fragment of 200 base pairs (*arrow*). R, virus-resistant cultivar; S, susceptible cultivar. M denotes DNA size marker ladder. *Source:* Courtesy of Kostya Kanyuka.

chromosome. The gene was later found to encode a host protein required for virus replication, and sequencing the gene from various barley lines identified variant alleles that differed in their specificity for particular strains of the viruses. Such sequence polymorphisms could therefore be used by breeders to select barley varieties with effective resistance to the diseases, depending on which virus strains were present in different geographic regions.

Genomics-Assisted Plant Breeding

The first complete plant genome from the model species *Arabidopsis thaliana* was published in 2000. This was a landmark for plant genetics as for the first time, the genetic blueprint for a higher plant could be studied and analyzed. The genomes of the important crop species are larger and more complex than that of *Arabidopsis* but with advances in sequencing technology, further plant genomes have been completed, including rice in 2006, maize in 2009, and bread wheat, a complex hexaploid species, in 2017. Reference genomes are now available for many crops, providing what has been described as an "instruction manual" for breeding better plants. This has added an extra dimension to the suite of tools available for crop improvement (Figure 12.3).

Whole-genome sequencing and resequencing data provide an unlimited resource for high-throughput genotyping technologies, such as DNA chip-based arrays, to detect polymorphisms. The genetic variation for important phenotypes such as disease and stress resistance can now be more easily mapped and associated with particular genes. Sequencing additional crop varieties, land races, and wild relatives can reveal the extent of genetic diversity available within the gene pool. Wider comparisons between different crop species can reveal whether sequences are conserved (a phenomenon known as synteny), and the extent to which genomes and genes have diverged during evolution. All this information is of potential value for breeding crops that are more resilient to disease and other stresses.

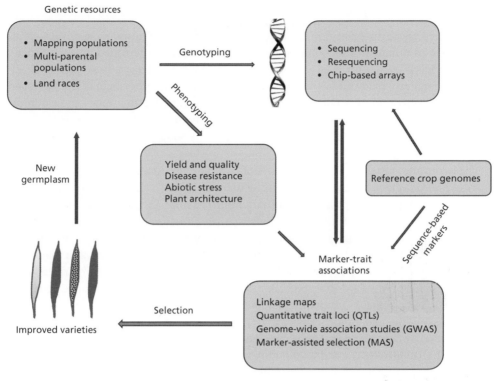

Figure 12.3 Flowchart for genomics-assisted plant breeding. *Source:* Adapted from Leng et al. (2017).

Resistance in the Field

Plant breeders are engaged in a constant drive to produce better crops showing improvements in yield, quality, or other agronomic characters including disease or stress resistance. In many respects, their efforts have been remarkably successful, leading to the so-called "Green Revolution" in which major food crops such as rice, wheat, and maize were progressively bred to give higher and higher yields. New, dwarf cultivars, that respond well to fertilizers and irrigation, mature earlier and can be cropped at high densities, replaced the older varieties. This intensification of agriculture, especially in developing countries, has raised some controversial issues, such as genetic uniformity and the loss of locally adapted crop types, but illustrates nevertheless the power of a sustained breeding program. In the UK, for instance, wheat yields rose by as much as 5% each year since 1950; about half of this increase could be attributed to improved cultivars and the remainder to improvements in the management of the crop, including disease control measures. There are increasing concerns, however, about the sustainability and environmental impact of agricultural systems based on monocultures of genetically uniform crops receiving high inputs of fertilizers and other agrochemicals, including pesticides. This has led to a renewed focus on breeding more resilient crops that can withstand biotic and abiotic stresses without resort to chemical inputs.

The Boom and Bust Cycle

As described above, breeders have succeeded in identifying and introducing novel sources of resistance to many diseases, but all too often the protection provided by such genetic resistance has proved short-lived in the field. The introduction of a new host gene for resistance was regularly followed by the appearance of a new pathotype (race) of the pathogen capable of overcoming the effects of the gene. This in turn led to a renewed search for alternative resistance genes which were again frequently countered by changes in the pathogen population. This scenario has been described as the boom and bust cycle (Figure 12.4) or the "breeder's treadmill."

The boom and bust cycle is a consequence of natural evolutionary processes. Widespread planting of a single host genotype containing one or a few resistance genes will inevitably favor pathogen genotypes possessing the matching virulence and, over time, lead to directional selection for such virulence. Yellow stripe rust, *P. striiformis*, on wheat in the UK serves as an example. During the 1980s, the winter wheat cultivar Slejpner was widely grown due to its high yield and other agronomic qualities. In 1989, there was a major yellow rust epidemic which was particularly severe on this cultivar, and crops not treated with fungicide suffered yield losses as high as 75%. Slejpner contains the resistance gene *Yr9*, and the epidemic coincided with the emergence and spread of a new race of the rust pathogen possessing the matching virulence *Yv9* (Figure 12.5). Annual surveys showed that by 1988, over 60% of the rust population sampled contained this virulence. A second race with combined virulence to *Yr9* and a second resistance gene, *Yr6*, found in another popular wheat cultivar, Hornet, subsequently emerged, so that by 1989 both Slejpner and Hornet were vulnerable to rust infection. This change was reflected in the relative disease ratings of the two cultivars, which slumped from a score of 9 (highly resistant) to 2 (highly susceptible) within 2–3 seasons (Figure 12.5).

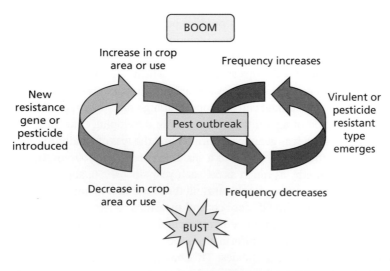

Figure 12.4 The boom and bust cycle of crop resistance. Note that a similar cycle can be shown for the introduction and use of single-site fungicides.

The capacity of pathogens such as *P. striiformis* to evolve new virulent forms is graphically demonstrated by the history of yellow stripe rust in Australia. The disease was unknown in the country until 1979, when a single race of the pathogen was detected, most likely introduced from Europe. During the following 10 years, regular surveys of pathogen isolates in Australia and New Zealand (to where the fungus is believed to have traveled by wind transport of spores) revealed the emergence of at least 15 pathogenic variants with different virulences. In most cases, each variant seemed to arise from a previous form by a single mutational event, so that a family tree of pathotypes could be constructed (Figure 12.6).

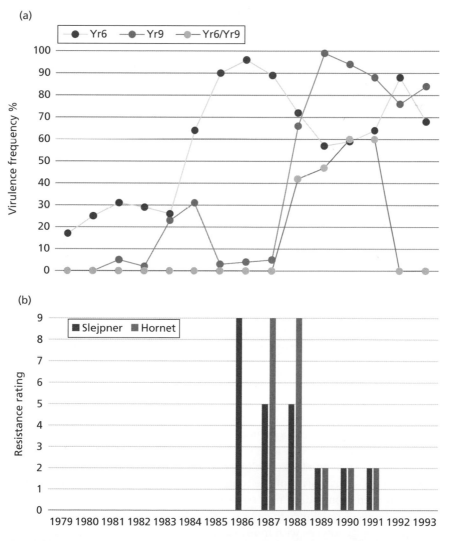

Figure 12.5 Annual changes in virulence frequencies (a) in yellow rust (*Puccinia striiformis*) populations on wheat crops in the UK, and corresponding cultivar resistance ratings (b). Increases in the frequency of virulence towards *R* genes *Yr9* and *Yr6* led to revision of the resistance ratings of wheat cultivars Slejpner (*Yr9*) and Hornet (*Yr6* and *Yr9*). *Source:* Data courtesy of the UK Cereal Pathogens Virulence Survey, National Institute of Agricultural Botany.

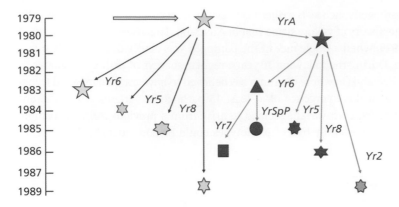

Figure 12.6 Evolution of virulence in the yellow rust (*Puccinia graminis*) population in Australia following the introduction of the pathogen in 1979. The proposed phylogenetic tree shows the year of detection of virulence to particular rust resistance (*Yr*) genes and apparent relationships between the different pathotypes. *Source:* After Wellings and McIntosh (1990).

Table 12.2 Yellow rust (*Puccinia striiformis* f.sp. *tritici*) race changes in the UK. More complex genotypes with virulence to multiple host *R* genes have emerged over time.

Year	Wheat variety	Key resistance gene combination[a]
1983	Slejpner	*Yr9*
1988	Hornet	*Yr6, Yr9*
1994	Brigadier	*Yr9, Yr17*
1996	Madrigal	*Yr6, Yr9, Yr17*
2000	Robigus	*Yr4, Yr9, Yr17, Yr32*
2008	Solstice	*Yr6, Yr9, Yr17, Yr32*
2011	Warrior	*Yr6, Yr7, Yr9, Yr17, Yr32* plus Cv, Spaldings Prolific

[a] Virulence scored in seedling tests.
Source: UK Cereal Pathogens Virulence Survey.

Such rapid evolution of novel virulence, in this case in the absence of any sexual process in the fungus, highlights the difficulties facing the plant breeder attempting to introduce lasting resistance into a crop.

More recently, the global population of yellow rust has further diversified with the emergence and migration of sexual forms of the fungus, as described earlier (see p.123). Plant breeders have continued to introduce new cultivars with combinations of major R genes, and more durable forms of resistance to counter these changes. In the UK and elsewhere, more complex pathogen genotypes with multiple virulences have arisen (Table 12.2), leading to further breakdown of previously resistant cultivars. In 2011, a new race of exotic origin (known as the Warrior race) appeared in the UK and rapidly increased in frequency, resulting in disease outbreaks on some previously resistant cultivars such as Claire. In this case, however, there was a reduction rather than complete breakdown of resistance, due to the presence of additional host genes conferring partial protection against this disease. The dynamics of the interaction has now become more complex due to the greater diversity of genotypes present in the pathogen population, and the efforts of breeders to broaden the basis of host resistance

Breaking the Boom and Bust Cycle

The phenomenon of boom and bust, typified by the rapid loss of effectiveness of a resistance gene in the field, is a legacy not only of breeding for particular types of plant resistance, but also how resistance genes have been deployed. For obvious reasons, breeders have usually selected single genes which have a major effect on disease development. Such genes are expressed in seedlings as well as adult plants, and can therefore be easily detected in progeny segregating in a breeding program. Usually, these *R* genes conform to the gene-for-gene model of host pathogen interaction, and most likely encode a pathogen recognition function. Failure of recognition, that is, susceptibility, can occur readily through loss of the corresponding avirulence (*Avr*) gene in the pathogen. Deployment of single *R* genes in monocultures on a large scale then exerts strong selection for any pathogen genotypes that can evade recognition.

Can the boom and bust cycle be delayed or broken to prevent selection for virulence? As shown in Table 12.3, there are several options based on either alternative breeding strategies or diversification of the resistance used in space and time. First, one can try to breed for types of resistance which are effective against all the genetic variants of the pathogen. Such resistance is often described as **race nonspecific** as opposed to **race specific**. An alternative and useful way of thinking about this is the concept of horizontal versus vertical resistance. Figure 12.7 illustrates diagrammatically the interaction of a host plant containing a single *R* gene with seven races of a pathogen designated by the *Avr* genes they contain. The host gene (*R1*) confers a high degree of resistance to races containing the corresponding *Avr* gene, but very little to others (Figure 12.7a). Such vertical resistance can be contrasted with the situation in Figure 12.7b where a lower degree of protection is expressed more or less equally against all the races. A variety of terms have been used to describe these two types of resistance, summarized in Table 12.4.

Table 12.3 Strategies to counter the boom and bust cycle

---

1) **Breeding approaches**
   - Breed for polygenic resistance
   - Pyramid *R* genes/gene stacking
   - Novel resistances from genetic modification or gene editing
2) **Diversify deployment of resistance**
   - Multiline cultivars
   - Variety mixtures
   - Spatial diversity between regions
   - Spatial diversity between fields
   - Temporal diversity between seasons
   - Intercropping

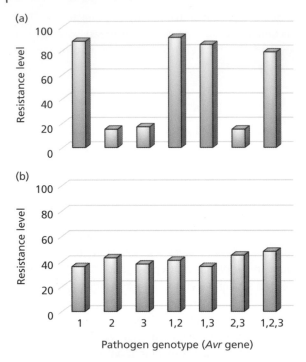

Figure 12.7 Vertical and horizontal resistance compared. A cultivar with a single gene (*R1*) for vertical resistance (a) shows a high level of resistance to pathogen genotypes containing the corresponding avirulence gene (*Avr1*) but little resistance to other genotypes, while a cultivar with horizontal resistance (b) shows an intermediate level of resistance to all pathogen races. *Source:* After Vanderplank (1963).

Table 12.4 Race-specific and race-nonspecific resistance compared

	Race specific	Race nonspecific
Genetic control	One or a few genes involved	Several or many genes involved
Description	Generally clear-cut, conferring a high level of resistance in seedlings and adult plants	Variable; seedlings less resistant, increases as plants mature
Mechanism	Usually a hypersensitive host reaction	Reduced rate and degree of infection, development and/or reproduction of pathogen
Efficiency	Highly effective toward specific pathogen races	Variable, but effective against all pathogen races
Stability	Liable to break down due to emergence of new pathogen races	Unaffected by changes in virulence of the pathogen
Commonly used approximate synonyms	Monogenic Major gene Vertical Qualitative	Polygenic Minor gene Horizontal Quantitative Adult plant Rate reducing

Source: Adapted from Hayes and Johnson (1971).

It is generally assumed that race-specific resistance is encoded by one or very few genes, whereas race nonspecific resistance is a more complex genetic trait based on multiple genes, hence the alternative terms monogenic versus polygenic. In many cases, however, the genetic basis of a particular phenotype, for instance adult plant or rate-reducing resistance, is not clearly defined, and hence these various types of resistance may not be directly comparable. Selecting for nonspecific resistance is a more demanding task for plant breeders, due to the partial effects of each gene, but techniques such as QTL mapping and genome-wide association studies can aid identification of genetic regions contributing to such resistance and these may be combined to produce agronomically useful levels of protection against disease.

In practical terms, what matters is not so much the nature or even the genetic basis of resistance, but whether it will remain effective over a long period. Hence the concept of **durable resistance**, defined simply as resistance which continues to provide control even after an extended period of exposure to the pathogen.

It is often assumed that resistance based on single genes of major effect is doomed to break down if deployed on a commercial scale over several seasons. In fact, there are examples of single genes which have remained effective year after year even when widely used. The *mlo* gene for resistance to barley powdery mildew, *Erysiphe graminis* f.sp. *hordei*, was introduced in 1979 but still gives satisfactory control of the disease throughout Europe. Interestingly, this gene was first discovered in mutant barley lines produced by exposure to X-rays, but was later detected occurring naturally in a few barley samples from Ethiopia. The gene *Sr26* transferred into wheat from a wild relative has provided durable resistance to stem rust, *Puccinia graminis*, in Australia since its introduction in 1969, and is also effective against new threatening races such as Ug99 currently spreading from Africa into the Middle East and Asia. Similarly, the *Lr34* gene in wheat for resistance to leaf rust, *Puccinia triticina*, first identified in South American cultivars, has remained effective for many years in several different countries.

Why certain genes prove to be durable in practice is not fully understood. One idea is that such genes are associated with resistance mechanisms which are different from the hypersensitive response activated by typical *R* genes, and cannot easily be overcome by changes in the pathogen. The *Mlo* gene, which encodes a transmembrane protein, triggers formation of a cell wall barrier, or papilla (see p.216), at sites of attempted penetration by the fungus. This mechanism appears to be effective against all genetic variants of the pathogen. The *Lr34* gene reduces infection and delays development of the pathogen without any associated host cell death. It is mainly effective in adult plants and also confers partial resistance to another rust (*P. striiformis*) as well as powdery mildew. *Lr34* encodes a transporter protein and while the resistance mechanism is not yet known, related genes have been implicated in the movement of toxic compounds into the apoplast.

Due to the recurrent problems with vertical (usually monogenic) resistance, the emphasis in resistance breeding in many crops shifted toward polygenic sources of resistance. In potatoes, for example, breeders started to evaluate potato lines which, while not fully resistant to infection, nevertheless withstand severe blight epidemics. The pathogen is able to infect the plant, and even to cause disease lesions, but the time from initial infection to the production of further spores is extended and pathogen reproduction is decreased. Such resistance is often described as **rate reducing** as it serves to delay epidemic development.

The genetic basis of this resistance is not fully understood, but it most likely involves several genes. Commercial exploitation of such lines has to date been limited as the resistance is often associated with undesirable characteristics, such as late maturity of the crop.

Effector-Guided Resistance Breeding

Ultimately, the durability or otherwise of single resistance genes will depend upon the frequency of mutations to virulence in the pathogen, and whether such mutations can survive and spread in the pathogen population. It is possible that mutations countering host resistance genes might in certain cases incur a fitness penalty in the pathogen. Such mutants would therefore fail to survive in competition with the remainder of the population. Evidence supporting this idea came initially from studies on bacterial pathogens. Resistance genes such as *Xa-21* from rice and *Bs2* from pepper are effective against all known races of the corresponding pathogens *Xanthomonas oryzae* and *X. campestris* pv, *vesicatoria* (Xcv); every race apparently contains a copy of the matching *Avr* gene. In the case of Xcv, loss of the gene concerned, *avrBs2*, leads to a severe reduction in the vigor of the bacterium. The gene product therefore appears to be essential for survival of the pathogen.

The discovery that bacterial, fungal, and oomycete pathogens produce a range of effector molecules that target components of the plant immune system (see Chapter 10), and that these effectors are themselves recognized by plant proteins encoded by R genes, has created new opportunities for breeding and deploying more durable forms of resistance. First, such effectors can be used as molecular probes to identify host targets and genes encoding recognition functions. Some of these may be novel sources of resistance that are of value to plant breeders. Second, analyzing the occurrence and distribution of genes encoding effectors in the pathogen population should help to predict the relative durability of the matching R gene. If, for instance, a particular avirulence (*Avr*) gene encoding a specific effector is present in all the isolates sampled, then the corresponding R gene is likely to be effective against the whole pathogen population. Conversely, if a specific avirulence is absent, or nonfunctional in some isolates, there is a risk that such pathogen genotypes will be selected and the resistance will eventually break down.

Effectors are chemically and functionally diverse but most are small secreted proteins, and some possess common features, such as the RXLR motif found in effectors from oomycete pathogens including *Phytophthora* and downy mildew species. This has enabled discovery and compilation of effector inventories and subsequent screening of pathogen populations to identify which effectors are present, and especially the "core" complement found in all isolates. The appropriate matching R genes can then be deployed, either singly or ideally in different combinations (see Figure 12.8).

Diversification of Resistance

Directional selection for virulence in the pathogen population can be reduced by changing the ways in which R genes are deployed (Table 12.3). The main aim is to achieve greater genetic diversity and thereby create a more complex genetic puzzle for the pathogen to overcome. Rather than the pathogen dictating events, directional selection for virulence is reduced and the effectiveness of certain *R* genes preserved. Various schemes have been proposed, summarized diagrammatically in Figure 12.8. First, one can combine several different major *R* genes in a single host genotype, a process known as **gene pyramiding** or stacking. There is a risk that more complex pathogen strains with multiple virulences (so-called super-races) might be selected, but this will depend on the fitness of the complex race, and the number and nature of the genes in the mix. As shown in Table 12.2, there is evidence for the gradual accumulation of virulence in some pathogen genotypes but how complex can it become before fitness is compromised?

Alternatively, one can introduce major *R* genes into a host genotype with partial resistance conferred by minor genes, the theory being that such genes might provide a degree of protection for the major gene and enhance its durability. Experimental support for this hypothesis is scarce but has been obtained for a few pathosystems, for instance stem canker of oilseed rape (canola) caused by the fungus *Leptosphaeria maculans*. Several major genes (designated *Rlm*) have been used in breeding resistance to this damaging disease, but these have often broken down in the field. One of these genes, *Rlm6*, was introduced into a cultivar with a fully susceptible genetic background, and another with a degree of quantitative resistance conferred by several minor genes. The cultivars were then deployed in a five-year field trial exposed to natural populations of the pathogen. The *Rlm6* resistance in the fully susceptible background proved ineffective after three seasons but when combined with the quantitative resistance, it was still effective after five years of selection. This confirmed that the presence of additional quantitative resistance due to minor genes can increase the durability of a major *R* gene.

Greater diversity can be created in space and time by growing mixtures of host genotypes containing different *R* genes, or varying the *R* genotypes grown in different fields or across regions, or alternating them between seasons. A multiline is a series of near-isogenic

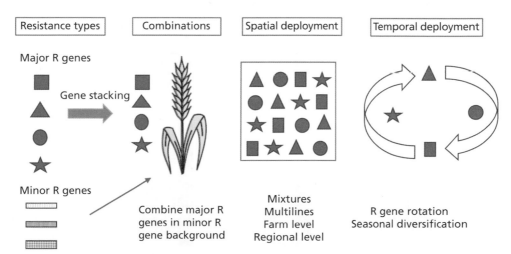

Figure 12.8 Options for the use and deployment of *R* genes in crops. Major genes can be combined in a single host variety (gene stacking), together with minor genes that may contribute to durability. Alternatively, genes can be deployed in variety mixtures, each component containing a different *R* gene, or diversified spatially between fields, farms or regions. Temporal diversification involves *R* gene rotation in successive crops or seasons. *Source:* Lucas (2017b).

(i.e., genetically identical) plant lines which differ in a single character, such as disease resistance. These lines can be grown together like a conventional crop and retain the agronomic advantages of a normal cultivar, such as uniformity, while confronting the pathogen with a mosaic of *R* genes. Breeding multilines by conventional crossing is a time-consuming task, but GM technology provides the option of introducing DNA cassettes with different combinations of genes, and also of varying the *R* genes used between seasons. The concept of cultivar mixtures is similar to multilines but easier to put into practice. In this case, one simply mixes together seed of several genetically distinct cultivars and grows them as a single crop. Each cultivar in the mixture contains one or more different *R* genes. Trials have shown that such mixtures often develop less disease than would be expected from levels of infection observed on the component cultivars grown alone (Figure 12.9).

Several mechanisms might account for this reduction in disease, including less efficient spread from plant to plant, and possible induction of host resistance by incompatible pathotypes attempting to infect the different hosts. Once again, one can vary the mixture from season to season and hence keep one step ahead of the pathogen. Cultivar mixtures have been used in cereal crops in Europe but there has been some reluctance on the part of growers to adopt them. The root of the problem is that many markets demand consistent quality, and variation in certain properties is not tolerated. Barley, for instance, is used in malting for beer production, and stringent standards are applied to the grain to be processed. Variety mixtures will usually vary in maturity date and hence quality at harvest, and meeting these standards is a challenge. Currently, the prospects for mixtures are more promising in alternative crops grown for nonfood uses, such as biomass. One option for reducing rust infection in short-rotation coppice species such as willows (see Chapter 1, Figure 1.4) is to grow mixtures that differ in their susceptibility to the prevailing pathogen races.

Diversification of resistance between farms and regions requires a high degree of coordination among farmers and advisors. In the USA, for instance, the normal annual pattern of

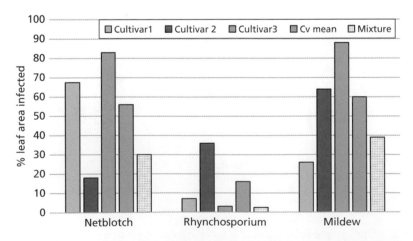

Figure 12.9 Amounts of foliar infection by three pathogens of barley on three cultivars grown separately or in a mixture. In each case, the level of infection in the mixture is lower than the expected mean based on scores for individual cultivars. *Source:* After Wolfe (1993), with kind permission from Kluwer Academic Publishers.

rust epidemics on cereals is for the disease to spread from the south to the north during the season. Early outbreaks arise from inoculum overwintering in the warmer southern states or Mexico, and spores are then carried northwards to infect later-maturing crops. In this case, it is possible to plant cultivars containing different *R* genes along the disease pathway. Selection for virulence on a particular host genotype does not then confer an advantage once the pathogen spreads to the next geographic zone.

All these strategies, one way or another, represent a move away from monoculture, and begin to recreate some of the diversity found in natural plant communities. The ultimate version of such diversification is intercropping, in which different crop species are grown together in a relatively small area. While such genetic mosaics may be desirable for limiting disease, they are not compatible with many of the labor-saving techniques of industrialized agriculture, and require more intensive management, as practiced by many smallholder farmers.

Biotechnology and Breeding for Disease Resistance

The advent of techniques for the stable genetic transformation of plants created apparently limitless opportunities for engineering novel pest- and disease-resistant crops. It is now possible to introduce genes from unrelated plants, or more exotic sources such as microorganisms or even animals. In theory, the biological barrier of genetic incompatibility no longer applied. In practice, there are constraints on the technology, both biological, such as the stability and efficacy of the transgenic trait, and political, due to the regulatory framework for GM crops imposed in many countries. Nonetheless, biotechnology has added a powerful new tool to the breeders' toolbox.

The basic principle of genetic engineering is straightforward. A piece of foreign DNA, usually described as a construct, is introduced into the plant cell and becomes integrated into the plant genome. There are several ways of getting foreign DNA into plant cells, including uptake into isolated protoplasts and biolistic methods where the DNA is coated on small particles which are then shot into a suitable tissue. The method of choice for the majority of crop species is, however, to utilize the natural capability of *Agrobacterium* to introduce the Ti plasmid into plant cells (see p.205). The novel gene is engineered into a modified plasmid which is then transferred by infection with the bacterium. Ideally, the introduced gene or genes will be stably integrated and expressed so that the desired resistance trait is present in the transformed plants. First-generation transgenic crops were usually transformed with a single gene fused to a constitutive promoter so that the gene product is produced all the time, which can incur a biosynthetic cost to the plant. Further tricks are required if one aims to express the transgene in a particular part of the plant or at a particular time, for instance when the plant is under attack by an invading pest or pathogen. This can be achieved by using a tissue-specific or inducible promoter sequence so that the gene is only switched on when necessary.

Table 12.5 lists some examples of the use of GM to increase the resistance of plants to pathogens. The first experiments were done with plants such as tobacco which is relatively easy to transform and regenerate from tissue culture. Several genes with antifungal activity were tested, including the enzymes chitinase and glucanase, which attack fungal cell walls, and small inhibitory proteins such as the defensins, initially isolated from seeds. The transgenic

Table 12.5 Some examples of genetically engineered plants with enhanced resistance to pathogens

Transgene	Plant transformed	Target pathogen(s)
Chitinase	Tobacco	*Rhizoctonia solani* (F)
Antifungal protein (defensin)	Tobacco	*Alternaria longipes* (F)
Chitinase + glucanase	Tobacco	*Cercospora nicotianae* (F)
Stilbene synthase (phytoalexin synthesis)	Tobacco	*Botrytis cinerea* (F)
NPR1 (regulator of systemic acquired resistance)	*Arabidopsis*	*Pseudomonas syringae* (B)
		Hyaloperonospora parasitica (O)
WRK45 (transcription factor)	Rice	*Magnaporthe oryzae* (F)
		Xanthomonas oryzae (B)
Pattern recognition receptor (*EFR*) from *Arabidopsis*	Tobacco and tomato	Several bacteria (*Ralstonia, Xanthomonas, Agrobacterium*)
Immune receptor *Ve1* from tomato	Tobacco and cotton	*Verticillium* spp. (F)
R gene *Bs2* from pepper	Tomato	*Xanthomonas* spp. (B)
R gene *Rpi-vnt1.1* from *Solanum venturii*	Potato	*Phytophthora infestans* (O)
Putative R gene (*RGA2*) from wild banana subspecies	Cultivated banana (Cavendish)	*Fusarium oxysporum* f.sp. *cubense* race 4 (F)
R gene pair (*RPS4* and *RRS1*) from *Arabidopsis*	*Brassica rapa* and *B. napus*	*Colletotrichum higginsianum, C. orbiculare* (F)
	Tomato, tobacco, and cucumber	*Ralstonia* and *Pseudomonas* (B)
Virus coat protein	Papaya	Papaya ringspot virus (PRSV)
Virus coat protein	Cassava	Cassava brown streak virus (CBSV) and cassava mosaic viruses
Virus replicase gene (*AC1*)	Common bean	Bean golden mosaic virus (BGMV)

B, bacterium; F, fungus; O, oomycete.

plants were generally less susceptible to fungal pathogens, but higher levels of resistance were obtained when more than one gene was introduced. A typical experiment is shown in Figure 12.10. Initially tobacco was transformed with separate constructs for chitinase and glucanase, and transformants expressing one or other enzyme were then crossed to give hybrids which contained both novel genes. These hybrids expressing both enzymes were significantly more resistant to the leaf spot pathogen *Cercospora nicotianae* than plants with only one transgene. An interesting exception to introduction of a gene with a directly inhibitory product is the case of stilbene synthase (Table 12.5). This enzyme catalyzes a key step in the synthesis of the phytoalexin resveratrol, which is found in plants such as grapevine but not in tobacco. However, the chemical precursors of resveratrol are present in tobacco, so if the gene coding for stilbene synthase is engineered into and constitutively expressed in tobacco, the phytoalexin is produced. When this was done, the transgenic plants were found to be more resistant to colonization by the fungus *Botrytis cinerea*. This is an example of metabolic engineering, in which the biosynthetic capacities of the plant are altered by diverting a pathway or otherwise changing its regulation.

Following these pioneering studies, other targets for GM have been explored, in particular genes involved in pathogen recognition or components of the plant innate immune response. As already discussed in Chapter 9, NPR1 is a key regulator of systemic acquired resistance. The *NPR1* gene was cloned from *Arabidopsis* and was then overexpressed in transgenic plants. The transformants showed elevated resistance to both a bacterial pathogen, *Pseudomonas syringae*, and the oomycete *Hyaloperonospora parasitica*. A similar approach was taken in rice plants with overexpression of a transcription factor, WRK45, that responds to the plant defense activators benzothiadiazole and probenazole. The transgenic plants were more resistant to two major rice pathogens, the rice blast fungus *Magnaporthe oryzae* and the bacterial blight bacterium *Xanthomonas oryzae*. These

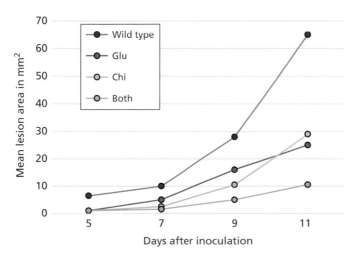

Figure 12.10 Rate of disease development of frog-eye of tobacco, a leaf spot disease caused by *Cercospora nicotianae*, in transgenic tobacco plants. The data show the mean lesion area in wild-type plants, compared with plants expressing either chitinase or glucanase, or both enzymes. *Source:* After Zhu et al. (1994).

cases show that genetic manipulation of genes regulating plant defense can result in broad-spectrum resistance to diverse pathogens.

Pattern recognition receptors upstream of the response pathway are also potential targets for intervention. The leucine-rich receptor kinase EFR recognizes a bacterial pathogen-associated molecular pattern (PAMP) known as elongation factor Tu (EF-Tu). The EFR gene from *Arabidopsis* was transferred to tobacco and tomato, two plants in a different family from the donor. The transgenic plants responded to several bacterial pathogens from different genera, demonstrating that pattern recognition receptors from one plant family can also function in unrelated host plants. In tomato, the immune receptor Ve1 confers race-specific resistance to the wilt pathogen *Verticillium* through recognition of a fungal effector molecule. When transferred to the closely related species tobacco and distantly related species cotton, the gene retained its ability to recognize *Verticillium* strains, suggesting that the defense signaling pathway mediated by theVe1 receptor is conserved between different plant species. There may therefore be potential for plant breeders to exploit such interactions in extending wilt resistance to further crops via a biotechnological approach.

Transgenic Resistance in Plant Breeding

All the examples discussed above could be described as "proof-of-concept" experiments demonstrating the feasibility of altering the disease reaction type of plants by introducing genes of various kinds by GM methods. As far as is known, they have not yet led to field trials or the development of commercial transgenic cultivars for use by farmers. To date, there are relatively few examples of GM crops resistant to pathogens in agriculture or horticulture, unlike the scenario with insect pests where crops expressing the *Bacillus thuringiensis* (Bt) toxin are now grown in many countries on an area recently estimated at 23 million hectares, including major commodities such as maize, soybean, cotton, and potatoes. This situation, however, may be about to change.

The resistance genes introduced into crop cultivars by conventional crossing and selection can also be manipulated by GM methods. The main advantage is that the gene or genes can be directly transferred into an elite cultivar without the need for laborious back-crossing. The phenomenon of boom and bust had discouraged the use of major *R* genes, but new insights into the genetics of virulence in pathogen populations suggested such genes might be recycled and deployed in more durable ways. The *Bs2* gene for resistance to bacterial spot disease caused by *Xanthomonas* species (Table 12.5) was originally identified in pepper and subsequently transferred to the related species tomato. Transgenic tomato lines expressing the *Bs2* gene were then grown in replicated field trials in Florida over three seasons. These lines had consistently lower levels of bacterial spot disease than standard commercial varieties, and also produced competitive yields. The *Bs2* gene is now being transferred into parental tomato lines and hybrids for use in breeding programs. Interestingly, the gene was initially selected for use on the basis that the corresponding pathogen effector that it recognizes is present in all six currently known races of the bacterial spot complex. Nonetheless, the recommendation for breeders is for the gene to be combined with additional sources of resistance to guard against possible future changes in the pathogen population.

The development of methods for the genetic transformation of major crops such as the cereals, brassicas, and potatoes has given plant scientists the opportunity to revisit wild relatives as sources of disease resistance. Research in the UK, Netherlands, and Belgium has been exploring this approach for controlling potato late blight (*P. infestans*). The blight-susceptible potato variety Desiree was transformed with *Resistance to P. infestans* (*Rpi*) genes obtained from the wild *Solanum* species *S. stoloniferum*, *S. venturii*, and *S. bulbocastanum*. In the UK trial, held over three seasons, disease development and crop yields were compared in plots exposed to natural *P. infestans* populations. In the final season, conditions were favorable for blight infection and disease severity on the Desiree control plots reached around 100%, while transgenic lines containing the *Rpi-vnt1-1* gene from *S. venturii* remained fully resistant to blight. The comparison of plot tuber yields showed a reduction of between 50% and 75% in the nontransgenic plots (Figure 12.11b).

Trials in Belgium compared blight severity on commercial varieties with transgenic lines containing single *Rpi* genes, or a combination of three stacked genes from the wild species listed above. Figure 12.11a shows disease progress curves for the 2012 season. The transgenic lines all delayed disease development, while those containing the *Rpi-vnt1-1* gene, either alone or stacked with other *Rpi* genes, provided the best control. Considering results from two trial seasons in both the Netherlands and Belgium, lines containing the three transgenes generally provided higher levels of resistance than those with single genes.

Overall, these promising results suggest that *R* genes from wild relatives are functional in transgenic lines. Furthermore, they can be combined to give durable control of late blight based on knowledge of the pathogen races predominating in a particular location or season.

Panama disease, caused by the vascular wilt fungus *Fusarium oxysporum* f.sp. *cubensis* (Foc), has been a major scourge of bananas since the first half of the twentieth century, when the disease devastated production of the main export cultivar Gros Michel in South and Central America. This cultivar was subsequently replaced by Cavendish, which is resistant to race 1 of Foc, the strain responsible for the initial epidemic. Since then, Cavendish has become the most important cultivar in global banana production and dominates the export market. In the 1990s, a new strain of Foc, virulent on Cavendish, emerged in South-East Asia and is now spreading internationally. The new strain, known as tropical race 4 (TR4), now poses a significant threat to commercial banana production worldwide. Conventional breeding of bananas by hybridization is complex as the commercial varieties are triploid.

Some variation in resistance to Foc TR4 has been found in plant populations regenerated from tissue culture, but insufficient to provide good protection against the disease. Researchers have therefore explored the possibility of introducing resistance using GM approaches. Two transgenes, one of them a putative NLR type resistance gene identified in a wild banana subspecies, were introduced into a Cavendish genotype. The transgenic bananas were then assessed in a field trial in the Northern Territory of Australia, where the TR4 race is now endemic. By the end of the three-year experiment, more than 60% of the susceptible control banana cultivars were either dead or showing disease symptoms, while 30% or fewer of plants from the 10 transgenic lines tested developed symptoms. Two lines remained completely disease free throughout the trial. Further field trials are now under way to identify transgenic lines suitable for progression to commercial release, while several *R* gene homologs found in Cavendish might also be good candidates for gene editing (see later in this chapter).

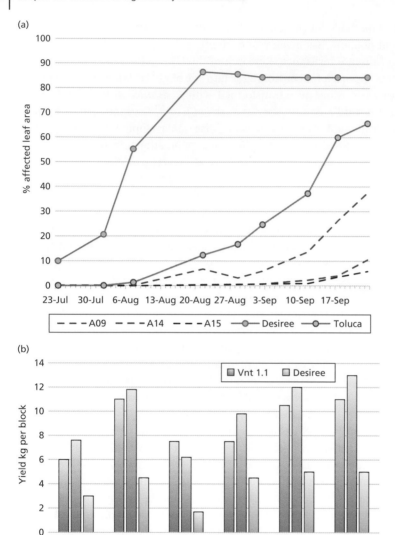

Figure 12.11   Control of potato late blight (*Phytophthora infestans*) in field trials with transgenic potato lines in Belgium (a) and the UK (b). (a) Disease progress curves on two nontransgenic cultivars, Desiree (susceptible) and Toluca (moderate resistance), and Desiree lines transformed with the *sto1* gene from *Solanum stoloniferum* (A09), the *vnt 1.1* gene from *S. venturii* (A15), and three stacked genes, *sto1*, *vnt 1.1*, and *blb3*, from *S. bulbocastanum* (A14). Plots were artificially inoculated with a mixture of *P. infestans* isolates as well as being exposed to natural inoculum. (b) Yield data from six plots comparing nontransgenic Desiree with a transgenic Desiree line containing the *vnt 1.1* gene. The plots were exposed to natural infection during the 2012 season when disease pressure was high due to unusually wet weather. *Source:* (a) adapted from Haesaert et al. (2015), data kindly provided by Geert Haesaert, Ghent University; (b) from Jones et al. (2014).

The examples above show that typical *R* genes can be successfully introduced into crops from wild relatives by GM methods, but can they be transferred and function effectively in unrelated plant families and species? This question has been partly answered by experiments in which two NB-LRR type *R* genes from *Arabidopsis* (*RPS4/RPS1*) that act together against several pathogens were introduced into two crop species from the same family as the donor (the Brassicaceae), and also into tobacco and tomato, from the unrelated family Solanaceae (Table 12.5). The two genes functioned correctly in *Brassica rapa* and *B. napus*, conferring resistance to a fungal pathogen, but also reduced infection by two unrelated bacterial pathogens in tomato, as well as recognizing the appropriate bacterial effectors in transgenic tobacco. In addition, the two genes controlled another fungus in transgenic cucumbers, from a further plant family. The hope now is that *R* gene-based immunity can be transferred directly to distantly related species using GM as a new strategy to control multiple diseases.

Transgenic Resistance to Viruses

While recent advances have raised hopes that GM crops resistant to fungal and bacterial pathogens will become available in the near future, transgenic crops with effective resistance to viruses have been in commercial use for at least 20 years. It has been known for a long time that infection of a plant with one virus can interfere with infection by a second, related virus, a phenomenon known as "cross-protection". The presence of the first virus somehow inhibits the replication cycle of the second. Starting from this observation, it was reasoned that expression of foreign viral genes in plant cells might have the same effect.

The first practical demonstration of the feasibility of this approach was in 1986, when it was shown that transgenic tobacco plants expressing the coat protein gene from tobacco mosaic virus (TMV) were resistant to infection by the virus. Since then, similar results have been obtained with a whole series of other plant viruses, including cucumber mosaic, potato viruses X and Y, and cassava brown streak and mosaic viruses. The term pathogen-derived resistance (PDR) was proposed to describe resistance based on genes taken from the virus. While most cases have utilized coat protein genes, it has been shown that other parts of the virus genome can also give effective protection, for instance the genes encoding the replicase function or movement protein. The mechanism of virus cross-protection was initially unknown, but it then transpired that it operates via posttranscriptional gene silencing, otherwise known as RNA interference (RNAi), a natural defense pathway targeting and degrading foreign nucleic acids.

Resistance based on expression of viral transgenes has proved highly effective in the control of some damaging virus diseases in the field. One of the best-known examples concerns papaya ringspot virus (PRSV) which was discovered infecting crops in Hawaii in 1992. Over the next few years, the disease became widespread, leading to many papaya plantations being abandoned. Production declined and the industry in Hawaii faced a potential economic disaster. Fortunately, research had already started on papaya transformed with a coat protein gene from PRSV, including testing in an initial field trial. Transgenic cultivars were released for commercial use in 1998 and gave highly effective resistance, even in conditions of high disease pressure (Figure 12.12). Subsequently, papaya production in PRSV-infected regions of Hawaii recovered. Some questions remain, however, over the longer-term durability of the resistance due to variation in virus populations, and take-up of GM papaya has been low in other countries mainly due to regulatory issues and concerns about GM produce in some global markets.

Figure 12.12 Aerial view of field trial of transgenic papaya resistant to papaya ringspot virus in Hawaii. The transgenic plants in the center of the plot are surrounded by nontransgenic susceptible plants severely infected with the virus. *Source:* Gonsalves et al. (2004).

Cassava production in large areas of Africa has been severely affected by epidemics of two virus diseases: cassava mosaic disease (CMD) and cassava brown streak disease (CBSD). These diseases have proved difficult to manage by selection of naturally resistant genotypes, due to the occurrence of a complex of viruses and the lack of good sources of resistance to CBSD. Transgenic resistance to both viruses has been developed and is now being evaluated in the field in Kenya and elsewhere. One advantage of the GM approach is that resistance can be introduced into the cassava varieties preferred by farmers. It is also possible to introduce transgenes into genotypes with some natural resistance. For instance, a cassava variety selected for resistance to CMD was transformed with a sequence of the coat protein gene from the cassava brown streak virus, giving good protection against both diseases.

One of the criticisms of GM technology is that it is owned by large corporations which control its availability. Some of the work on cassava described above was funded through charities with the aim of helping subsistence farmers to manage diseases that threaten their food security. Another example is transgenic control of the bean golden mosaic virus (BGMV) infecting common beans in Latin America. No bean genotypes have adequate resistance to the virus and control has relied on use of insecticides to control the whitefly vector. In public sector research funded by the Brazilian agricultural agency Embrapa, a construct containing part of the AC1 gene involved in virus replication was introduced into a number of common bean lines. One showed high levels of resistance even when large numbers of the virus-carrying vector were present. After further development, the transgenic beans were approved by the Brazilian biosafety authority in 2011, and seed was distributed to smallholder family farmers.

Host-Induced Gene Silencing (HIGS)

In another recent breakthrough, it has been shown that small RNA molecules produced in plant cells can be taken up by invading pests and pathogens, leading to silencing of any genes homologous with the RNA sequence. The concept of host-induced and spray-induced gene silencing (HIGS and SIGS) has already been introduced in Chapter 11, as potential novel routes for delivery of molecules to inhibit invading pathogens.

RNA silencing is typically initiated by introduction of long double-stranded RNAs (dsRNA) into the cell (see Chapter 11, Figure 11.4). These RNA molecules form a hairpin structure that is then cut into small interfering RNA molecules (siRNA). During HIGS, these are taken up by cells of the invading organism, for instance fungal hyphae, where they are recognized as foreign RNA and bound by proteins into an RNA-induced silencing complex (RISC). Any fungal messenger RNA containing a complementary sequence to the introduced RNA is then also bound and degraded, leading to silencing of the endogenous gene. A proposed alternative pathway is for the dsRNA itself to transfer into the invading cell where it is complexed by its own endogenous silencing system.

One of the first studies demonstrating the feasibility of HIGS was with transgenic cereals engineered to express dsRNA targeting a virulence factor in the powdery mildew fungus *Blumeria graminis*, leading to reduced mildew infection. Subsequently, HIGS has been shown to operate in a series of plant pathogens, including those causing difficult-to-control plant diseases such as the vascular wilts and head blight of cereals. For instance, with *Fusarium graminearum*, dsRNA complementary to genes encoding an enzyme required for sterol biosynthesis (CYP51) was expressed in transgenic barley. When these plants were inoculated with the fungus, growth was restricted to the initial infection sites. One interesting aspect of this study is that the CYP51 enzyme is also the target of the well-known azole class of fungicides, described in the previous chapter. In fact, *Fusarium* contains three genes encoding CYP51 enzymes, and all three were effectively silenced. In a parallel study, an enzyme involved in chitin synthesis, an essential component of fungal cell walls, was targeted, again conferring high levels of resistance to head blight. HIGS has also been shown to function against oomycete pathogens such as *P. infestans* and downy mildew of lettuce, *Bremia lactucae*.

In summary, HIGS is now emerging as a promising new strategy for boosting plant resistance against pathogens by silencing genes that are essential for growth or pathogenicity. However, there may be some limitations to its application. As HIGS usually involves delivery via a transgenic route, it is regarded as a further type of GM technology, so will be subject to the same regulatory hurdles. Some crop species remain difficult to transform. There may also be potential problems with silencing of nontarget genes or effects on beneficial relationships such as mycorrhizas. Nonetheless, once fully evaluated, it should provide an additional option for the management of currently intractable plant diseases.

Future Prospects for Crop Resistance to Pathogens

Recent advances in genetics and genomics have provided new technologies for gene discovery and use in breeding programs, as well as insights into pathogen populations that should inform decisions on how resistance genes are deployed in the field. At the same time, improved understanding of the recognition and response systems involved in plant defense is suggesting new ways of manipulating these to combat plant disease. There is now the real prospect of engineering receptor proteins and *R* genes to create novel specificities or extend the spectrum of pathogens controlled. The latest promising technology is gene editing in which a DNA sequence can be inserted, deleted, modified, or replaced in a specific location in the genome using nuclease enzymes as "molecular scissors." There are several ways of doing this but the best known are transcription activator-like effector nucleases (TALENs), based on natural bacterial proteins with DNA-binding domains fused to a cutting enzyme, and clustered, regularly interspaced short palindromic repeats (CRISPRs) that originate from virus sequences that insert into genomes, together with a nuclease enzyme, Cas9. The CRISPR-Cas9 system can be programmed using a guide RNA to target any DNA sequence for cleavage. There are many potential applications for this technology, including engineering different components of the plant immune system, as well as altering genes conferring susceptibility to necrotrophic pathogens.

While the technological options for crop improvement continually advance, there are constraints on the use of biotechnology, ranging from societal mistrust of GM to issues of cost and availability that may limit adoption in developing countries. In Europe, in particular, public skepticism together with an adverse regulatory environment has so far prevented widespread adoption of GM crops. It is possible that gene editing might be viewed differently as the engineered gene(s) are already present in the plant genome. From a crop protection perspective, it is hoped that these technologies can be applied by plant breeders as they promise to diversify the types and sources of resistance available, as well as accelerating our ability to counter evolutionary changes in pathogen populations. This can make an important contribution to the current quest for global food security.

The rapid progress taking place in genome sequencing and bioinformatics will also impact on conventional plant breeding by providing detailed catalogs of *R* genes in crops and their relatives, as well as improved markers for tracking genes in breeding lines. Corresponding analysis of field populations of pathogens will identify effector repertoires and aid targeted deployment of resistant crop varieties. This approach is likely to be part of the wider agenda to create more diverse and resilient agricultural systems.

Conclusion

The ultimate goal of plant breeders is to deliver durable disease control via the seed, without the need for crop protection chemicals or other inputs. How realistic is this goal? Irrespective of the route by which resistance genes are introduced or engineered into a crop, once deployed in the field they will be subject to the same evolutionary pressures as before. It is prudent, therefore, to integrate cultivar resistance with a range of other disease control measures to ensure that it remains effective for as long as possible. The remaining chapters focus on such alternative cultural and biological approaches and their combined use.

Further Reading

Books

Collinge, D.B. (ed.) (2016). *Plant Pathogen Resistance Biotechnology*. Hoboken, New Jersey: Wiley.

Reviews and Papers

Andolfo, G., Iovieno, P., Frusciante, L. et al. (2016). Genome-editing technologies for enhancing plant disease resistance. *Frontiers in Plant Science* 7: 1813. https://doi.org/10.3389/fpls.2016.01813.

Boyd, L.A., Ridout, C., O'Sullivan, D.M. et al. (2013). Plant-pathogen interactions: disease resistance in modern agriculture. *Trends in Genetics* 29: 233–240.

Cao, H., Li, X., and Dong, X.N. (1998). Generation of broad-spectrum disease resistance by overexpression of an essential regulatory gene in systemic acquired resistance. *Proceedings of the National Academy of Sciences United States of America* 95: 6531–6536.

Collinge, D.B., Lund, O.S., and Thordal-Christensen, H. (2008). What are the prospects for genetically engineered, disease resistant plants? *European Journal of Plant Pathology* 121: 217–231.

Dangl, J.L., Horvath, D.M., and Staskawicz, B.J. (2013). Pivoting the plant immune system from dissection to deployment. *Science* 341: 746–751.

Ellis, J.G., Lagudah, E.S., Spielmeyer, W. et al. (2014). The past, present and future of breeding rust resistant wheat. *Frontiers in Plant Science* 5: 641. https://doi.org/10.3389/fpls.2014.00641.

Haesaert, G., Vossen, J.H., Custers, R. et al. (2015). Transformation of the potato cultivar Desiree with single or multiple resistance genes increase resistance to late blight under field conditions. *Crop Protection* 77: 163–175. https://doi.org/10.1016/j.cropro.2015.07.018.

Jones, J.D.G., Witek, K., Verweij, W. et al. (2014). Elevating crop disease resistance with cloned genes. *Philosophical Transactions of the Royal Society B-Biological Sciences* 369: 20130087. https://doi.org/10.1098/rstb.2013.0087.

Lacombe, S., Rougon-Cardoso, A., Sherwood, E. et al. (2010). Interfamily transfer of a plant pattern-recognition receptor confers broad-spectrum bacterial resistance. *Nature Biotechnology* 28: 365–369. https://doi.org/10.1038/nbt.1613.

Lius, S., Manshardt, R.M., Fitch, M.M.M. et al. (1997). Pathogen-derived resistance provides papaya with effective protection against papaya ringspot virus. *Molecular Breeding* 3: 161–168. https://doi.org/10.1023/A:100961450.

Mikaberidze, A., McDonald, B.A., and Bonhoeffer, S. (2015). Developing smarter host mixtures to control plant disease. *Plant Pathology* 64: 996–1004. https://doi.org/10.1111/ppa.12321.

Mundt, C.C. (2002). Use of multiline cultivars and cultivar mixtures for disease management. *Annual Review of Phytopathology* 40: 381–410. https://doi.org/10.1146/annurev.phyto.40.011402.113723.

Rimbaud, L., Papaix, J., Barrett, L.G. et al. (2018). Mosaics, mixtures, rotations or pyramiding: what is the optimal strategy to deploy major gene resistance? *Evolutionary Applications* 11 (10): 1791–1810. https://doi.org/10.1111/eva.12681.

Scott, P., Thomson, J., Grzywacz, D. et al. (2016). Genetic modification for disease resistance: a position paper. *Food Security* 8: 865–870.

Thrall, P.H., Oakeshott, J.G., Fitt, G. et al. (2011). Evolution in agriculture: the application of evolutionary approaches to the management of biotic interactions in agro-ecosystems. *Evolutionary Applications* 4: 200–215.

Wiesner-Hanks, T. and Nelson, R. (2016). Multiple disease resistance in plants. *Annual Review of Phytopathology* 54: 229–252. https://doi.org/10.1146/anurev-phyto-080615-100037.

Wulff, B.B.H. and Moscou, M.J. (2014). Strategies for transferring resistance into wheat: from wide crosses to GM cassettes. *Frontiers in Plant Science* 5: 692. https://doi.org/10.3389/fpls.2014.00692.

Zhu, Q., Maher, E.A., Masoud, S., Dixon, R.A., Lamb, C.J. (1994). Enhanced protection against fungal attack by constitutive co-expression of chitinase and glucanase genes in transgenic tobacco. *BioTechnology* 12: 807–812.

13

Biological Control of Plant Disease

The future of biological control will be limited only by our imagination.

(R.J. Cook, 1984)

Throughout most of the long history of agriculture, prior to the discovery of chemical pesticides and understanding of the genetic basis of heredity, crops were grown without any specific strategy for disease control. Practices evolved, however, which tended to restrict the losses caused by disease; these included time-honored rotations using different sequences of crops. Also, the crops themselves were often genetically diverse and consisted of a mixture of genotypes with different properties, including differences in susceptibility to disease. This type of small-scale mixed cropping is still practiced in many developing countries, for instance with legume crops in many parts of Africa.

The demands of an ever-increasing world population have inevitably led to larger and more intensive agricultural regimes. The development of highly effective pesticides, which seemed to offer instant answers to the threat of disease, shifted attention to nonbiological methods of control. The application of genetic principles to plant breeding further transformed agricultural practice. Selection for increased yield and responsiveness to fertilizers led to greater uniformity and the development of agricultural systems relying on routine inputs of chemicals. These approaches have been highly successful in boosting crop production and maintaining a reliable supply of the major commodities, but there is now concern over the costs and environmental side effects of such intensive regimes. Hence there is renewed interest in developing sustainable systems of production, with reduced inputs, which maintain some of the biodiversity of the agricultural landscape. This shift has refocused attention on the application of ecological principles to disease control.

Cultural Practices and Disease Control

Understanding the effects of cultural practices on the incidence of disease is vital if we are to develop more natural control measures. A whole range of agronomic factors influence disease, including sowing date, cultivations, nutrients and water, soil organic matter, management of crop residues such as straw, and crop rotations. Altering cultural practices can

Plant Pathology and Plant Pathogens, Fourth Edition. John A. Lucas.
© 2020 John Wiley & Sons Ltd. Published 2020 by John Wiley & Sons Ltd.
Companion website: www.wiley.com/go/Lucas PlantPathology4

both increase or decrease disease risk. Sowing date provides a good example. In northern Europe, there has been a trend toward growing winter cereals, sown in the autumn, as the final yields exceed those of spring-sown crops. Further yield increases may be gained by sowing early in the autumn, for instance in September. However, early-sown crops are more at risk of infection by certain pathogens such as barley yellow dwarf virus (Figure 13.1). This is because the early crop is emerging during a period when aphid vectors which may be carrying the virus are still active (see Chapter 4, Figure 4.12). Later sown crops avoid this risk period. The potential advantages of early sowing must therefore be balanced against the increased risk of yield penalties or control costs due to virus disease. In some regions, intensive cereal production has actually created a situation in which late-harvested crops overlap with the next crop emerging, so that host plants are continuously available. This so-called "green bridge" can encourage transfer of biotrophic pathogens such as powdery mildew, *Blumeria graminis*, or rust fungi, which require living host tissue to develop.

Crop Rotation

Varying the type of crop grown on a particular plot of land has long been known to influence the amount of disease present; indeed, this is arguably the oldest form of disease control. Figure 13.2 shows yield data for two varieties of pea grown on the same land with a four-, five-, or six-year break between crops. Rotations less than six years in duration gave reduced yield, primarily due to infection by a mixture of fungal pathogens attacking the roots and stem base of the plants.

Continuous cropping with a single species of host plant will, almost always, lead to a build-up of pathogens adapted to that crop, especially soil-borne pathogens. Black root rot, caused by the fungus *Thielaviopsis basicola*, is an important disease in cotton crops grown in California. The pathogen survives in soil mainly as chlamydospores, and there is a direct relationship between the amount of inoculum present and disease severity (Figure 13.3). Soil samples from fields cropped continuously with cotton for three or four years contained on average more than 100 propagules per gram of soil, whereas soils rotated to other crops

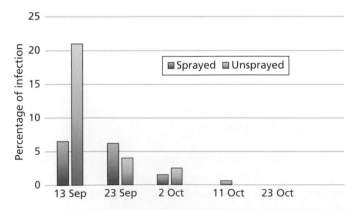

Figure 13.1 Effect of sowing date on the incidence of barley yellow dwarf virus (BYDV) in an autumn-sown winter cereal crop the following season (sampled in May). The sprayed crop was treated with an insecticide at the end of October. *Source:* Plumb (1986).

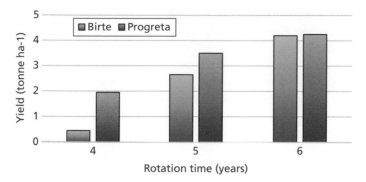

Figure 13.2 Effect of rotation on the yield of two pea cultivars. *Source:* After Rogers-Lewis (1985); © Crown Copyright.

Figure 13.3 Relationship between inoculum density of *Thielaviopsis basicola* in soil and disease severity in cotton seedlings. Bars indicate standard errors. *Source:* After Holz and Weinhold (1994).

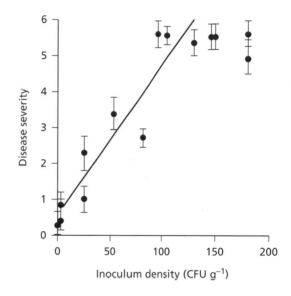

contained less than 20 propagules per gram. This difference in pathogen inoculum levels is critical in terms of symptom severity in the crop, with severe infection in the former case compared with only trace symptoms in the latter.

Previous crop history also affects the incidence of stem base- and root-infecting pathogens of cereal crops, such as eyespot disease, which is generally more severe in continuous wheat cultivation. Populations of the take-all fungus, *Gaeumannomyces graminis*, can be substantially reduced by a 2–3-year rotation to grass leys. The eradication of *Gaeumannomyces* is associated with an increase in the population of a fungal antagonist, *Phialophora radicicola*, which is especially common on the surfaces of grass roots. Many of the beneficial effects of crop rotation on disease levels may operate via similar mechanisms of microbial antagonism, which raises the possibility of enhancing such processes by means of appropriate cultural practices (see p.344).

Clearly, rotation is only feasible as a disease management tool where the pathogen concerned is unable to survive for extended periods away from the host. Fungi with very durable survival structures may persist at damaging levels beyond any economically acceptable rotation scheme. The best examples are fungal species forming sclerotia. The vascular wilt pathogen *Verticillium dahliae* survives as microsclerotia, and the minimum period required to effectively reduce inoculum between susceptible crops, such as potatoes, is 5–10 years. The shorter 2–3-year rotations favored by growers are in this case ineffective. Other sclerotial fungi are even better survivors; the white rot fungus of onion, *Sclerotium cepivorum*, can persist in soil for at least 20 years (see p.36), so that once land becomes infested it is, for all practical purposes, unusable for this crop.

Soil Amendments

Crop rotation as a means of disease control relies on natural factors to gradually reduce the level of inoculum in soil. Can anything be done to accelerate this process, and to reduce the viability of inoculum? Chemical treatment of soil with fumigants such as methyl bromide is one possibility, but in addition to the expense of the treatment, these toxic sterilants kill nontarget organisms and pollute the atmosphere, so their use is now restricted. A more benign method is needed. One promising approach is to cover the soil with materials which raise the surface temperature above ambient. Traditionally, such **mulches** have used compost, sawdust, or other waste products. These organic treatments accelerate crop growth by several means, including increased water retention and nutrient availability as well as effects on soil temperature.

More recently, plastic sheeting has replaced such traditional materials, a method described as **soil solarization.** The usual procedure is to overlay the soil for a period of one or two months prior to planting of the crop. Both transparent and black polyethylene sheets can be used, and when laid over soil during periods of high air temperature, these treatments can reduce or eradicate a variety of pathogens and pests, including soil-borne fungi and nematodes, as well as weed seeds. The method has proved most successful in regions with hotter climates, such as southern Europe, Israel, and South Africa. This is because the soil temperature needs to rise above a certain critical threshold before pathogen propagules and weed seeds are actually killed. In the central valley of California, for instance, surface soil layers may reach 40 °C during summer, and deeper layers 30 °C, but solarization can raise these temperatures by around 10 °C. Most pathogen propagules are unable to withstand temperatures of 50 °C for more than a few hours, and longer periods at lower temperatures may also be lethal. Best results are achieved if the soil is moist, so solarization is often combined with an irrigation treatment (see below).

Soil solarization compares favorably with chemical fumigation as a means of controlling soil-borne diseases. Figure 13.4 shows results from a trial in Spain in which soil was artificially infested with the vascular wilt fungus *Fusarium oxysporum*, and then fumigated or solarized for one or two months. Levels of pathogen inoculum declined in all treatments, but the highest yields of a subsequent melon crop were obtained on soil which had been solarized for two months. Solarization can also control diseases caused by sclerotial pathogens such as onion white rot, *S. cepivorum*. Two to three months treatment can reduce sclerotial populations in surface soil to undetectable levels, with major effects on epidemic development, even in highly susceptible crops such as garlic (Figure 13.5).

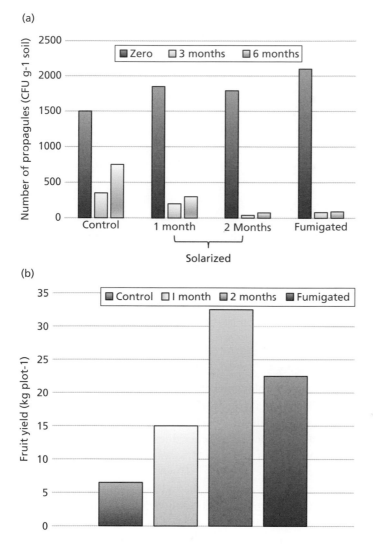

(a)

(b)

Figure 13.4 Effects of soil solarization and fumigation on (a) numbers of propagules of *Fusarium oxysporum* in soil before treatment, and three or six months after treatment, and (b) yield of a subsequent melon crop. CFU, colony-forming units. *Source:* After Gonzâlez-Torres et al. (1993).

The effectiveness of soil solarization can be increased by irrigating the soil prior to covering it, and also by addition of a carbon source such as ethanol, agricultural by-products like molasses or manure, or residues from legume or cruciferous crops. This approach to preplanting control of weeds, pests, and pathogens has been described as **anaerobic soil disinfestation** (ASD). The mechanisms of ASD are not yet fully understood, but a combination of changes in the soil microbial community, production of volatile organic compounds, and the creation of lethal anaerobic conditions has been suggested. For instance, cruciferous crop residues are known to break down to release toxic sulfur-containing chemicals such as

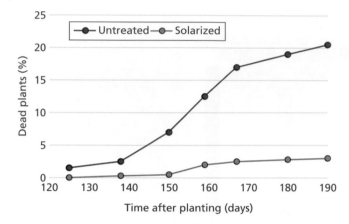

Figure 13.5 Effect of soil solarization on the disease progress curves of garlic white rot in field naturally infested with *Sclerotium cepivorum*. *Source:* Redrawn from Basallote-Ureba and Melero-Vera (1993) with kind permission of Elsevier Science Ltd.

isothiocyanates and sulfides. Such soil amendments alone can reduce levels of viable pathogen inoculum, but combination with solarization has been shown to enhance efficacy.

Tillage

Traditionally, soils have been cultivated by plowing between crops, to bury crop residues and with them any pathogens that survive on crop debris. More recently, there has been a trend to reduce or completely remove such tillage, an approach described as **no-till farming**. The following crop is then directly drilled into the residues of the previous one. The advantages of no-till include preservation of soil structure, water conservation, and less time and energy expended on cultivations.

There are concerns that retention of crop residues on the soil surface might increase the risk of carry-over of pathogen survival structures and inoculum between seasons. However, there does not appear to be a consistent trend in the incidence of soil-borne and straw-borne diseases in no-till systems as opposed to cultivation. Some have been shown to increase, while others decline. This is most likely due to the development of antagonistic microbial communities on straw or other crop residues which reduce pathogen survival. In fact, when straw burning was banned in cereal crops in the UK, contrary to expectations, some straw-borne diseases, such as eyespot (*Oculimacula* spp.), actually increased in incidence. One possible explanation for this is the removal by burning of natural biological agents able to partially control the pathogen. Such examples illustrate the complexity of microbial interactions in agricultural systems and the need to better understand the overall ecology of the crop.

Thermotherapy

The sensitivity of many pathogens to high temperatures has also been put to good use in the treatment of seeds or propagation material to eradicate disease agents prior to planting. The optimum combination of time and temperature depends on the type of plant material and its relative ability to withstand heat without a reduction in viability. The fireblight bacterium *Erwinia amylovora* is unable to withstand temperatures of 45–50 °C for much more than one hour, and moist and dry heat treatments may therefore be used to control the disease during propagation of apple and pear shoots. Many other bacterial diseases are seed borne, and treatment with hot air or water can be a relatively cheap and efficient way to disinfect seed stocks. Thermotherapy can also cure plant tissues of virus infection, and is especially effective when combined with tissue culture, for instance by maintaining excised shoot tips or buds for a period at a temperature around 50 °C during micropropagation.

Biological Control

The fact that cultural practices such as crop rotation can reduce disease indicates that pathogen populations may be regulated by natural means, rather than by human intervention. During dispersal, survival, and the early stages of infection on the surface of the host, pathogens are exposed to other organisms which may affect their growth or viability. The starting point for biological control is to identify these natural constraints on pathogens and exploit them wherever possible to limit disease outbreaks.

The term **biological control** has been defined in a broad sense as the reduction of the damaging effects of an organism by one or more other organisms. This concept includes both direct and indirect effects, due to either introduced antagonists or manipulation of existing populations to reduce disease (Table 13.1). Agronomic practices such as soil amendment with organic matter to boost microbial populations to suppress pathogens or pests are therefore covered by this definition. More usually, however, biocontrol refers to the targeted introduction of a biological agent to reduce the population of a pest, pathogen, or weed.

It should be noted that biological agents are also a valuable source of chemical products with antimicrobial activity. These are described as **biochemical pesticides** to distinguish them from use of the agent itself as a control method.

Table 13.1 Approaches used for biological control of plant pathogens

Direct
Introduction of microbial antagonists able to suppress activity of the pathogen
Indirect
Manipulation of the microbial community (microbiome) by crop practices, cultural measures, soil supplements or other means
Host mediated
Induction of localized or systemic plant resistance
Virus cross-protection

Classic Biocontrol

The earliest examples of successful biological control almost all involved pests or weeds. The first breakthrough in the field concerned control of the cotton cushion scale insect in the USA. This pest was first recorded in 1868 in California where it spread rapidly through the newly established citrus groves. Entomologists suspected that the pest had come from Australia on imported nursery stock, and a search was made there for natural predators. Two of these, a lady beetle and a fly, were selected, imported, and released in California in 1888, and within a few months the problem of cotton cushion scale had been solved.

The principle involved in this example is simple yet extremely effective. Pests almost invariably have natural enemies somewhere in their geographic range, and introduction of such enemies can reduce the size of the pest population. There are now numerous successful case histories where natural predators or parasites have been released to control insects or weeds. But there are biological constraints to this classic, inundative approach. Potential natural control agents may fail to establish in the environment, or to multiply to levels which significantly reduce the pest or pathogen population. The dynamics of predator–prey relationships, in which there is often a cyclical equilibrium between the two organisms, means that the degree of control achieved may be less than with a chemical agent. It is worth noting that many of the most successful biological control systems concern protected environments such as glasshouses, where conditions can be manipulated to favor the introduced antagonist. The use of chemical pesticides to control many glasshouse pests has now been almost completely replaced by natural enemies. Control in field crops is more difficult, and requires a good understanding of the population biology and ecology of both the target pest or pathogen and the biocontrol agent (BCA).

The majority of commercial biocontrol products currently on the market are used for the management of insect pests. Microbial pesticides based on the toxin-producing bacterium *Bacillus thuringiensis* (Bt) represent a large proportion of these, and Bt toxins are also now widely used in transgenic crops to control lepidopteran pests. Other biological insecticides include entomopathogenic fungi, viruses, and nematodes. Most existing and potential biological products for weed control are based on plant-pathogenic fungi and hence are known as **mycoherbicides**. To date, there are relatively few examples of biological agents developed for the control of plant pathogens. Some examples of commercial products are shown in Table 13.2.

There have been numerous demonstrations of promising control of plant diseases by BCAs in experimental systems, but scaling this up has often given disappointing results. Why is this? The following discussion will focus on the search for effective BCAs for plant pathogens, and consider both the problems and prospects for commercialization of such agents.

Microbial Antagonism

Studies on several soil-borne plant pathogens have shown that other microorganisms resident in the soil can have a large influence on the incidence and severity of disease. It has been known for many years that certain soils can reduce disease caused by vascular wilt fungi, such as *Fusarium*, and that the suppressive properties of such soils can be removed

Table 13.2 Some examples of commercial biocontrol products for plant diseases

Biocontrol agent	Target pathogens or disease	Commercial product[a]
Bacteria		
Agrobacterium radiobacter	Crown gall	NoGall™
Bacillus subtilis	Root-infecting fungi	Kodiak®
	Seedling diseases	Quantum 4000®
	Botrytis cinerea	Serenade®
Pseudomonas fluorescens	Fireblight (*Erwinia amylovora*)	BlightBan®
Pseudomonas chlororaphis	Cereal foliar pathogens	Cedomon®
Streptomyces griseoviridis	Seedling diseases, damping-off	Mycostop®
Streptomyces lydicus	Foliar and soil-borne pathogens	Actinovate®
Fungi		
Phlebiopsis gigantea	*Heterobasidion annosum* rot	RotStop®
Coniothyrium minitans	*Sclerotinia* spp.	Contans®
Trichoderma harzianum	Root-infecting and other fungi	RootShield® PlantShield®
Trichoderma virens	Soil-borne fungal pathogens	SoilGard®

[a] Tradenames may vary between countries and be subject to change.

by sterilization. It is also possible to convert a "conducive" soil into a "suppressive" one by mixing in a proportion of the suppressive type. This evidence suggests that living organisms present in some soils are capable of reducing the viability or infectivity of plant pathogens.

Further indirect evidence for such microbial antagonism comes from surveys of disease in continuous monoculture. For instance, the normal pattern of take-all (*G. graminis*) in successive wheat crops is an increase in severity over the first few seasons, followed by a decline. Logic suggests that in the absence of any crop rotation, the disease should become more and more prevalent as inoculum levels continue to build up in the soil. But instead, the amount of disease actually falls (Figure 13.6). Similar patterns of incidence have been reported with other soil-borne fungal pathogens. The simplest explanation for this phenomenon is that as the pathogen population increases, there is a corresponding increase in the amount or activity of antagonistic microorganisms.

The exact mechanism of such antagonism is still debated. It is known that certain types of bacteria found in the root zone, or rhizosphere, for instance fluorescent pseudomonads, can interfere with infection of roots by fungi, and over time these types may come to predominate. Alternatively, there may be increasing microbial competition for the nutrient substrates on which the pathogen survives between crops, or even an increase in the level of predation or parasitism of the pathogen by other living agents, including viruses. Bacteriophages, for example, can infect bacteria and cause lysis of cells. It has recently been demonstrated that treatment with mixtures of phages can reduce the incidence and severity of black-leg disease of potatoes caused by the bacterial rot pathogen *Dickeya solani*

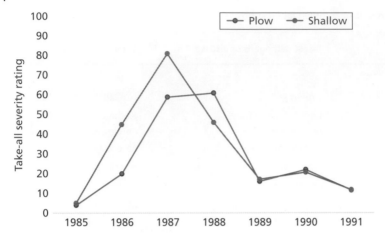

Figure 13.6 Severity of take-all disease in continuous winter wheat crops at Rothamsted sown after plowing or shallow cultivation. After an initial increase year on year, the disease naturally declines. *Source:* Adapted from Jenkyn et al. (1994).

in both lab assays and the field. Such **phage therapy** may therefore provide an alternative biocontrol option for bacterial diseases.

Overall, disease suppression may well be due to a combination of several of these processes. But the important conclusion is that such natural antagonism might be exploited to control disease, by managing the crop to either maximize the impact of resident antagonists or to isolate and identify the most effective agents for use in specific biocontrol programs.

Selection of BCAs for Plant Disease Control

There is little difference, in principle, between screening chemicals or testing biological agents for disease control. Both selection processes require an appropriate assay system to detect activity against the target pathogen or pathogens. Random screening of large numbers of microorganisms for suppressive effects on plant disease has often been used to select the most promising strains for further development. But the chances of success can be improved by adopting a more rational approach based on ecology.

The argument is as follows. To be effective, a BCA must be able to colonize a particular habitat, or occupy a specific niche, in sufficient numbers to interfere with the growth or survival of the target pathogen. Rather than introducing a randomly selected microbial antagonist, it would be better to introduce one known to be adapted to the habitat concerned. Hence, the best place to look for potential BCAs is in the specific environment in which they are to be used. For instance, if the target is a pathogen which infects plant roots, the logical place to look for an antagonist is in the rhizosphere. Any microorganism isolated from this habitat is likely to have biological and physiological attributes enabling successful multiplication in the root zone, and is therefore "rhizosphere competent." Similar concepts apply to leaf surfaces or other substrates such as straw where a particular stage in the pathogen life cycle might be disrupted. Screening for BCAs should therefore focus on specific natural sources chosen to optimize the chances of isolating strains with the correct biological properties.

Production and Formulation of BCAs

An important difference between a chemical and a BCA is that the latter is a living organism and usually needs to be metabolically active to be effective. One of the most difficult steps in the commercialization of a BCA is to produce, package, and deliver sufficient quantities of the agent in a viable and stable form. Variation is a normal feature of microorganisms but each batch of a BCA needs to have similar activity. This can be a particular problem in scaling up production, as microbial populations can change their properties during growth in fermenters. The BCA then has to be harvested and distributed in a formulation which ensures viability. To be of practical use, a BCA needs to have a reasonable shelf-life so that it can be stored for a period without significant loss of activity. It then needs to be applied to the crop, or into soil, in a way which ensures that the antagonist grows and persists in the environment for sufficient time to exert control.

Several different approaches have been used to solve these various problems, including freeze-drying microbial cultures, or mixing cells with inert carriers such as clay or talc. Alternatively, biomass may be immobilized or encapsulated in an alginate polymer. A food source such as wheat bran can be added, which not only acts as a carrier but also releases nutrients promoting growth of the microorganism once applied. There are a variety of application methods, for instance as liquid sprays or drenches, seed dressings, and pelleted formulations which slowly release the BCA into the environment. As with chemicals, the choice depends upon the target pathogen, and the mode of action of the BCA.

Table 13.2 lists several BCAs which have been used, or are being used, on a commercial scale to control plant pathogens. There are many more examples which have shown promise in experimental trials, but have yet to make a significant impact in the marketplace.

Some Case Histories

Crown gall is a plant tumor caused by the soil-borne bacterium *Agrobacterium tumefaciens*; galls occur on the roots or often at the crown of the plant between root and shoot (see Chapter 8, Figure 8.14). The pathogen has a wide host range but the disease is especially important in stone fruits such as peaches, on grapevines, and woody ornamentals such as roses. A highly effective biocontrol method has been developed using a related, nonpathogenic bacterium, *Agrobacterium radiobacter* (now reclassified as *Rhizobium rhizogenes*), to treat roots during transplanting. The BCA is commercially available (Table 13.2) and gives cheaper, more effective control than antibacterial chemicals.

Not all strains of *A. radiobacter* are able to protect plants from the disease. Effective strains possess two important properties. First, they are able to colonize host roots to a higher population density than ineffective strains. Second, biologically active strains produce an antibiotic, agrocin, which is toxic to *A. tumefaciens*. This molecule is a "rogue" nucleotide (Figure 13.7) which interferes with DNA synthesis. Agrocin production is encoded by a plasmid which also carries genes for insensitivity to the toxin. As plasmids are often able to transfer between bacterial strains and even species, one concern with the use of this BCA is the possibility that the genes for agrocin insensitivity might "jump" into the pathogen, rendering it immune to the toxin. Work in Australia has therefore concentrated on engineering *A. radiobacter* strains carrying modified plasmids which are unable to transfer between bacteria. This was done by deleting the region of the agrocin plasmid encoding transfer functions. Such *Tra⁻* strains are now registered for use in the field.

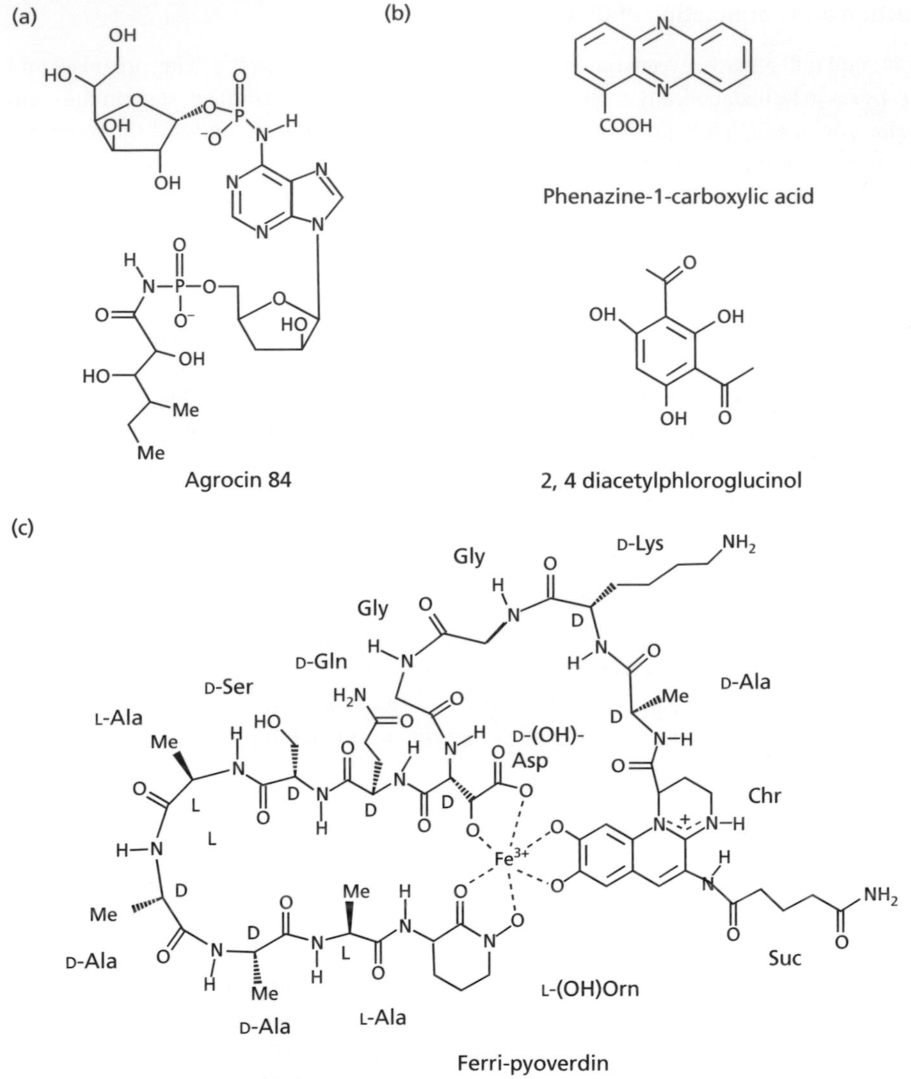

(a)

(b)

Agrocin 84

Phenazine-1-carboxylic acid

2, 4 diacetylphloroglucinol

(c)

Ferri-pyoverdin

Figure 13.7 Some compounds produced by bacterial antagonists which play a role in biological control. (a) The bacteriocin agrocin. (b) The antibiotics phenazine and 2,4-diacetylphloroglucinol. (c) The siderophore pyoverdin. *Source:* After Mohn et al. (1994).

The soil-borne fungus *Heterobasidion annosum* is a serious problem in forestry plantations where it causes root and butt rot of conifers (Figure 13.8). The fungus is not an aggressive pathogen of intact trees but instead exploits cut tree stumps to establish a food base from which it can spread along the roots and infect adjacent hosts. As forestry practice involves frequent thinning and felling of trees, this provides the pathogen with ample opportunity to gain access. Treating stumps with chemicals only delays infection, as the concentration of fungicide eventually declines to an ineffective level. Instead, a biological solution was sought by looking for microbial antagonists able to colonize cut timber and prevent the

Figure 13.8 Damage caused to conifer timber due to infection by butt rot, *Heterobasidion annosum*. Note extensive rot of heartwood. *Source:* Courtesy of Steve Gregory, Forestry Commission.

pathogen from becoming established. The most effective agent proved to be another wood-rotting basidiomycete, *Phlebiopsis gigantea*. Spores of this fungus are painted or sprayed onto cut stumps or can now be applied automatically by timber-harvesting machines. Around 10^6 spores are used per square meter of stump surface, which ensures rapid colonization of the timber by the antagonist, thereby occupying the infection court and denying access to the pathogen. In the UK, this treatment has provided cost-effective and environmentally safe control of the disease in pine plantations for many years, while in Scandinavia strains of the BCA have been developed for use on Norway spruce.

This example of protection of an infection court in a woody host has parallels in several other plant diseases. For instance, many pathogens, such as *Nectria galligena* which causes silverleaf of fruit trees, gain entry to the host via pruning wounds. These sites can be treated with a fast-growing antagonist, such as formulations of the saprophytic fungus *Trichoderma*, to prevent infection by the pathogen.

Soil-borne fungal pathogens which infect seeds and roots are a serious constraint to agricultural production as they affect crop establishment, leading to patchy growth and delayed development. Examples include damping-off diseases, caused by *Pythium* species and *Rhizoctonia solani*, and take-all of cereals, caused by *G. graminis*. One feature common to the infection cycle of these diverse pathogens is the need to colonize the zone surrounding seeds or roots prior to penetration of host tissue. Opportunities exist, therefore, to interfere with this step by introducing aggressive microbial competitors or manipulating resident microbial populations to reduce infection.

It has already been noted that the decline of take-all disease in cereal monocultures coincides with changes in the microbiology of the rhizosphere. Bacteria isolated from this zone (often described as "rhizobacteria") have been intensively studied as potential BCAs for take-all and several seedling diseases. The most promising candidate strains are usually

Figure 13.9 Protection of pea seedlings against preemergence damping-off caused by *Pythium ultimum*. The same number of seeds were sown in each tray. Treatments (from left to right) are compost + pathogen; compost + pathogen + *Pseudomonas fluorescens* drench; compost + pathogen + fungicide (metalaxyl) drench; compost without pathogen. *Source:* Courtesy of Donna Murray.

isolates of *Pseudomonas fluorescens*, a group of bacteria which are well adapted to growth in the rhizosphere. Figure 13.9 shows the effects of a soil drench with *P. fluorescens* on the emergence of pea seedlings in soil infested by *Pythium*, compared with a fungicide treatment, under controlled environment conditions. Several strains of *P. fluorescens* have now also been tested in plot and field trials for the control of soil-borne fungal pathogens with varying degrees of success, although to date few commercial products based on this bacterium have been launched.

An interesting alternative to *Pseudomonas* spp. are species of *Bacillus* which, unlike most bacteria, produce resistant endospores which can be kept as a dry formulation for long periods without losing viability. This property might also aid use in situations exposed to drying, such as on leaves and other aerial plant surfaces. Formulations of *Bacillus* cells have been microencapsulated with various materials to improve shelf-life and efficacy when used as sprays. Encapsulation with a mixture of maltose-dextrin and gum Arabic improved survival rates and stability of the BCA during storage, and gave good control of *Rhizoctonia* rot of tomatoes in field trials (Figure 13.10). The BCA treatments compared favorably with control provided by a fungicide spray, suggesting that such formulations might have the potential to be successful commercial products.

Mode of Action of BCAs against Plant Pathogens

The examples discussed above almost all involve protection of an infection court by prior treatment with a microbial antagonist. Effective suppression of the pathogen requires the establishment of a metabolically active threshold population of the BCA at the infection site. There are also examples of BCAs which act against pathogens in soil or on plant debris, affecting the survival of resting structures or propagules, rather than preventing infection.

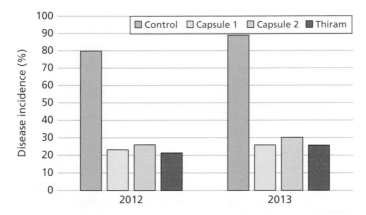

Figure 13.10 Control of *Rhizoctonia* rot of tomatoes in the field by sprays of microencapsulated *Bacillus subtilis* at two concentrations compared with a fungicide spray (Thiram). Data show disease incidence for the different treatments in two seasons. *Source:* Ma et al. (2015).

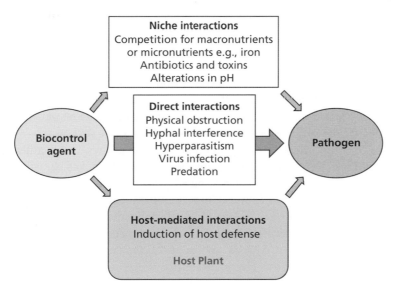

Figure 13.11 Modes of action of biocontrol agents.

The exact mechanism of antagonism is often not known. There are, however, a number of possible modes of action (Figure 13.11).

At one extreme, suppression might simply be due to occupation of a particular niche by the BCA, leading to physical exclusion of the pathogen or to competition for essential nutrients. It was originally thought that control of *Heterobasidion* by *Phlebiopsis* was due to this type of effect. Now it is suspected that more direct interactions between the hyphae of the two fungi may occur, in a type of territorial combat known as **hyphal interference**. A more clearly characterized mode of action is production of a toxic or inhibitory compound by the BCA, such as the example of agrocin mentioned earlier. Antibiotics have been shown to be important in the suppression of root diseases such as cereal take-all (*G. graminis*) and damping-off, caused by *Pythium* spp. Strains of the root-colonizing bacterium *P. fluorescens* produce several antimicrobial metabolites including 2,4-diacetylphloroglucinol (DPAG) and phenazine-1-carboxylic acid (PCA) (Figure 13.7). DPAG is highly active against *G. graminis* and strains producing this antibiotic are consistently found on roots of wheat plants growing in soils developing take-all decline. The population densities of DPAG producers are high enough to account for suppression of the disease. Production of phenazine has been linked to the suppression of seedling damping-off; mutants which have lost the ability to produce this antibiotic are less effective in controlling the disease. This correlation has been confirmed by introduction of a functional gene for biosynthesis of phenazine which restored biocontrol activity. Furthermore, enhancing expression of the gene cluster responsible for phenazine production increased the activity of the bacterium in suppressing damping-off disease. These studies confirmed that antibiotics play a key role in the biocontrol of some soil-borne pathogens.

Siderophores are low molecular weight compounds which have a high affinity for iron and aid transport into cells. These chemicals are efficient scavengers of iron and may thus mop up all of the available supply in the immediate environment. Many pathogens require iron as an essential mineral nutrient for growth, and in some cases iron is required for virulence. Hence, production of a siderophore by a BCA may reduce the growth of a pathogen or its ability to attack the host. Fluorescent pseudomonads produce several siderophores, such as the pigmented compound pyoverdin (Figure 13.7), and the most convincing evidence that these contribute to disease control again comes from studies on nonproducing mutants which are less effective than wild-type strains.

Direct parasitism or predation of the pathogen can also occur and may be particularly important in reducing the viability of spores or survival structures such as sclerotia. Fungal pathogens forming sclerotia are difficult to control as these propagules persist for long periods in soil. Several fungi which invade sclerotia and act as parasites have now been identified. One of the most interesting is *Sporidesmium sclerotivorum*, which is an obligate parasite of sclerotia of several important pathogens, including species of *Sclerotinia*, *Sclerotium*, and *Botrytis cinerea*. Multicellular conidia of the parasite are stimulated to germinate by chemicals diffusing from nearby sclerotia, and germ tubes infect the sclerotia, causing eventual lysis. Spores of this parasite added in sufficient quantities to soil have been shown to give good control of diseases such as lettuce drop, caused by *Sclerotinia minor*. Unfortunately, the BCA is difficult to produce on a large scale in pure culture. Another sclerotial parasite, *Coniothyrium minitans*, has demonstrated biocontrol potential and has been developed as a commercial product (Table 13.2).

Parasitic microorganisms which infect other parasites are usually described as **hyperparasites** (Figure 13.11) or in the case of fungi infecting other fungi, **mycoparasites**. These agents themselves have diverse host–parasite relationships, ranging from necrotrophy to biotrophy, and may infect different stages in pathogen life cycles. Some are close relatives of the pathogens themselves; for instance, parasitic species of *Pythium* such as *P. nunn* and *P. oligandrum* coil around and lyze hyphae of the damping-off pathogen *P. ultimum*. Figure 13.12 shows parasitism of the downy mildew *Plasmopara*, an oomycete pathogen of grapevines, by a species of the fungus *Fusarium*. The diversity of interactions occurring in natural environments means that it is usually possible to find an organism

(a)

(b)

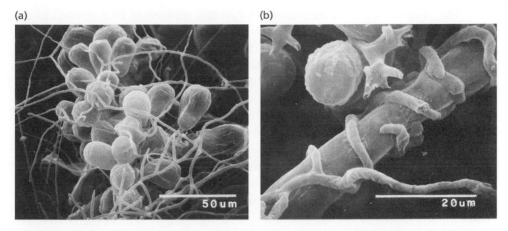

Figure 13.12 An example of mycoparasitism. Infection of the grapevine downy mildew pathogen *Plasmopara viticola* by *Fusarium proliferatum*. (a) Thin hyphae of the mycoparasite overgrowing sporangia of the downy mildew. (b) Parasite hyphae curling around and penetrating a sporangiophore. *Source:* Falk et al. (1996).

parasitic on the target pathogen or pest. Hence fungi attacking other fungi, oomycetes, nematodes, and insects are all being studied as potential BCAs.

Predation is an important mode of action in the biocontrol of insect pests but has not been exploited to any significant extent in the suppression of pathogens. Fungal spores are subject to predation in soils, for instance by large, mobile amoebae which are able to penetrate even highly resistant spores and destroy them. These protozoa, known as vampyrellids, contact spores by means of thin pseudopodia, drill a hole in the spore wall, enter, and consume the spore contents. Such natural predation is no doubt important in reducing pathogen inoculum levels in soil but so far, these antagonists have proved difficult to produce in culture.

Hypovirulence

A unique form of natural biocontrol has been observed in the pandemic of the highly destructive chestnut blight disease, caused by *Cryphonectria parasitica*. This fungus infects via wounds, causing aggressive lesions, or cankers, which girdle the stem, leading to the death of shoots above the infection site. The disease was first recorded in the USA in 1904 and despite attempts to eradicate it, spread to destroy most of the native chestnuts in the eastern states. The pathogen is believed to have originated from Asia, and the severity of the epidemic is consistent with a "new encounter" disease (see p.21), in which a pathogen infects a previously unexposed and highly susceptible host population.

When, in 1938, the disease was recorded in Italy, it was feared that European chestnut trees would suffer a similar fate. The initial European outbreak was also severe, but subsequently many infected trees began to show signs of recovery, with spontaneous healing of cankers. Significantly, strains of the fungus isolated from such cankers were found to be less virulent than the original pathogen. Furthermore, if these **hypovirulent** isolates were co-inoculated with highly virulent strains, the resulting cankers also healed. The most intriguing observation, however, was that virulent strains could be converted to the hypovirulent phenotype by hyphal contact and fusion in culture. Some transmissible factor moved from the hypovirulent strain into the more aggressive one. The agent(s) responsible were shown to be cytoplasmic and were subsequently identified as double-stranded (ds) RNA molecules. Several different-sized dsRNAs have been isolated from hypovirulent strains of the fungus, and the larger of these have sequence homology with certain plant virus genomes. It seems likely that hypovirulence is therefore a type of virus infection.

The natural transfer of hypovirulence between strains of *C. parasitica* suggests that the disease might be managed simply by introducing hypovirulent isolates into the pathogen population. In Europe, this has to some extent occurred naturally but the situation in the USA is more complex, due to the greater genetic variation between *Cryphonectria* strains in that region. For transfer to occur, hyphae must fuse and this is often limited by natural compatibility barriers. To overcome this problem, there is now interest in engineering replicating forms of the dsRNA into a range of *C. parasitica* strains representing the different compatibility types. In this way, more efficient spread of hypovirulence should occur and the epidemic should be restricted.

Induction of Host Resistance

An alternative, and quite different, mode of action of BCAs is disease suppression through the induction of host defense mechanisms. It has been known for many years that inoculation of plants with avirulent strains of pathogens, or agents causing necrosis, can trigger both local and systemic plant resistance (see p.233). More recently, it has been shown that plant growth-promoting rhizobacteria applied to seeds or roots can induce a systemic resistance response expressed against pathogens infecting aerial tissues. Examples include treatment with *Pseudomonas* spp. increasing resistance to the vascular wilt fungus *F. oxysporum*

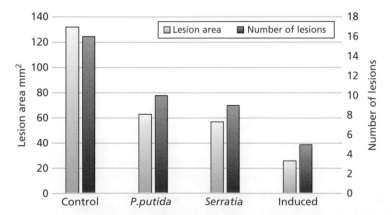

Figure 13.13 Severity of angular leaf spot, caused by the bacterium *Pseudomonas syringae* pv. *lacrymans* on cucumber plants grown from seeds treated with the plant growth-promoting rhizobacterium *Pseudomonas putida* and *Serratia marcescens*, both inducing systemic resistance. Control shows values for unprotected plants, induced shows results for plants in which systemic acquired resistance (SAR) was switched on by prior inoculation of the first leaf with the pathogen. *Source:* Data from Liu et al. (1995).

f.sp. *dianthi* in carnation, as well as to leaf pathogens such as *Colletotrichum orbiculare* and bacterial blight, *Pseudomonas syringae* pv. *phaseolicola*, in bean and *Ps. syringae* in cucumber (Figure 13.13). This resistance is associated with the induction of pathogenesis-related (PR) proteins, as seen in a typical systemic acquired resistance (SAR) response (see p.234). Presumably, the colonizing BCA produces a signal molecule(s) which activates the SAR pathway. Other rhizobacteria, for instance *P. fluorescens* strains producing the polyketide antibiotic DAPG (Figure 13.7) trigger the induced systemic resistance (ISR) pathway mediated by ethylene and jasmonic acid. Experiments with *Arabidopsis* and resistance to the foliar pathogen *Ps. syringae* suggest that the antibiotic itself might be a key determinant of the induced response. It seems likely, therefore, that different BCAs may operate through different pathways to influence host defense.

Developing More Effective Biocontrol Agents

The main obstacle to more widespread acceptance of biological control as a genuine alternative to chemicals has been the inconsistent performance of most BCAs in the field. Growers are reluctant to adopt methods which occasionally fail to benefit the crop. Generally, it is possible to find natural antagonists of pathogens which work well under certain conditions. The challenge therefore is to improve the performance of these agents over a range of conditions, either by devising more effective methods of formulation and delivery or by improving the properties of the BCA itself.

To be effective, a BCA has to function in a complex and variable environment which includes interactions with other organisms. Considering control of a root-infecting pathogen as an example, it is clear that the BCA must be able to enter the root zone and establish a threshold population in competition with the indigenous microflora. Properties aiding this process may include motility, chemotaxis toward chemical signals from the root, attachment to the surface, and ability to utilize nutrients present in root exudates. Interference with the pathogen might involve one or more of the mechanisms discussed above, such as nutrient competition, production of antibiotics or siderophores, direct parasitism, or induction of host defense. The genetic control of these different traits is likely to be complex, so that manipulation by mutation or, more precisely, genetic engineering may be difficult. Nonetheless, there are several examples of increasing antibiotic production as well as transfer of the ability to produce antibiotics such as phenazine and 2,4-DAPG to nonproducing strains, thereby enhancing their biocontrol activity.

Other traits that might be manipulated in this way include the production of lytic enzymes such as chitinases and glucanases which degrade fungal cell walls, as well as the insect exoskeleton. Sequencing the genomes of effective BCAs can identify features contributing to colonization of plant surfaces and microbial antagonism, such as gene clusters involved in the biosynthesis of inhibitory metabolites. It should therefore be possible to engineer strains combining traits for establishment, growth, and survival on or in plants with those conferring antimicrobial properties. Eventually, this might lead to improved BCAs giving more consistent control performance.

Given the potential regulatory hurdles with genetically modified organisms, an alternative approach is to co-formulate mixtures of BCAs with different modes of action. There is already some evidence that mixtures can perform better than individual agents, especially if the antagonistic properties are combined with other beneficial traits such as growth promotion or induction of plant defense. Such cocktails can also combine strains which vary in environmental tolerance, thereby extending the range of conditions under which the formulation will perform. There is also the possibility of co-formulating BCAs with chemical agents such as fungicides to provide complementary modes of action, and to reduce reliance on chemistry alone. If the BCA is itself a fungus, fungicide-resistant strains might be used or developed. Hence there are a number of permutations for integrating use of BCAs with other methods for disease management.

The Future of Biocontrol

In some respects, the record of biocontrol of plant diseases has been a disappointment. Few commercial products have reached the market and much of the early promise has not been achieved in practical terms. Is this a fair assessment?

To a large extent, the problems of biocontrol are due to two factors. First, we still have only a partial understanding of microbial ecology and the factors leading to sustained performance of a BCA in a natural environment. This is a state of knowledge problem. Second, the expectation that natural control agents will substitute for chemicals in terms of instant results is unreasonable. There needs to be a more realistic approach in which these agents are seen as part of an integrated strategy for disease management.

It should also be noted that the classic concept of biocontrol, involving introduction of a specific antagonist, is now being superseded by a wider vision. During their co-evolution with microorganisms, green plants have recruited various beneficial species as partners in mutualistic symbioses. Well-known examples include nitrogen-fixing bacteria such as *Rhizobium* and mycorrhizal fungi. Both of these are already used in agriculture and forestry as inoculants to improve plant performance. The discovery of endophytes and their beneficial effects on plant health should provide further opportunities to exploit microbes in disease control. Next-generation sequencing of the plant microbiome, the whole community of plant-associated microorganisms, is now providing insights into the diversity of species present, and an improved understanding of their functional properties should follow.

Rather than focusing on individual agents, it may be possible to assemble or manipulate a community of species or strains with multiple modes of action. Furthermore, bioactive molecules produced by endophytes and other members of the plant microbiome are a potential source of novel classes of chemicals for disease control. Hence, methods for improving plant health based on chemical, biological, and genetic approaches are converging, and the future will see increasing integration of these once separate areas of discovery.

Further Reading

Books

Walters, D. (ed.) (2009). *Disease Control in Crops: Biological and Environmentally-Friendly Approaches*. Chichester: Wiley Blackwell.

Reviews and Papers

Bakker, P.A.H.M., Berendsen, R.L., Doornbos, R.F. et al. (2013). The rhizosphere revisited: root microbiomics. *Frontiers in Plant Science* 4: 165. https://doi.org/10.3389/fpls.2013.00165.

Kim, Y.C. and Anderson, A.J. (2018). Rhizosphere pseudomonads as probiotics improving plant health. *Molecular Plant Pathology* 19 (10): 2349–2359. https://doi.org/10.1111/mpp.12693.

Klein, E., Katan, J., and Gamliel, A. (2011). Soil suppressiveness to Fusarium disease following organic amendments and solarization. *Plant Disease* 95 (9): 1116–1123. https://doi.org/10.1094/PDIS-01-11-0065.

Le Cocq, K., Gurr, S.J., Hirsch, P.R. et al. (2017). Exploitation of endophytes for sustainable agricultural intensification. *Molecular Plant Pathology* 18 (3): 469–473. https://doi.org/10.1111/mpp.12483.

Liu, K., McInroy, J.A., Hu, C.-H. et al. (2018). Mixtures of plant-growth-promoting rhizobacteria enhance biological control of multiple plant diseases and plant-growth promotion in the presence of pathogens. *Plant Disease* 102 (1): 67–72. https://doi.org/10.1094/PDIS-04-17-0478-RE.

Lugtenberg, B.J., Caradus, J.R., and Johnson, L.J. (2016). Fungal endophytes for sustainable crop production. *FEMS Microbiology Ecology* 92: fiw194. https://doi.org/10.1093/femsec/fiw194.

Mazzola, M. and Freilich, S. (2017). Prospects for biological soilborne disease control: application of indigenous versus synthetic microbiomes. *Phytopathology* 107: 256–263. https://doi.org/10.1094/PHYTO-09-16-0330-RVW.

Mendes, R., Garbeva, P., and Raaijmakers, J.M. (2013). The rhizosphere microbiome: significance of plant beneficial, plant pathogenic, and human pathogenic microorganisms. *FEMS Microbiology Reviews* 37 (5): 634–663. https://doi.org/10.1111/1574-6976.12028.

Nuss, D.L. (1992). Biological control of chestnut blight: an example of virus-mediated attenuation of fungal pathogenesis. *Microbiology and Molecular Biology Reviews* 56 (4): 561–576.

Schlatter, D., Kinkel, L., Thomashow, L. et al. (2017). Disease suppressive soils: new insights from the soil microbiome. *Phytopathology* 107: 1284–1297. https://doi.org/10.1094/PHYTO-03-17-0111-RVW.

Shlevin, E., Gamliel, A., Katan, J. et al. (2018). Multi-study analysis of the added benefits of combining soil solarization with fumigants or non-chemical measures. *Crop Protection* 111: 58–65. https://doi.org/10.1016/j.cropro.2018.05.001.

Strauss, S.L. and Kluepfel, D.A. (2015). Anaerobic soil disinfestation: a chemical-independent approach to pre-plant control of plant pathogens. *Journal of Integrative Agriculture* 14 (11): 2309–2318. https://doi.org/10.1016/S2095-3119(15)61118-2.

Timmusk, S., Behers, L., Muthoni, J. et al. (2017). Perspectives and challenges of microbial application for crop improvement. *Frontiers in Plant Science* 8: 49. https://doi.org/10.3389/fpls.2017.00049.

Weller, D.M., Mavrodi, D.V., van Pelt, J.A. et al. (2012). Induced systemic resistance in *Arabidopsis thaliana* against *Pseudomonas syringae* pv. *tomato* by 2, 4-diacetylphloroglucinol-producing *Pseudomonas fluorescens*. *Phytopathology* 102 (4): 403–412. https://doi.org/10.1094/PHYTO-08-11-0222.

14

An Integrated Approach to Disease Management

A sustainable system of crop production is one that may be used continuously for many years, is soundly based on the potential and within the limitations of a particular region, does not unduly deplete its resources or degrade its environment, makes the best use of energy and materials, ensures good and reliable yields, and benefits the health and wealth of the local population ...

(R.K.S. Wood, 1919–2017)

From an historical perspective, many of the problems encountered in crop protection have arisen from a "quick-fix" approach, in which a single strategy has been used, often intensively, to control a disease or pest. The boom and bust cycle seen with many *R* genes and agrochemicals is a direct consequence of this simplistic approach. What is required is a more lasting solution, using a combination of different control strategies. The overall aim is to develop sustainable systems of disease and pest management, based on a sound understanding of the whole crop ecosystem. Ideally, such a holistic approach will not only maintain the efficacy of host resistance and agrochemicals, but also bring other benefits such as reduced environmental impacts and lower control costs.

Much of the early progress toward an integrated approach was made by entomologists attempting to manage insect pests. The concept arose largely from problems with insecticides, including the rapid development of resistance and a growing awareness of the environmental impact associated with their use. It was realized that chemicals used to control pests often affected nontarget species, including natural enemies of the pest. Hence the emphasis shifted to a more ecological approach, in which insecticides are used sparingly and the natural constraints on pest populations are enhanced rather than suppressed. **Integrated pest management** (IPM) views the crop and its environs as a single system, and incorporates information on pest ecology and behavior, as well as natural predators, parasites, and other factors regulating pest populations. It also defines action thresholds at which pesticides might need to be strategically deployed. Finally, an IPM system must take account of economic and sociological factors, so that it is sustainable within the overall agricultural context.

Plant Pathology and Plant Pathogens, Fourth Edition. John A. Lucas.
© 2020 John Wiley & Sons Ltd. Published 2020 by John Wiley & Sons Ltd.
Companion website: www.wiley.com/go/Lucas_PlantPathology4

Integrated Control of Plant Disease

Initially, there was less emphasis on IPM to control plant diseases for several reasons. Use of host genetic resistance was a relatively successful strategy, albeit short-lived, in a series of cases. A range of effective fungicides was available to manage many of the most important pathogens of arable crops. Also, there were fewer obvious biological options for controlling pathogens, compared with the natural enemies predating insects. Biological control agents effective against foliar pathogens were not readily available. Finally, it should be noted that many of the traditional practices for limiting disease losses, such as crop rotation, already included an element of integrated control.

The current imperative to adopt a more holistic approach to disease management originates from several sources. There is mounting pressure to reduce the overall use of pesticides in agricultural systems. Repeated use of single-site fungicides has often selected pathogen genotypes able to resist the previously effective chemical. Likewise, resistance genes selected by breeders when deployed in the field have frequently proved vulnerable to the emergence of virulence in pathogen populations. There is now much greater awareness of the dynamic nature of pathogen populations and the need to deploy more complex evolutionary hurdles for them to overcome. Hence the shift toward more sustainable crop protection programs which, ideally, combine cultural, chemical, biological and host genetic factors to control plant diseases.

To date, there are relatively few examples of fully integrated disease management systems, but increased efforts are being made to put together packages of measures which not only reduce reliance on chemical agents but also offer other advantages to growers, such as reduced costs, and environmental benefits. Figure 14.1. outlines some of the key principles of an IPM program in a stepwise form, with associated actions, starting from prevention of disease in the first place, followed by effective monitoring and disease surveillance, leading to decisions to either intervene or take no action. Should disease occur, there are a number of possible interventions, including use of pesticides, but a key aim is to reduce their use wherever possible.

Finally, it is important to conduct a detailed evaluation of the process, to estimate costs and benefits, environmental aspects, and in the longer term any effects that might carry over between seasons. This should allow the program to be refined and objective standards to be set.

It should be noted that in subsistence agriculture, the need for low-cost, integrated approaches to disease control has always been preeminent as inputs of agrochemicals are either unavailable or unaffordable.

Integration of Fungicides with Host Resistance

One salutary fact in plant pathology is that some well-known historical scourges are still with us today. The potato late blight pathogen *Phytophthora infestans*, which caused destruction and famine more than 150 years ago, continues to damage potato crops throughout the world. Single-gene, qualitative resistance has not yet provided a lasting

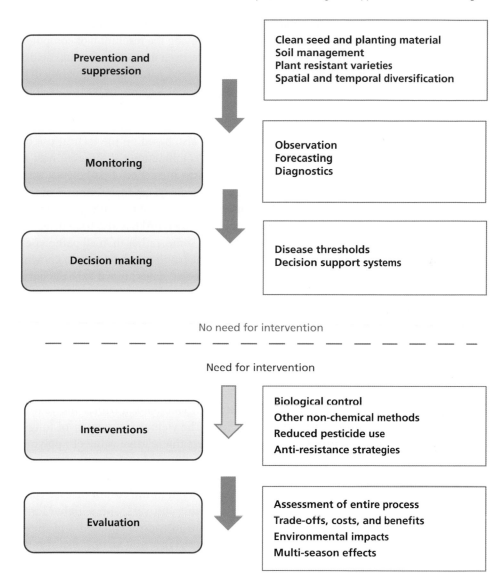

Figure 14.1 Principles of integrated pest management, with sequential actions. *Source:* Adapted from Barzman et al. (2015).

solution and the pathogen developed resistance to some of the most effective fungicides used for control of the disease, such as metalaxyl.

However, cost-effective control can be achieved by combining several different measures. For example, it has been demonstrated that polygenic, quantitative resistance to late blight can significantly reduce fungicide use compared with the amount necessary to protect susceptible cultivars. Further improvements in the efficiency of fungicide use may be achieved if spray treatments are applied according to a disease forecasting system such as Blitecast and BlightPro (see p.106). Consequent reductions in the number of spray applications not only reduce costs but also limit mechanical damage to the crop and reduce fungicide residues. Further refinements to the system include improving the resolution of computer models so that disease risk can be assessed on a local scale, even down to individual fields. This requires on-farm monitoring of the main environmental parameters regulating disease epidemics. There is also the prospect of stacking different *R* genes for late blight control using biotechnological approaches (see p.330) which may introduce more durable resistance into the crop.

Control of lettuce downy mildew, *Bremia lactucae*, is also based on the tactical use of host resistance combined with fungicides. In this case, combinations of different host *R* genes have helped to limit losses, deployed according to the incidence of pathotypes with corresponding virulence to these genes. The fungicide metalaxyl gave good control of the disease but as with potato late blight, fungicide-insensitive strains of the pathogen are now common. Reduced sensitivity to an alternative compound fosetyl-Al has also been detected in California. However, resistance to metalaxyl occurs in some, rather than all pathotypes of *B. lactucae*, and provided the genetic composition of the pathogen population is known, the fungicide can be used in combination with host resistance genes known to be effective against metalaxyl-resistant pathotypes. The success of this strategy depends on careful monitoring of the pathogen population from season to season, to detect any changes in virulence or fungicide sensitivity which might increase the vulnerability of the lettuce cultivars planted. As a fall-back, fungicides with an alternative mode of action, including multisite inhibitors, can be applied.

Recent surveys of fungicide use on cereal crops in Europe have shown wide variations in the amounts applied in different countries. This reflects the relative prevalence of the main foliar diseases such as Septoria tritici leaf blotch (*Zymoseptoria tritici*) and yellow and brown rusts (*Puccinia striiformis* and *P. triticina*). The yield gain from use of fungicides also varies between countries and therefore affects the relative value of applying them. The greatest responses to a fungicide program are obtained with the most susceptible cultivars, while the number of applications and dose of fungicide can be reduced on more resistant cultivars. Economic analysis has shown that the cost of disease control by fungicides can be approximately halved by growing cultivars with a higher level of resistance. At present, cultural and biological options to control epidemics of these foliar pathogens are limited, so a combination of host resistance and tactical use of fungicides is recommended.

Prolonging the Effectiveness of Host Resistance and Fungicides

There are several strategies for delaying the evolution of virulence to resistant crop cultivars based on either combining different *R* genes in a single cultivar or diversifying the deployment of *R* genes in space and time. Alternatively, fungicides may be used to slow the rate of multiplication of the pathogen population. In principle, this should equally affect both virulent and avirulent genotypes of the pathogen and hence reduce directional selection for virulence. Strategic use of fungicides can therefore extend the durability of host resistance.

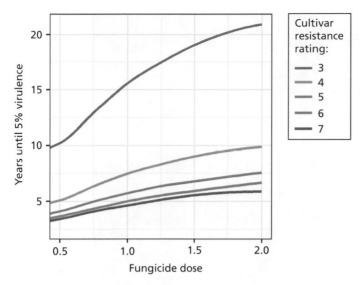

Figure 14.2   Extending the durability of cultivar resistance with fungicides. The model shows the effect of fungicide dose, measured as time to the emergence of virulence in 5% of the pathogen population on different crop cultivars varying in their resistance rating (from 3 = low resistance to 7 = high resistance). More resistant cultivars require lower fungicide doses but exert higher selection pressure for virulence. *Source:* Adapted from Carolan et al. (2017), courtesy of Joe Helps.

Figure 14.2 shows the output from an epidemiological model based on potato late blight, and the combination of host cultivars with different resistance ratings with varying doses of fungicide. The model measures the time taken from introduction of the resistant cultivar to emergence of virulence to the cultivar, measured as the moment at which 5% of the pathogen population is virulent. For practical purposes, this will be the point at which the resistance begins to lose its effectiveness. The model predicts that cultivars with higher resistance ratings exert greater selection for virulence, and that use of a fungicide can prolong the effectiveness of the resistance, depending on dose rate. Higher doses of fungicide delay the emergence of virulence more than lower doses.

The same principles can be applied to the sustainability of fungicides when applying single-site compounds that are vulnerable to resistance development. Use of resistant crop cultivars in combination with such chemicals reduces selection for fungicide resistance by slowing down the rate of epidemic growth and permitting lower doses of fungicide to be used. So, it should be possible to determine optimum combinations of cultivars and fungicides to maximize the durability of disease control.

Integration of Cultural and Biological Measures

Many *Phytophthora* species are soil-borne pathogens which cause root and crown rot diseases of important crops such as soybean, peppers, citrus, and avocado. *Phytophthora* root rot (PRR) caused by *P. cinnamomi* occurs in avocado (*Persea americana*) groves in many parts of the world. In plantations established on shallow desert soils in southern California,

Figure 14.3 Avocado grove in southern California affected by *Phytophthora* root rot. Tree at left shows reduced vigor and thinning of the crown, while more severely affected trees have been cut back. *Source:* Photo by John Lucas.

the disease can result in a severe decline, with many trees being killed outright (Figure 14.3). As PRR occurs in more than 60% of groves in this region, it poses a serious threat to avocado production.

The initial response to the problem of PRR was to expand new plantings onto virgin land free from the pathogen. However, this provided only a short-term solution, and current attempts to manage the disease are based on a series of different measures (Figure 14.4). The avocado cultivars originally planted are highly susceptible to PRR, and hence there has been a systematic program to find more resistant rootstocks, by screening germplasm from the center of genetic diversity of *Persea* in Central America. Several new sources have been discovered, including wild species such as *Persea schiedeana*. Rootstocks developed from such material do not completely prevent infection but reduce the severity of symptoms caused by PRR. Several routine precautions and cultural measures can also limit the spread and effects of this disease. Rigorous hygiene in the nurseries supplying seedlings, combined with systematic checks and quarantine measures, can prevent distribution of the disease to new areas. It has also been shown that careful management of the soil and water regime in avocado groves can reduce infection.

It is wellknown that many soil-borne *Phytophthora* species are favored by irrigation (Figure 14.5) which encourages the production and dispersal of zoospores. Planting trees on mounds, which aids drainage, together with precise regulation of irrigation to prevent waterlogging, is beneficial. Most of the soils used for avocado production are relatively low in organic matter, and mulching or manuring can improve soil structure and may also boost populations of potentially antagonistic soil microorganisms. Certain soils, for instance in Australia, have been shown to be naturally suppressive to *P. cinnamomi*, and

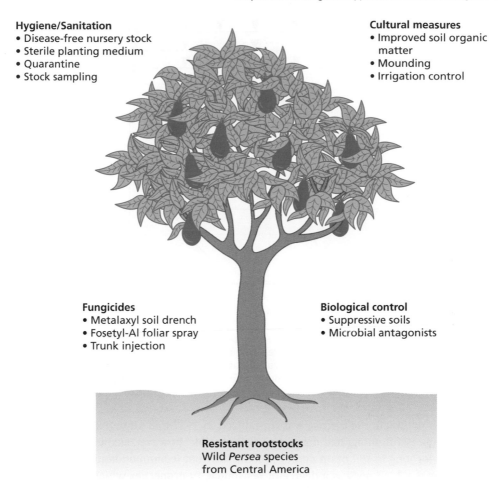

Hygiene/Sanitation
• Disease-free nursery stock
• Sterile planting medium
• Quarantine
• Stock sampling

Cultural measures
• Improved soil organic matter
• Mounding
• Irrigation control

Fungicides
• Metalaxyl soil drench
• Fosetyl-Al foliar spray
• Trunk injection

Biological control
• Suppressive soils
• Microbial antagonists

Resistant rootstocks
Wild *Persea* species
from Central America

Figure 14.4 Strategies for integrated control of *Phytophthora* root rot of avocado (*Persea americana*). *Source:* Based on Coffey (1987).

such biological control might be supplemented by use of appropriate microbial antagonists capable of colonizing the rhizosphere.

Finally, at least two classes of fungicides, the acylalanides and phosphonates, are active against *P. cinnamomi* and have been used to control PRR in the field. Metalaxyl is usually applied as a soil drench but fosetyl-Al, being phloem mobile, can be used as a foliar spray or introduced into individual trees by trunk injection (see Chapter 11, Figure 11.5). The latter method is very efficient, with two injections a year giving good protection of large trees. The various strategies for integrated control of PRR in avocado are summarized in Figure 14.4.

Good crop hygiene and cultural practices combined with host resistance and the strategic use of chemicals have also been employed to control the bacterial disease citrus canker *Xanthomonas axonopodis* pv. *citri*. The stringent quarantine measures applied to prevent introduction of this pathogen have been described earlier (see p.131–2).

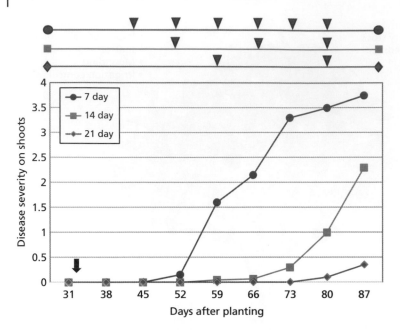

Figure 14.5 Effect of irrigation on severity of root rot of squash (*Cucurbito pepo*) caused by *Phytophthora capsici*. Plots with infested soil were irrigated at 7-, 14- or 21-day intervals. The arrow on the *x*-axis indicates day when soil was inoculated, and the arrows above the graph indicate the times of the different irrigation treatments. *Source:* After Café-Filho et al. (1995).

In the south of Brazil, the disease was introduced in 1957 and attempts to eradicate it proved unsuccessful. Integrated management is possible, however, as the pathogen does not survive for long in soil or on nonhost plants. Current recommendations are to plant citrus cultivars which possess a degree of resistance, in properties which are free from the disease or in which eradication measures have been applied at least one year previously. The health of nursery stock is carefully monitored to ensure that diseased plants are not distributed. These hygiene measures are combined with the use of windbreaks to reduce dispersal of the bacterium, and applications of copper-based bacteriocides to trees each year during the new flush of growth.

Making Decisions

Practical crop protection requires the grower to take decisions such as which cultivar to plant, and when or whether to apply agrochemicals or some other control measure. In situations where an obviously destructive disease outbreak is occurring, such as rust epidemics on cereals or late blight in a potato crop, the decision is usually simple. But for many diseases, the critical factors may not be obvious and the final outcome may be far from clear. There is a need for additional information on which to base a course of action. Improving the accuracy of decision making is therefore an important aspect of integrated disease control. Hence the ongoing interest in the development of **decision support systems** (DSS) to help growers make the correct choices for pest and disease management. Such systems derive from long-established principles of disease forecasting (see p.104), but differ in the scale and complexity of data they can handle and the active participation of growers in the decision process. An appropriate analogy is with financial markets where investment decisions are made based on share indices and predictions of future events. Furthermore, as field crops are exposed to simultaneous attack by several different pathogens or pests, any management system must take these threats into account within an integrated scheme of control measures.

Initial forecasting systems were mainly based on integrating disease risk factors such as the occurrence of weather favorable for infection, pathogen reproduction or vector migration on a regional basis, with alerts issued to growers. With the advent of web-based systems, growers can now input information on the location of their farm, crop cultivar (and hence resistance rating), sowing date, and other information relevant to disease risk. Outputs include definition of pest or disease thresholds at which action needs to be taken. Advice on appropriate spray treatments and a cost–benefit analysis of fungicide use can be incorporated to allow economic comparisons to be made. As more data are gathered, season by season, there is also the opportunity to refine disease simulation models to more accurately predict the outcome of a particular course of action. Adoption of such systems has been shown to reduce fungicide use when compared with calendar spray regimes. For instance, use of a decision support system for control of fruit rots of strawberries caused by the fungi *Colletotrichum acutatum* and *Botrytis cinerea* in the south-eastern USA was shown to provide protection equivalent to that achieved with a calendar-based program while applying only half the number of sprays.

Practical experience with DSS has shown that their adoption by farmers is often low. Many of the forecasting systems require regular inputs by growers or advisors and are seen as too time-consuming compared with more routine strategies. In many cases, there is also a risk-averse culture with a reluctance to rely on predictions, often based on weather variables, rather than an insurance approach with regular preventive treatments. There is the additional problem that farmers usually have to deal with a range of biotic threats that vary season by season and the DSS may not cover all the relevant problems. To help overcome these limitations, wider consultation and more effective knowledge exchange between scientists and end-users are required during development of a DSS to ensure that the system fits with the wider demands of farm management.

The Impact of New Technology – Digital Agriculture

As new technologies for the detection and diagnosis of disease come on stream, coupled with more sophisticated communication networks, real-time monitoring of crop health, increased computing power, and robotics, the practice of crop protection is likely to undergo major changes. Figure 14.6 shows the potential integration of some of the new sensing technologies with computer models based on increasingly large datasets, providing a range of site-specific outputs to growers and advisors. In addition to collecting on-farm meteorological data, inputs might include remote sensing by satellites and drones or in-field biosensors detecting air-borne pathogen inoculum. Hand-held devices to help identify

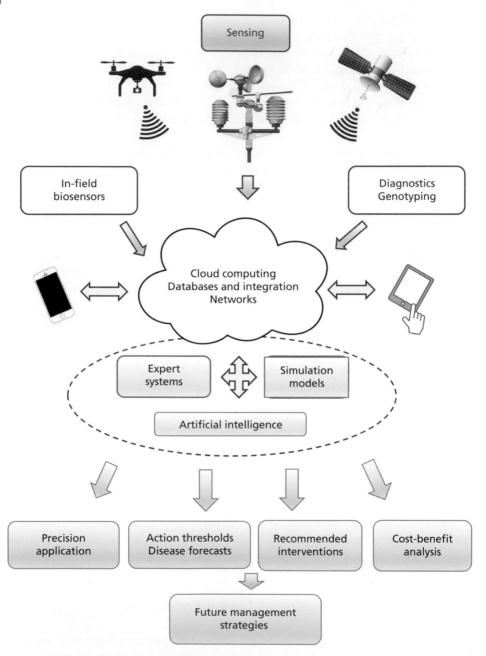

Figure 14.6 Digital agriculture and potential applications in crop protection.

diseases and also genotype pathogens for virulence, fungicide resistance, and other properties can provide real-time information analogous to the point-of-care diagnostics now being used in medicine. The large amounts of information collected and stored in the Cloud can be continually added to existing datasets to provide a valuable resource for further analysis.

For example, techniques of artificial intelligence (AI) such as machine learning might be employed to identify recurrent patterns and improve the identification of pests, diseases, and weeds. Such artificially trained systems can achieve high levels of accuracy, saving staff time and resources spent on conventional diagnosis. Access to more data can also continually refine simulation models and improve the accuracy of disease forecasts.

The outputs from an integrated digital agriculture system (Figure 14.6) include generating action thresholds to inform decisions on disease and pest control, or alternatively to guide automated machinery for precision interventions, such as applying pesticides when and where required. Driverless tractors or robots guided by a global positioning system (GPS) are already being used on farms and in glasshouses, for instance to target agrochemicals or to harvest produce, and the trend toward greater automation will continue. The overall aim is to make the practice of farming more efficient and productive, with fewer inputs and less impact on the environment. However, the extent to which such systems will be adopted will depend on how well they perform in practice, as well as socioeconomic factors and the type of crops grown.

The Impact of Biotechnology and Genomics

Another series of technological advances that will provide new opportunities for disease management are now arising from genomics and techniques for modifying crops by transformation or gene editing (Table 14.1). The availability of full-genome sequences of major crop species is already providing more rapid and precise ways of identifying novel sources of genetic resistance to pathogens, within both the gene pool of the crop and its wild rela-

Table 14.1 Some applications of genomics and gene editing in disease management

The host plant

- Discovery of new *R* genes by sequence capture and enrichment
- Identifying genes involved in quantitative disease resistance
- Creating new combinations (stacking) of resistance genes
- Engineering novel *R* gene specificities
- Manipulating genes involved in immune response pathways

The pathogen

- Designing more specific molecular diagnostics
- Rapid characterization of new invasive pathogens
- Defining effector repertoires in pathogen populations
- Field pathogenomics to track changes in virulence
- Identifying mutations determining resistance to fungicides
- Exploring variation in fungicide targets to counter resistance
- Discovering novel targets for intervention by plant breeding or chemistry

The microbiota

- Characterizing microbial communities influencing disease development
- Identifying genetic traits important for biocontrol of plant pathogens

tives. The fact that many *R* genes possess conserved features aids the identification of previously unknown genes of potential value in resistance breeding. Genome-wide searches have, for instance, shown that most plants possess hundreds of NLR-encoding genes exhibiting extensive diversity. Using a technique known as sequence capture, combined with association genetics, functional genes for resistance to black stem rust *Puccinia graminis* have recently been identified in the wild grass *Aegilops tauschii*, one of the progenitors of cultivated wheat. Such approaches should be applicable to any crop species where germplasm collections are available. Other families of genes involved in host–pathogen interaction, such as pattern recognition receptors, can also be explored.

Together, these represent a rich source of natural genetic variation that should broaden the options for creating more durable resistance by stacking combinations of genes detecting disease agents. In the longer term, it may also be possible to engineer new recognition specificities through gene editing, thereby further extending the spectrum of pathogens and variants which can be controlled.

One obstacle to the wider use of GM and gene editing, however, is the regulatory framework in many countries, especially in Europe, that places tight restrictions on the deployment of genetically modified crops. There is a need to develop a forward-looking system based on scientific evidence rather than ideology.

Further advances are being made based on the comparative genomics of plant pathogens and other microbes associated with plants. These include more accurate discrimination between species and pathotypes, together with identification of specific sequences for the design of molecular diagnostic assays. Genomic approaches have proved of particular value in characterizing new invasive pathogens, such as the ash dieback fungus *Hymenoscyphus fraxineus*, that has recently spread across Europe from its center of origin in Asia. Much lower levels of genetic diversity are found in the European population than in the pathogen's native range, suggesting a bottleneck was passed through during its westward spread. Tighter biosecurity measures are therefore recommended to prevent the introduction of further diversity that might exacerbate an already destructive pandemic. Another example of the rapid deployment of genomic resources to analyze a disease epidemic concerns a sudden outbreak of blast on wheat crops in Bangladesh in 2016.

The fungus *Magnaporthe oryzae* is best known as a destructive pathogen of rice and some wild grasses, but more recently has emerged as a serious threat to wheat in parts of South America. Disease symptoms are typically seen on the ears of infected plants, with premature bleaching, shriveled grain, and eventual loss of the whole ear. Wheat-infecting isolates of *M. oryzae* have been shown to be genetically distinct from rice-infecting isolates of the fungus. Wheat blast had not been reported in Asia until early 2016 when severe outbreaks occurred on crops in several regions of Bangladesh. In response to this emergency, samples were collected from the field and the transcriptome of infected and uninfected tissues was sequenced. The sequences obtained were then mapped to the genomes of wheat and *Magnaporthe grisea*. This analysis confirmed that the causal agent was indeed the wheat blast fungus but in addition, the isolates responsible were closely related to those found in Brazil, indicating that the pathogen was most likely introduced from South America. This discovery again confirmed the need for more stringent biosecurity in international trade networks, but also suggested that measures taken to limit losses from blast in Brazil might also be effective in Asia.

Similar techniques can be used to characterize effector repertoires and virulence in pathogen populations, informing choice of crop cultivars that might counter any changes and also mutations likely to affect sensitivity to fungicides (Table 14.1). It is possible to explore such variation to more effectively manage resistance, and in the longer term to identify new targets for intervention in the infection process. Recent advances in the analysis of the plant microbiome should clarify interactions affecting crop growth or disease suppression, and identify novel candidates for potential use as biocontrol agents.

Designing Future Farms

Agriculture in the twenty-first century faces major challenges. On the one hand, there is the need to produce more food and outputs such as biofuels and other renewable feed-stocks for industry. At the same time, this has to be achieved without requiring additional land, water, and nutrient inputs, while reducing harmful by-products such as greenhouse gases and maintaining biodiversity. Preventing losses due to pests, weeds, and diseases is an important part of this challenge. There is now an ongoing debate about how to farm more efficiently and sustainably without damaging or degrading natural ecosystems.

Part of the solution should be provided by the new technologies described in the previous sections. More resilient crop varieties that require fewer inputs of fertilizer and other agrochemicals, including pesticides, are likely to be produced using advanced breeding methods. As more plant genomes are sequenced, new sources of resistance to pests and pathogens will be discovered to further expand the gene pool available to breeders. Novel approaches for editing and stacking resistance genes will, subject to regulatory constraints, provide more durable protection and hence reduce reliance on pesticides. But there is also a requirement to design crop production systems that exert less selection for pathogen genotypes able to invade and exploit the available host plants. Greater diversification of crops and cultivars by various means to create genetic mosaics should reduce the risk of epidemic disease or sudden pest outbreaks. This might be managed at different scales ranging from individual fields to farms to the wider landscape or production region. Diversification in time can be achieved by greater use of seasonal rotations, with either different crop species or cultivars possessing different combinations of *R* genes.

There is no single or simple answer to the challenges currently facing agriculture, and different solutions may be found depending on the crop, available technology, and socio-economic circumstances. There is pressure to move away from the industrialized farming model based on uniform monocultures, but intensive production systems will still be required if future demands for food security are to be met. Organic farms generally use fewer inputs and provide environmental benefits in terms of soil quality and biodiversity but are, on average, 20% less productive than conventional farms. There is a trend toward smaller-scale local production units around major cities, as well as novel solutions such as vertical farms on urban buildings that require less land and use hydroponics to supply water and nutrients. Integration of crop production with aquaculture is being explored as one way of conserving water and recycling nutrients that are often lost to rivers and coastal waters. Systems based on combinations of annual crops, farm forestry, and livestock production are also being used in several countries as a means to exploit more natural fertilizers and also to provide the farmer with a range of commodities for sale.

At the opposite extreme are industrial-scale glasshouses covering many hectares with high levels of environmental control and automation, combined heat and power generation, and advanced technologies such as robotic handling and harvesting. One such complex in southern England provides around 10% of all the tomatoes, peppers, and cucumbers consumed annually in the country. Under the strictly maintained light levels and heat and carbon dioxide concentrations, tomatoes can be harvested every single day of the year. The crop protection problems in these very different systems will vary, although the degree of environmental control and surveillance in modern glasshouses, coupled with the use of biological agents, should reduce dependence on pesticides.

The Technology Gap

While these technological advances promise to change the ways in which we farm, and to provide new solutions for the threats posed by pests and diseases, we must not lose sight of the fact that for many crops, in less developed regions of the world, productivity remains poor and limited by lack of knowledge and resources. This dilemma is exacerbated by the continual growth in human populations in such regions and environmental degradation such as deforestation, salination, and soil erosion. Here, the real challenge is to boost the yield and reliability of staple crops through improvements in agronomy and a better understanding of the main constraints on productivity. The scale of losses due to pests and pathogens remains unacceptably high, especially if postharvest losses are included. For some tropical crops, the principal disease problems are poorly defined, and basic information on pathogen epidemiology is still lacking. In this situation, improved diagnosis and advice on low-cost control measures are required. There is currently a shortage of scientists skilled in knowledge transfer and the practical application of new technologies in the field. However, the examples described above using genomic techniques to identify and track emerging diseases show that these approaches are not necessarily confined to more developed countries.

Even if improved productivity can keep pace with population growth, there are other uncertainties, such as the likely impact of climate change which is already leading to more extreme temperatures, wide fluctuations in rainfall, and additional stresses on crops. Alterations in global climate patterns will affect the distribution of pathogens and their vectors, and hence the incidence and severity of disease. The predicted outcome of such change is largely a matter of speculation but one thing is certain. New problems will continue to arise, and the plant pathologist will still have an important role to play in limiting the threat posed by plant disease.

Further Reading

Reviews and Papers

Arora, S., Steuernagel, B., Gaurav, K. et al. (2019). Resistance gene cloning from a wild crop relative by sequence capture and association genetics. *Nature Biotechnology* 37: 139–143. https://doi.org/10.1038/s41587-018-0007-9.

Barzman, M., Bàrberi, P., Birch, A.N.E. et al. (2015). Eight principles of integrated pest management. *Agronomy for Sustainable Development* 35 (4): 1199–1215. https://doi.org/10.1007/s13593-015-0327-9.

Carolan, K., Helps, J., van den Berg, F. et al. (2017). Extending the durability of cultivar resistance by limiting epidemic growth rates. *Proceedings of the Royal Society B* 284: ii. https://doi.org/10.1098/rspb.2017.0828.

Fedoroff, N.V., Battisti, D.S., Beachy, R.N. et al. (2010). Radically rethinking agriculture for the 21st century. *Science* 327 (5967): 833–834. https://doi.org/10.1126/science.1186834.

Islam, M.T., Croll, D., Gladieux, P. et al. (2016). Emergence of wheat blast in Bangladesh was caused by a South American lineage of *Magnaporthe oryzae*. *BMC Biology* 14: 84. https://doi.org/10.1186/s12915-016-0309-7.

Jørgensen, L.N., Hovmøller, M.S., and Hansen, J.G. (2014). IPM strategies and their dilemmas including an introduction to www.eurowheat.org. *Journal of Integrative Agriculture* 13 (2): 265–281. https://doi.org/10.1016/S2095-3119(13)60646-2.

Liakos, K.G., Busato, P., Moshou, D. et al. (2018). Machine learning in agriculture: a review. *Sensors* 18 (8): 2674. https://doi.org/10.3390/s18082674.

Michelmore, R., Coaker, G., Bart, R. et al. (2017). Foundational and translational research opportunities to improve plant health. *Molecular Plant-Microbe Interactions* 30 (7): 515–516. https://doi.org/10.1094/MPMI-01-17-0010-CR.

Newlands, N.K. (2018). Model-based forecasting of agricultural crop disease risk at the regional scale, integrating airborne inoculum, environmental, and satellite-based monitoring data. *Frontiers in Environmental Sciences* 6: 63. https://doi.org/10.3389/fenvs.2018.00063.

Pertot, I., Caffi, T., Rossi, V. et al. (2017). A critical review of plant protection tools for reducing pesticide use on grapevine and new perspectives for the implementation of IPM in viticulture. *Crop Protection* 97: 70–84. https://doi.org/10.1016/j.cropro.2016.11.025.

Pretty, J. (2018). Intensification for redesigned and sustainable agricultural systems. *Science* 362: eaav0294. https://doi.org/10.1126/science.aav0294.

Smith, V.H., McBride, R.C., and Shurin, J.B. (2015). Crop diversification can contribute to disease risk control in sustainable biofuels production. *Frontiers in Ecology and the Environment* 13 (10): 561–567. https://doi.org/10.1890/150094.

Appendix 1

Annotated List of Pathogens and the Diseases they Cause

Main diseases considered in the text. These are arranged according to the type of pathogen and within each group on an alphabetic basis. Note: Latin names subject to taxonomic revision.

Pathogen	Host	Disease	Notes
FUNGI			
Alternaria solani	Potato and tomato	Early blight	Common in warmer countries.
Armillaria mellea	Trees	Butt and root rot	Common in plantations and gardens.
Bipolaris maydis *(Cochliobolus heterostrophus)*	Maize	Leaf blight	Cause of southern corn leaf blight, now controlled by resistant varieties.
Bipolaris sorokiana *(Cochliobolus sativus)*	Cereals	Root rot and leaf spot	Wide host range, common as seedling disease.
Blumeria graminis	Cereals and grasses	Powdery mildew	Prevalent in Europe, control by host resistance and systemic fungicides.
Botrytis cinerea	Vegetables and vines	Gray mold	Opportunistic pathogen of a wide range of fruits and vegetables. Control by fungicides but resistance is common.
Botrytis fabae	Broad beans (*Vicia*)	Chocolate spot	Aggressive pathogen in wet conditions.
Cercospora nicotianae	Tobacco	Frog-eye	Favored by warm weather.
Cladosporium fulvum	Tomato	Leaf mold	Occurs as many races. Well defined gene-for-gene system.
Claviceps purpurea	Cereals and grasses	Ergot	Replaces grain with black sclerotia containing alkaloids.

(Continued)

Plant Pathology and Plant Pathogens, Fourth Edition. John A. Lucas.
© 2020 John Wiley & Sons Ltd. Published 2020 by John Wiley & Sons Ltd.
Companion website: www.wiley.com/go/Lucas_PlantPathology4

Pathogen	Host	Disease	Notes
Cochliobolus (Helminthosporium) carbonum	Maize	Leaf spot	Pathogenic strains produce a host specific toxin (HC-toxin) inactivated by resistant maize genotypes.
Cochliobolus (Helminthosporium) victoriae	Oats	Victoria blight	Susceptible cultivars sensitive to host-specific toxin victorin.
Colletotrichum acutatum	Strawberry, lupin, others	Black spot and anthracnose	Worldwide distribution, spread by vegetative propagation. Notifiable disease in UK.
Colletotrichum circinans	Onions	Smudge disease	Attacks white-skinned varieties.
Colletotrichum gloeosporioides	Fruits and plantation crops	Postharvest rots and anthracnose of leaves	Common in wet tropics.
Colletotrichum graminicola	Maize and other cereals	Leaf blight and stalk rot	Infects all aerial parts of plant, especially the stalk. Common in no-till cultivation.
Colletotrichum higginsianum	Many hosts in Brassicaceae including *Arabidopsis*	Anthracnose	Model hemibiotrophic pathogen.
Colletotrichum lindemuthianum	Bean	Anthracnose	Occurs as races differentiated on bean cultivars.
Colletotrichum orbiculare	Cucumber, melon	Anthracnose	Common in humid regions. Control by cultural methods and fungicides.
Cronartium ribicola	Pines	White blister rust	Alternate host is *Ribes* spp.
Cryphonectria parasitica	Chestnut	Chestnut blight	Causes cankers and death of trees. Serious in N. America and Europe.
Epichloe typhina	Grasses	Choke disease	Systemic infection which inhibits flowering.
Erysiphe cichoracearum	Wide host range including cucurbits and *Arabidopsis*	Powdery mildew	White powdery lesions on leaves
Fusarium culmorum	Cereals	Foot rot	One of a complex of pathogens which infect the stem base.
Fusarium graminearum	Wheat and other cereals	Head blight	One of several *Fusarium* species infecting the ear causing contamination by mycotoxins.

Pathogen	Host	Disease	Notes
Fusarium oxysporum	Many hosts	Vascular wilt	Soil-borne pathogen occurring as distinct form species infecting different hosts.
e.g. f.sp. *cubense*	Banana	Panama disease	Destructive disease controlled by host resistance, but new races occur.
f.sp. *lycopersici*	Tomato	Wilt	Worldwide occurrence, controlled by soil sanitation and resistant cultivars.
Fusarium solani f.sp. *pisi* (*Nectria haematococca*)	Pea	Foot and root rot	Other form species attack beans and other legumes.
Fusarium virguliforme	Soybean	Sudden death syndrome	Controlled by soil sanitation measures.
Gaeumannomyces graminis	Cereals	Take-all	Soil-borne pathogen infecting roots.
Gibberella fujikuroi	Rice	Foolish seedling or bakanae	Seed or soil-borne disease.
Hemileia vastatrix	Coffee	Leaf rust	Most destructive disease of coffee, now worldwide.
Heterobasidion annosum	Conifers	Root and butt-rot	Common in plantations, controlled using antagonistic fungus.
Hymenoscyphus fraxineus	Ash trees	Dieback	Invasive pathogen has recently spread throughout Europe.
Leptosphaeria maculans	Oilseed rape (canola)	Stem canker	Managed by cultivar resistance and autumn fungicide sprays.
Magnaporthe oryzae	Rice and grasses (*M. grisea*)	Rice blast	Species complex, major rice pathogen
Melampsora lini	Flax	Rust	Widespread disease, best known as basis of gene for gene theory.
Melampsora spp.	Willow	Rusts	*M. epitea* and several other species infecting biomass willows.
Microcyclus ulei	Rubber	Leaf blight	Endemic in S. America, but not in Malaysia, from which it is excluded by quarantine.
Microsphaera alphitoides	Oak	Powdery mildew	Can be serious on young trees in nurseries.
Monilinia fructicola	Apple and plum	Brown rot	Postharvest disease of fruit.

(Continued)

Pathogen	Host	Disease	Notes
Moniliophthora (Crinipellis) perniciosa	Cocoa	Witch's broom	Damaging disease in S. America.
Mycosphaerella fijiensis	Bananas	Black sigatoka	Worldwide spread, now dominant disease of banana replacing yellow sigatoka, *M. musicola.*
Nectria galligena	Apples and pears	Canker	Sanitation important, plus fungicides and biocontrol.
Nectria haematococca	Peas	Foot and root rot	See *Fusarium solani* f.sp. *pisi.*
Oculimacula yallundae and *O. acuformis*	Wheat and other cereals	Eyespot	Infect stem base causing lodging.
Ophiostoma novo-ulmi	Elm	Dutch elm	New aggressive strain responsible for pandemic in northern hemisphere
Parastagonospora nodorum	Wheat	Leaf and glume blotch	Late-season infections attack ears causing severe lesions.
Penicillium digitatum	Citrus fruit	Green mold	Postharvest rot, infects via wounds.
Penicillium expansum	Apples and pears	Blue mold	Postharvest soft rot.
Penicillium italicum	Citrus fruit	Blue mold	Postharvest rot occurs with green mold (above).
Phakopsora pachyrhizi	Soybean	Asian soybean rust	Major pathogen of soybeans has now spread to western hemisphere
Pilgeriella anacardii	Cashew	Black mold	Infects dwarf cashew trees during rainy season.
Plasmodiophora brassicae	Brassica	Clubroot	Causes galls on roots. Persists in soil for long periods.
Polymyxa graminis *P. betae*	Cereals Sugarbeet		Chytrid fungi which infect host roots. Unimportant as pathogens but vectors of serious virus diseases, e.g., rhizomania.
Puccinia coronata	Oats and barley	Crown rust	Controlled by resistant cultivars and fungicides.
Puccinia graminis	Cereals	Black stem rust	Form species on different hosts, e.g., f.sp. *tritici* (wheat) and f.sp. *hordei* (barley).

Pathogen	Host	Disease	Notes
Puccinia striiformis	Cereals	Yellow stripe rust	Sporadic occurrence, usually in cool conditions. Has now migrated worldwide.
Puccinia triticina	Wheat and other cereals	Brown rust/leaf rust	Mainly controlled by host resistance.
Pyrenophora avenae	Oats	Seedling blight, leaf stripe	Seed-borne disease effectively controlled by seed dressing with fungicides.
Pyrenophora teres	Barley	Net blotch	Seed-borne disease, occurs as net and spot forms.
Pyrenophora tritici-repentis	Wheat	Tan spot	Survives on stubble, infection from air-borne ascospores.
Ramularia collo-cygni	Barley	Leaf spot	Emerging disease in NW Europe and New Zealand
Rhizoctonia cerealis	Cereals	Sharp eyespot	Survives on infected stubble, part of stem base disease complex.
Rhizoctonia solani	Potato	Stem canker	Also affects tubers, causing black scurf.
Rhynchosporium commune	Barley	Leaf blotch/scald	Splash-dispersed pathogen prevalent in wet conditions.
Rigidoporus microporus	Trees, especially rubber	White root disease	Major disease of rubber plantations, spreads by rhizomorphs infecting roots.
Sclerotinia (Clarireedia) homeocarpa	Turf grasses	Dollar spot	Problem on golf courses, managed by nutrient regime and fungicides.
Sclerotinia minor	Lettuce	Lettuce drop	Favored by wet conditions, survives as sclerotia.
Sclerotinia sclerotiorum	Numerous	White mold, stem rot	Survives in soil as sclerotia.
Sclerotium cepivorum	Onion and garlic	White rot	Survives in soil as sclerotia.
Sclerotium rolfsii	Numerous	Root and collar rot	Very wide host range, especially on vegetable crops in warm climates.
Septoria lycopersici	Tomato and potato	Leaf spot	Potato form restricted to S. America.
Sphaerotheca fuliginea	Cucurbits	Powdery mildew	Important in glasshouses.

(Continued)

Pathogen	Host	Disease	Notes
Taphrina deformans	Peach	Leaf curl	Can also infect fruits, control by fungicide spray in autumn or spring before bud break.
Thielaviopsis basicola	Cotton and tobacco	Black root rot	Soil-borne pathogen with wide host range, including vegetables and ornamentals.
Tilletia indica	Wheat	Karnal bunt	Originally in Asia, has now spread to Mexico.
Tilletia tritici syn. *Tilletia caries*	Wheat, rye and grasses	Common bunt	Devastating disease now controlled by seed treatment.
Uromyces appendiculatus syn. *Uromyces fabae*	Bean	Rust	One of a number of *Uromyces* species causing rust on legumes.
Uromyces vicia fabae	Faba bean	Rust	Common disease on faba (broad) bean.
Ustilago maydis	Maize	Corn smut	Forms galls on aerial parts of plant, especially on ears.
Ustilago nuda	Barley	Loose smut	Controlled by treating seeds with fungicides.
Venturia inaequalis	Apple	Scab	Almost all commercial cultivars are susceptible, hence control relies on fungicides.
Verticillium albo-atrum	Hop, alfalfa, tomato and others	Wilt	Soil-borne pathogen, difficult to control.
Verticillium dahliae	Numerous, including cotton and potatoes	Wilt	Produces microsclerotia. More common in warmer regions.
Zymoseptoria tritici	Wheat	Septoria tritici leaf blotch	Occurs worldwide. Has increased in importance in recent years.

OOMYCETES

Pathogen	Host	Disease	Notes
Albugo candida	Brassicas	White blister	Obligate pathogen occurring on many crucifer species.
Bremia lactucae	Lettuce	Downy mildew	Important disease controlled by host resistance and fungicides.
Hyaloperonospora aradopsidis	*Arabidopsis*	Downy mildew	Model pathosystem for study of host–pathogen interactions.

Pathogen	Host	Disease	Notes
Hyaloperonospora parasitica	Brassica	Downy mildew	Occurs on many crucifer species.
Peronospora pisi (P. viciae)	Pea and other legumes	Downy mildew	Transmitted by seed- and soil-borne oospores.
Peronospora hyoscyami f.sp. *tabacina*	Tobacco	Blue mold	Destructive disease controlled by host resistance and fungicides.
Phytophthora alni	Alder	Decline disease	Recently emerged in Europe through hybridization between two other species.
Phytophthora cinnamomi	Trees and shrubs	Root rot	Many woody hosts attacked, including avocado and Jarrah dieback, a devastating disease of native Australian forests.
Phytophthora erythroseptica	Potato	Pink rot	Occurs in waterlogged soils.
Phytophthora infestans	Potato	Late blight	Affects haulm and tubers, control mainly by fungicide sprays based on disease forecasts.
Phytophthora megasperma	Vegetable crops and trees	Root rot	Wide host range, but host-adapted forms occur, e.g., *P. megasperma* var. *sojae* on soybean.
Phytophthora ramorum	Oak, larch, and other forest trees and shrubs	Sudden oak death	Invasive species spreading in California and Europe.
Phytophthora parasitica	Tobacco	Black shank	Infects roots and stem base.
Plasmopara viticola	Vines	Downy mildew	European grape varieties are most susceptible, control by fungicides.
Pseudoperonospora cubensis	Cucurbits	Downy mildew	Particularly important in glasshouse crops. Control by fungicides.
Pseudoperonospora humuli	Hops	Downy mildew	Outbreaks in wet conditions, control by resistant cultivars and fungicides.
Pythium species	Numerous	Damping off	Seedling diseases prevalent in wet soils.

(Continued)

Pathogen	Host	Disease	Notes
BACTERIA			
Agrobacterium tumefaciens	Numerous dicotyledons	Crown gall	Plant tumor, worldwide distribution, can be controlled by bacterial antagonist.
Agrobacterium rhizogenes	Numerous dicotyledons	Hairy root	Closely related to crown gall.
Candidatus Liberibacter asiaticus	Citrus	Huanglongbing (greening)	Most destructive disease of citrus worldwide. Phloem-limited bacterium spread by insect vectors.
Dickeya solani	Potato	Black leg	Emerging pathogen similar to *Pectobacterium* but more aggressive.
Erwinia amylovora	Pears and apples	Fireblight	Control by sanitation and copper sprays.
Pectobacterium atrosepticum	Potato	Black leg	Attacks stem base and tubers, causing necrosis.
Pectobacterium carotovorum subsp. *carotovorum*	Numerous vegetables	Soft rot	Often a postharvest pathogen, control by sanitation and reducing humidity.
Erwinia chrysanthemi	Vegetables and ornamentals.	Soft rot	Now reclassified as *Dickeya dadantii*
Erwinia tracheiphila	Cucurbits	Bacterial wilt	Spread by insect vectors, may be controlled by insecticides.
Pantoea stewartii	Maize	Stewart's wilt	Indigenous to the Americas. Transmitted by flea beetles.
Ralstonia solanacearum	Banana, tomato, potato, tobacco	Bacterial wilt Moko disease (banana)	Mainly important in the tropics, control by resistant varieties.
Pseudomonas syringae	Numerous	Diverse	Important pathogen existing as numerous pathovars adapted to different hosts.
Ps. syringae pv. *actinidiae*	Kiwifruit	Bacterial canker	Recently emerged aggressive strain threatening kiwi production worldwide.
Ps. syringae pv. *phaseolicola*	Bean	Halo blight	Seed-borne disease spread by rain splash.
Ps. syringae pv. *savastanoi*	Olives, oleander and privet	Knot and canker	Enters by pruning wounds, leaf scars and wounds. Control by sanitation and copper sprays.

Pathogen	Host	Disease	Notes
Ps. syringae pv. *tabaci*	Tobacco	Wildfire	Angular lesions turning to scorch symptoms. Most serious in tropical countries.
Streptomyces scabiei	Potato	Common scab	Favored by low soil moisture during tuber development and high pH.
Xanthomonas axonopodis pv. *citri*	Citrus	Canker	Serious disease usually controlled by quarantine and eradication measures.
Xanthomonas axonopodis pv. *vesicatoria*	Tomato and pepper	Bacterial spot	Control by sanitation, seed certification, cultural measures and copper sprays.
Xanthomonas campestris	Numerous	Diverse	Occurs as pathovars adapted on different hosts.
Xanthomonas campestris pv. *manihotis*	Cassava	Bacterial blight	Occurs in Asia, Africa and Latin America. Often spread by cuttings.
Xanthomonas campestris pv. *oryzae*	Rice	Leaf blight	Important in Far East. Bacterial blight of rice was the first disease in which the genomes of both the pathogen and host were sequenced.
Xylella fastidiosa	Many woody hosts including citrus, grapevine and olives	Citrus variegated chlorosis, Pierce's disease of grapevine, olive quick decline	Limited to xylem tissues, spread by vegetative propagation and probably insect vectors.
PHYTOPLASMAS			
Candidatus Phytoplasma palmae	Coconut	Lethal yellowing	High mortality, control by resistant varieties and hybrids.
Spiroplasma kunkelii sp.	Maize	Corn stunt	Transmitted by leafhopper vectors, controlled by resistant cultivars.
Spiroplasma citri	Citrus	Stubborn disease	Occurs in phloem tissues, spread by leafhopper vectors.
VIRUSES			
Cassava mosaic, including African (ACMV) and East African (EACMV) viruses	Cassava	Mosaic (CMD)	Widespread in Africa, transmitted by the whitefly *Bemisia tabaci*.

(Continued)

Pathogen	Host	Disease	Notes
Cassava brown streak virus (CBSV)	Cassava	Brown streak (CBSD)	At least two virus strains spreading in East Africa.
Barley yellow dwarf virus (BYDV)	Cereals	Yellow dwarf	Spread by aphid vectors.
Barley yellow mosaic virus (BaYMV)	Barley	Yellow mosaic	Spread by the soil-borne plasmodiophorid *Polymyxa graminis*.
Barley mild mosaic virus (BaMMV)	Barley	Mild mosaic	Spread by *Polymyxa graminis*.
Beet necrotic yellow vein virus (BNYVV)	Sugar beet	Rhizomania	Spread by another root-invading plasmodiophorid *Polymyxa betae*.
Cauliflower mosaic virus (CaMV)	Cruciferae, Solanaceae and others	Mosaic, stunting and abnormal growth	First DNA plant virus discovered.
Citrus tristeza virus (CTV)	Citrus	Decline, yellows and stem pitting	Most damaging virus of citrus worldwide, spread by vegetative propagation.
Cocoa swollen shoot virus (CSSV)	Cocoa	Swollen shoot	Serious in West Africa. Spread by scale-insect vectors, control by eradication.
Cucumber mosaic virus (CMV)	Numerous	Mosaic and stunting	Worldwide distribution. Has the widest host range of any plant virus.
Grapevine fanleaf virus (GFLV)	Vine	Fanleaf and mosaic	Transmitted by nematode vectors and vegetative propagation, control by cloning virus-free stocks.
Grapevine leaf roll virus (GLRV)	Vine	Leaf roll	Causes characteristic red pigmentation of leaves. Increasing problem in USA and other countries.
Papaya ringspot virus (PRSV)	Papaya	Ring spot	Tropical countries including Hawaii where it was controlled by transgenic resistance.
Potato leaf roll virus (PLRV)	Potato	Leaf roll	Transmitted by aphids in a persistent manner.
Potato virus X (PVX)	Potato	Mild mosaic, mottle	Often asymptomatic, but may cause severe disease when another virus is present, e.g., potato virus Y.
Potato virus Y (PVY)	Potato	Yield loss plus necrotic ringspot of tubers	Spread by aphids or in tubers.

Pathogen	Host	Disease	Notes
Prunus necrotic ringspot virus (PNRV)	Stone fruit trees	Diverse	Transmitted by budding, grafting, seed and pollen. Control by using virus-free nursery stock.
Raspberry bushy dwarf virus (RBDV)	*Rubus* spp.	Bushy dwarf	Transmitted by seed and pollen, control by using resistant cultivars.
Raspberry ringspot virus (RRV)	Raspberry, tomato, gooseberry and many perennial plants		Nematode and seed transmitted, control involves eradicating nematodes.
Tobacco mosaic virus (TMV)	Numerous	Mosaic	Wide host range but important only in Solanaceae. Often spread by contact.
Tobacco rattle virus (TRV)	Wide host range >400 species	Stunt	Transmitted by root-infecting nematodes.
VIROIDS			
Coconut cadang cadang viroid (CCVd)	Coconut	Translation – dying dying	Occurs in Philippines, transmitted mechanically and probably by insects.
Potato spindle tuber viroid (PSTVd)	Potato and tomato	Spindle tuber	Spread by contact, insects and seed. Control by planting viroid-free stock.
ANGIOSPERMS			
Arceuthobium spp.	Conifers	Dwarf mistletoe	Reduces tree growth and wood quality. Spreads by explosive discharge of seeds.
Orobanche spp.	Sunflower, tobacco, tomato and faba bean	Broomrape	Root parasite, seeds survive in infested soil for long periods.
Striga hermonthica	Maize, rice, sorghum, millet and sugarcane	Witchweed	Root parasite, control difficult. Quarantine, sanitation, catch crops and resistant cultivars may be used.
Viscum album	Angiosperm trees	European mistletoe	Damaging on fruit trees. e.g., apple.

Appendix 2

Reference Sources for Figures

Ali, S., Gladieux, P., Leconte, P. et al. (2014). Origin, migration routes and worldwide population genetic structure of the wheat yellow rust pathogen *Puccinia striiformis* f.sp. *tritici. PLoS Pathogens* 10 (1): e1003903. https://doi.org/10.1371/journal.ppat.1003903.

Ayres, P.G. and Jones, P. (1975). Increased transpiration and the accumulation of root absorbed 86Rb in barley leaves infected by *Rhynchosporium secalis* (leaf blotch). *Physiological Plant Pathology* 7: 49–58.

Bailey, J.A. and Deverall, B.J. (1971). Formation and activity of phaseollin in the interaction between bean hypocotyls (*Phaseolus vulgaris*) and physiological races of *Colletotrichum lindemuthianum. Physiological Plant Pathology* 1: 435–449.

Barzman, M., Bàrberi, P., Birch, A.N.E. et al. (2015). Eight principles of integrated pest management. *Agronomy for Sustainable Development* 35 (4): 1199–1215. https://doi.org/10.1007/s13593-015-0327-9.

Basallote-Ureba, M.J. and Melero-Vara, J.M. (1993). Control of garlic white rot by soil solarization. *Crop Protection* 12: 219–223.

Blanc, S., Drucker, M., and Uzest, M. (2014). Localizing viruses in their plant hosts. *Annual Review of Phytopathology* 52: 403–425. https://doi.org/10.1146/annurev-phyto-102313-045920.

Bonas, U., Schulte, R., Fenselau, S. et al. (1991). Isolation of a gene cluster from *Xanthomonas campestris* pv. *vesicatoria* that determines pathogenicity and the hypersensitive response on pepper and tomato. *Molecular Plant–Microbe Interactions* 4: 81–88.

Bowyer, P., Clarke, B.R., Lunness, P. et al. (1995). Host range of a plant pathogenic fungus determined by a saponin detoxifying enzyme. *Science* 267: 371–374. DOI: 10.1126/science.7824933.

Bowyer, P., Mueller, E., and Lucas, J. (2000). Use of an isocitrate lyase promoter-GFP fusion to monitor carbon metabolism in the plant pathogen *Tapesia yallundae* during infection of wheat. *Molecular Plant Pathology* 1 (4): 253–262.

Bracker, C.E. (1968). Ultrastructure of the haustorial apparatus of *Erysiphe graminis* and its relationship to the epidermal cell of barley. *Phytopathology* 58: 12–30.

Braun, E.J. and Howard, R.J. (1994). Adhesion of fungal spores and germlings to host plant surfaces. *Protoplasma* 181: 202–212.

Brown, I.R. and Mansfield, J.W. (1988). An ultrastructural cytochemistry and quantitative analyses, of the interactions between pseudomonads and leaves of *Phaseolus vulgaris* L. *Physiological and Molecular Plant Pathology* 33: 351–376.

Bunders, J. (1988). Appropriate biotechnology for sustainable agriculture in developing countries. *Trends in Biotechnology* 6: 173–180.

Café-Filho, A.C., Duniway, J.M., and Davis, R.M. (1995). Effects of the frequency of furrow irrigation on root and fruit rots of squash caused by *Phytophthora capsici*. *Plant Disease* 79: 44–48.

Campbell, W.P. and Griffiths, D.A. (1974). Development of endoconidial chlamydospores in *Fusarium culmorum*. *Transactions of the British Mycological Society* 63: 221–228.

Cardoso, J.E., Felipe, E.M., Cavalcante, M.J.B. et al. (2000). Precipitação pluvial e progresso da antracnose e do mofo-preto do cajueiro (*Anacardium occidentale*). *Summa Phytopathologica. Jaboticabal* 26 (4): 413–416.

Carolan, K., Helps, J., van den Berg, F. et al. (2017). Extending the durability of cultivar resistance by limiting epidemic growth rates. *Proceedings of the Royal Society B* 284: 20170828. DOI: https://doi.org/10.1098/rspb.2017.0828.

Carver, T.L.W. and Thomas, B.J. (1990). Normal germling development by *Erysiphe graminis* on cereal leaves freed of epicuticular wax. *Plant Pathology* 39: 367–375.

Chowdhury, J., Henderson, M., Schweizer, P. et al. (2014). Differential accumulation of callose, arabinoxylan and cellulose in nonpenetrated versus penetrated papillae on leaves of barley infected with *Blumeria graminis* f. sp. *hordei*. *New Phytologist* 204: 650–660. https://doi.org/10.1111/nph.12974.

Coffey, M.D. (1975). Ultrastructural features of the haustorial apparatus of the white blister fungus *Albugo candida*. *Canadian Journal of Botany* 53: 1285–1299.

Coffey, M.D. (1976). Flax rust resistance involving the K gene: an ultrastructural survey. *Canadian Journal of Botany* 54: 1443–1457.

Coffey, M.D. (1987). Phytophthora root rot of avocado. An integrated approach to control in California. *Plant Disease* 71: 1046–1052.

Coley-Smith, J.R., Mitchell, C.M., and Sansford, C.E. (1990). Long-term survival of sclerotia of *Sclerotium cepivorum* and *Stromatinia gladioli*. *Plant Pathology* 39: 58–69.

Cramer, H.H. (1967). *Plant Protection and World Crop Production*. Leverkusen: Bayer.

Daniels, A., Lucas, J.A., and Peberdy, J.F. (1991). Morphology and ultrastructure of W and R pathotypes of *Pseudocercosporella herpotrichoides* on wheat seedlings. *Mycological Research* 95: 385–397.

Dixon, R.A. and Harrison, M.J. (1990). Activation and structure of organization of genes involved in microbial defense in plants. *Advances in Genetics* 28: 165–217.

Dodds, P.N. and Rathjen, J.P. (2010). Plant immunity: towards an integrated view of plant–pathogen interactions. *Nature Reviews Genetics* 11: 539–548. https://doi.org/10.1038/nrg2812.

Eckert, J.W. (1977). *Antifungal Compounds, Vol. I, Discovery, Development and Uses* (eds. M.R. Siegel and H.D. Sisler), 269–352. New York: Marcel Dekker.

Esau, K. (1968). *Viruses in Plant Hosts*. Milwaukee, Wisconsin: University of Wisconsin Press.

Evans, E. (1977). Efficient use of systemic fungicides. In: *Systemic Fungicides* (ed. R.W. Marsh), 198–212. London: Longman.

Falk, S.P., Pearson, R.C., Gadoury, D.M. et al. (1996). *Fusarium proliferatum* as a biocontrol agent against grape downy mildew. *Phytopathology* 86: 1010–1017.

Faris, J.D., Liu, Z., and Xu, S.S. (2013). Genetics of tan spot resistance in wheat. *Theoretical and Applied Genetics* 126: 2197–2217. https://doi.org/10.1007/s00122-013-2157-y.

Flores, R., Gago-Zachert, S., Serra, P. et al. (2014). Viroids: survivors from the RNA world? *Annual Review of Microbiology* 68: 395–414.

Fotopoulos, V., Gilbert, M.J., Pittman, J.K. et al. (2003). The monosaccharide transporter gene AtSTP4 and the cell wall invertase Atβfruct1 are induced in Arabidopsis during infection with the fungal biotroph *Erysiphe cichoracearum*. *Plant Physiology* 132: 821–829. https://doi.org/10.1104/pp.103.021428.

Fraaije, B.A., Cools, H.J., Fountaine, J. et al. (2005). Role of ascospores in further spread of QoI-resistant cytochrome b alleles (G143A) in field populations of *Mycosphaerella graminicola*. *Phytopathology* 95: 933–941. https://doi.org/10.1094/phyto-95-0933.

Friesen, T.L., Stukenbrock, E.H., Liu, Z. et al. (2006). Emergence of a new disease as a result of interspecific virulence gene transfer. *Nature Genetics* 38: 953–956. https://doi.org/10.1038/ng1839.

Gan, P., Ikeda, K., Irieda, H. et al. (2013). Comparative genomic and transcriptomic analyses reveal the hemibiotrophic stage shift of *Colletotrichum* fungi. *New Phytologist* 197: 1236–1249. https://doi.org/10.1111/nph.12085.

Gao, X., Huang, Q., Zhao, Z. et al. (2016). Studies on the infection, colonization, and movement of *Pseudomonas syringae* pv. *actinidiae* in kiwifruit tissues using a GFPuv-labeled strain. *PLoS One* 11 (3): e0151169. https://doi.org/10.1371/journal.pone.0151169.

Georgopoulos, S.G. and Skylakakis, G. (1986). Genetic variability in the fungi and the problem of fungicide resistance. *Crop Protection* 5 (5): 299–305. https://doi.org/10.1016/0261-2194(86)90107-9.

Giraldo, M.C. and Valent, B. (2013). Filamentous plant pathogen effectors in action. *Nature Reviews Microbiology* 11 (11): 800–814. https://doi.org/10.1038/nrmicro3119.

Gonsalves, D., Gonsalves, C., Ferreira, S. et al. (2004). Transgenic virus resistant papaya: from hope to reality for controlling Papaya Ringspot Virus in Hawaii. http://www.apsnet.org/edcenter/apsnetfeatures/Pages/PapayaRingspot.aspx.

González-Torres, R., Meléo-Vara, J.M., Gómez-Vázquez, J., and Jiménez-Diaz, R.M. (1993). The effects of soil solarization and soil fumigation on fusarium wilt of watermelon grown in plastic houses in South-Eastern Spain. *Plant Pathology* 42: 858–864.

Greaves, D.A., Hooper, A.J., and Walpole, B.J. (1983). Identification of barley yellow dwarf virus and cereal aphid infestations in winter wheat by aerial photography. *Plant Pathology* 32: 159–172.

Green, C.F. and Ivins, J.D. (1984). Late infestations of take-all (*Gaeumannomyces graminis* var *tritici*) on winter wheat (*Triticum aestivum* cv. Virtue): yield, yield components and photosynthetic potential. *Field Crops Research* 8: 199–206.

Grimmer, M.K., van den Bosch, F., Powers, S. et al. (2014). Evaluation of a matrix to calculate fungicide resistance risk. *Pest Management Science* 70 (6): 1008–1016. https://doi.org/10.1002/ps.3646.

Haesaert, G., Vossen, J.H., Custers, R. et al. (2015). Transformation of the potato variety Desiree with single or multiple resistance genes increases resistance to late blight under field conditions. *Crop Protection* 77: 163–175. https://doi.org/10.1016/j.cropro.2015.07.018.

Hamer, J.E., Howard, R.J., Chumley, F.G., and Valent, B. (1988). A mechanism for surface attachment of spores of a plant pathogenic fungus. *Science* 239: 288–290. https://doi.org/10.1126/science.239.4837.288.

Hammond-Kosack, K.E. and Jones, J.D.G. (2015). Responses to plant pathogens. In: *Biochemistry and Molecular Biology of Plants* (eds. B.B. Buchanan, W. Gruissem and R.L. Jones), 984–1050. Rockville, Maryland: American Society of Plant Physiologists.

Hargreaves, J.A., Mansfield, J.W., and Rossall, S. (1977). Changes in phytoalexin concentrations in tissues of the broad bean plant (*Vicia faba* L.) following inoculation with species of *Botrytis*. *Physiological Plant Pathology* 11: 227–242.

Harrison, N.A., Myrie, W., Jones, P. et al. (2002). 16S rRNA interoperon sequence heterogeneity distinguishes strain populations of palm lethal yellowing phytoplasma in the Caribbean region. *Annals of Applied Biology* 141: 183–193. https://doi.org/10.1111/j.1744-7348.2002.tb00211.x.

Hartung, J.S., Beretta, J., Brlansky, R.H. et al. (1994). Citrus variegated chlorosis bacterium: axenic culture, pathogenicity, and serological relationships with other strains of *Xylella fastidiosa*. *Phytopathology* 84 (6): 591–597.

Hayes, J.D. and Johnson, T.D. (1971). Breeding for disease resistance. In: *Diseases of Crop Plants* (ed. J.H. Western), 62–88. London: Macmillan.

Hedrick, S.A., Bell, J.N., Boller, T., and Lamb, C.J. (1988). Chitinase cDNA cloning and mRNA induction by fungal elicitor, wounding and infection. *Plant Physiology* 86: 182–186.

Hewitt, H.G. and Ayres, P.G. (1975). Changes in CO_2 and water vapour exchange rates in leaves of *Quercus robur* infected by *Microsphaera alphitoides* (powdery mildew). *Physiological Plant Pathology* 7: 127–137.

Hickey, E.L. and Coffey, M.D. (1977). A fine structural study of the pea downy mildew fungus *Peronospora pisi* in its host *Pisum sativum*. *Canadian Journal of Botany* 55: 2845–2858.

Hirano, S.S., Rouse, D.I., Clayton, M.K., and Upper, C.D. (1995). *Pseudomonas syringae* pv. *syringae* and bacterial brown spot of snap bean: a study of epiphytic bacteria and associated disease. *Plant Disease* 79: 1085–1093.

Holtz, B.A. and Weinhold, A.R. (1994). *Thielaviopsis basicola* in San Joaquin valley soils and the relationship between inoculum density and disease severity of cotton seedlings. *Plant Disease* 78: 986–990.

Honegger, R. (1985). Scanning electron-microscopy of the fungus-plant cell interface. A simple preparative technique. *Transactions of the British Mycological Society* 84: 530–533.

Hooker, W.J. (1956). Foliage fungicides for potatoes in Iowa. *American Potato Journal* 33: 47–52.

Hubbard, A., Lewis, C.M., Yoshida, K. et al. (2015). Field pathogenomics reveals the emergence of a diverse wheat yellow rust population. *Genome Biology* 16: 23. https://doi.org/10.1186/s13059-015-0590-8.

James, W.C. (1971). An illustrated series of assessment keys for plant diseases and their preparation and usage. *Canadian Plant Disease Survey* 51 (2): 39–65.

Jeffree, C.E., Blake, E.A., and Holloway, P.A. (1976). Origins of the fine structure of plant epicuticular waxes. In: *Microbiology of Aerial Plant Surfaces* (eds. C.H. Dickinson and T.F. Preece), 119–158. London: Academic Press.

Jenkyn, J.F., Gutteridge, R.J., and Jalaluddin, M. (1994). Straw disposal and cereal diseases. In: *Ecology of Plant Pathogens* (eds. J.P. Blakeman and B. Williamson), 285–300. Wallingford, UK: CAB International.

Jones, J.D.G. and Dangl, G. (2006). The plant immune system. *Nature* 444: 323–329. https://doi.org/10.1038/nature05286.

Jones, J.D.G., Witek, K., Verweij, W. et al. (2014). Elevating crop disease resistance with cloned genes. *Philosophical Transactions of the Royal Society B* 369: 20130087. DOI: https://doi.org/10.1098/rstb.2013.0087.

Jones, S., Baizan-Edge, A., MacFarlane, S. et al. (2017). Viral diagnostics in plants using next generation sequencing: computational analysis in practice. *Frontiers in Plant Science* 8: 1770. https://doi.org/10.3389/fpls.2017.01770.

Karisto, P., Walter, A., Hund, A. et al. (2018). Ranking quantitative resistance to Septoria tritici blotch in elite wheat cultivars using automated image analysis. *Phytopathology* 108: 568–581. https://doi.org/10.1094/phyto-04-17-0163-r.

Keon, J., Antoniw, J., Carzaniga, R. et al. (2007). Transcriptional adaptation of *Mycosphaerella graminicola* to programmed cell death (PCD) of its susceptible wheat host. *Molecular Plant–Microbe Interactions* 20 (2): 178–193. https://doi.org/10.1094/mpmi-20-2-0178.

Koczan, J.M., McGrath, M.J., and Zhao, Y. (2009). Contribution of *Erwinia amylovora* exopolysaccharides amylovoran and levan to biofilm formation: implications in pathogenicity. *Phytopathology* 99 (11): 1237–1244. https://doi.org/10.1094/PHYTO-99-11-1237.

Kuck, K.H. and Russell, P.E. (2006). FRAC: combined resistance risk assessment. *Aspects of Applied Biology* 78: 3–10.

Lacerda, A.F., Vasconcelos, E.A.R., and Pelegrini, P.B. (2014). Antifungal defensins and their role in plant defense. *Frontiers in Microbiology 5: article*: 116. https://doi.org/10.3389/fmicb.2014.00116.

Large, E.C. (1952). Interpretation of progress curves for potato blight and other plant diseases. *Plant Pathology* 1: 109–117.

Latunde-Dada, A.O. and Lucas, J.A. (2001). The plant defence activator acibenzolar-S-methyl primes cowpea (*Vigna unquiculata* [L.] Walp.) seedlings for rapid induction of resistance. *Physiological and Molecular Plant Pathology* 58: 199–208. https://doi.org/10.1006/pmpp.2001.0327.

Latunde-Dada, A.O., Bailey, J.A., and Lucas, J.A. (1997). Infection process of *Colletotrichum destructivum* O'Gara from lucerne (*Medicago sativa* L.). *European Journal of Plant Pathology* 103: 35–41.

Leng, P.-F., Lübberstedt, T., and Xu, M.-L. (2017). Genomics-assisted breeding – a revolutionary strategy for crop improvement. *Journal of Integrative Agriculture* 16 (12): 2674–2685. https://doi.org/10.1016/S2095-3119(17)61813-6.

Liu, L., Kloepper, J.W., and Tuzun, S. (1995). Induction of systemic resistance in cucumber against bacterial angular leaf spot by plant growth-promoting rhizobacteria. *Phytopathology* 85: 843–847.

Livne, A. and Daly, J.M. (1966). Translocation in healthy and rust-affected beans. *Phytopathology* 56: 170–175.

Lucas, J.A. (2017a). Resistance management: we know why but do we know how? In: *Modern Fungicides and Antifungal Compounds*, vol. VIII (eds. H.B. Deising, B. Fraaije, Mehl et al.), 3–14. Braunschweig: Deutsche Gesellschaft.

Lucas, J.A. (2017b). Fungi, food crops, and biosecurity: advances and challenges. In: *Advances in Food Security and Sustainability*, vol. 2 (ed. D. Barling), 1–40. Burlington, Virginia: Academic Press.

Lucas, J.A., Hawkins, N.J., and Fraaije, B.A. (2015). The evolution of fungicide resistance. *Advances in Applied Microbiology* 90: 29–92. https://doi.org/10.1016/bs.aambs.2014.09.001.

Ma, X., Wang, X., Cheng, J. et al. (2015). Microencapsulation of *Bacillus subtilis* B99-2 and its biocontrol efficiency against *Rhizoctonia solani* in tomato. *Biological Control* 90: 34–41. https://doi.org/10.1016/j.biocontrol.2015.05.013.

Machado, A.K., Brown, N.A., Urban, M. et al. (2017). RNAi as an emerging approach to control fusarium head blight disease and mycotoxin contamination in cereals. *Pest Management Science* 74: 790–799. https://doi.org/10.1002/ps.4748.

Magyarosy, A.C., Schürmann, P., and Buchanan, B.B. (1976). Effect of powdery mildew infection on photosynthesis of leaves and chloroplasts of sugar beets. *Plant Physiology* 57: 486–489.

Mahlein, A.-K. (2016). Plant disease detection by imaging sensors – parallels and specific demands for precision agriculture and plant phenotyping. *Plant Disease* 100 (2): 241–251. https://doi.org/10.1094/PDIS-03-15-0340-FE.

Melchers, L.S. and Hooykaas, P.J.J. (1987). Virulence of *Agrobacterium*. In: *Oxford Surveys of Plant Molecular and Cell Biology* (ed. B.J. Miflin), 167–220. Oxford: Oxford University Press.

Melotto, M., Underwood, W., Koczan, J. et al. (2006). Plant stomata function in innate immunity against bacterial invasion. *Cell* 126 (5): 969–980. https://doi.org/10.1016/j.cell.2006.06.054.

Mohn, G., Koehl, P., Budzikiewicz, H. et al. (1994). Solution structure of pyoverdin GM-II. *Biochemistry* 33: 2843–2851.

Money, N.P., Caesar-TonThat, T.C., Frederick, B. et al. (1998). Melanin synthesis is associated with changes in hyphopodial turgor, permeability, and wall rigidity in *Gaeumannomyces graminis* var. *graminis*. *Fungal Genetics and Biology* 24: 240–251. https://doi.org/10.1006/fgbi.1998.1052.

Monteiro, F. and Nishimura, N.T. (2018). Structural, functional, and genomic diversity of plant NLR proteins: an evolved resource for rational engineering of plant immunity. *Annual Review of Phytopathology* 56: 243–267. https://doi.org/10.1146/annurev-phyto-080417-045817.

Mundt, C.C. and Leonard, K.J. (1985). A modification of Gregory's model for describing plant disease gradients. *Phytopathology* 75: 930–935.

Oerke, E.C. (2006). Crop losses to pests. *Journal of Agricultural Science* 144: 31–43.

Oerke, E.C., Dehne, H.-W., Schonbeck, F., and Weber, A. (1994). *Crop Production and Crop Protection*. Amsterdam: Elsevier.

Owera, S.A.P., Farrar, J.F., and Whitbread, R. (1983). Translocation from leaves of barley infected with brown rust. *New Phytologist* 94: 111–123.

Paul, N.D. and Ayres, P.G. (1986a). The impact of a pathogen, *Puccinia lagenophorae* on populations of groundsel, *Senecio vulgaris*, overwintering in the field. 1. Mortality, vegetative growth and the development of size hierarchies. *Journal of Ecology* 74: 1069–1084.

Paul, N.D. and Ayres, P.G. (1986b). The impact of a pathogen, *Puccinia lagenophorae*, on populations of groundsel, *Senecio vulgaris*, overwintering in the field. 2. Reproduction. *Journal of Ecology* 74: 1085–1094.

Petre, B. and Kamoun, S. (2014). How do filamentous pathogens deliver effector proteins into plant cells? *PLoS Biology* 12 (2): e1001801. https://doi.org/10.1371/journal.pbio.1001801.

Piasecka, A., Jedrzejczak-Rey, N., and Bednarek, P. (2015). Secondary metabolites in plant innate immunity: conserved function of divergent chemicals. *New Phytologist* 206 (3): 948–964. https://doi.org/10.1111/nph.13325.

Plumb, R.T. (1986). A rational approach to the control of barley yellow dwarf virus. *Journal of the Royal Agricultural Society of England* 147: 162–171.

Politis, D.J. (1976). Ultrastructure of penetration by *Colletotrichum graminicola* of highly resistant oat leaves. *Physiological Plant Pathology* 8: 117–122. https://doi.org/10.1016/0048-4059(76)90044-8.

Ramond, E., Maclachlan, C., Clerc-Rosset, S. et al. (2016). Cell division by longitudinal scission in the insect endosymbiont *Spiroplasma poulsonii*. *MBio* 7 (4): e00881–e00816. https://doi.org/10.1128/mBio.00881-16.

Read, N.D., Kellock, L.J., Knight, H. et al. (1992). Contact sensing during infection by fungal pathogens. In: *Perspectives in Plant Cell Recognition* (eds. J.A. Callow and J.R. Green), 137–192. Cambridge, UK: Cambridge University Press.

Rogers-Lewis, D. (1985). Dried peas – review of trials 1980–84. In: *Terrington Experimental Husbandry Farm Annual Review*. 32–37, UK: MAFF.

Saijo, P.Y., Loo, E.P., and Yasuda, S. (2018). Pattern recognition receptors and signaling in plant–microbe interactions. *Plant Journal* 93: 592–613. https://doi.org/10.1111/tpj.1380.

Saile, E., McGarvey, J.A., Schell, M.A. et al. (1997). Role of extracellular polysaccharide and endoglucananase in root invasion and colonization of tomato plants by *Ralstonia solanacearum*. *Phytopathology* 87: 1264–1271.

Scheffer, R.P. and Yoder, O.C. (1972). Host-specific toxins and selective toxicity. In: *Phytotoxins in Plant Diseases* (eds. R.K.S. Wood, A. Ballio and A. Graniti), 251–272. London: Academic Press.

Scholes, J.D. and Rolfe, S.A. (2009). Chlorophyll fluorescence imaging as tool for understanding the impact of fungal diseases on plant performance: a phenomics perspective. *Functional Plant Biology* 36: 880–892. https://doi.org/10.1071/FP09145.

Scholthof, K.B.G., Scholthof, H.B., and Jackson, A.O. (1993). Control of plant-virus diseases by pathogen-derived resistance in transgenic plants. *Plant Physiology* 102: 7–12.

Seck, M., Roelfs, A.P., and Teng, P.S. (1988). Effects of leaf rust (*Puccinia recondita tritici*) on yield of four isogenic wheat lines. *Crop Protection* 7: 39–42.

Sels, J., Janick, M., De Coninck, B.M.A. et al. (2008). Plant-related (PR) proteins: A focus on PR peptides. *Plant Physiology and Biochemistry* 46: 941–950. https://doi.org/10.1016/j.plaphy.2008.06.011.

Showalter, A.M., Bell, J.N., Cramer, C.L. et al. (1985). Accumulation of hydroxyproline-rich glycoprotein mRNAs, in response to fungal elicitor and infection. *Proceedings of National Academy of Sciences USA* 82: 6551–6556.

Sikora, E.J., Allen, T.W., Wise, K.A. et al. (2014). A coordinated effort to manage soybean rust in North America: a success story in soybean disease monitoring. *Plant Disease* 98: 864–875. https://doi.org/10.1094/PDIS-02-14-0121-FE.

Slusarenko, A.J., Croft, K.P., and Voisey, C.R. (1991). Biochemical and molecular events in the hypersensitive response of beans to *Pseudomonas syringae* pv. *phaseolicola*. In: *Biochemistry and Molecular Biology of Plant–Pathogen Interactions* (ed. C.J. Smith), 127–143. Oxford: Clarendon Press.

Smedegaard-Petersen, V. (1984). The role of respiration and energy generation in diseased and disease-resistant plants. In: *Plant Diseases: Infection, Damage and Loss* (eds. R.K.S. Wood and G.J. Jellis), 73–85. Oxford: Blackwell Scientific Publications.

Smith, V.L., Campbell, C.L., Jenkins, S.F. et al. (1988). Effects of host density and number of disease foci on epidemics of southern blight of processing carrots. *Phytopathology* 78: 595–600.

Soylu, S., Brown, I., and Mansfield, J.W. (2005). Cellular reactions in Arabidopsis following challenge by strains of *Pseudomonas syringae*: from basal resistance to compatibility. *Physiological and Molecular Plant Pathology* 66: 232–243. https://doi.org/10.1016/j.pmpp.2005.08.005.

Spanu, P.D., Abbott, J.C., Amselem, J. et al. (2010). Genome expansion and gene loss in powdery mildew fungi reveal tradeoffs in extreme parasitism. *Science* 330: 1543–1546. https://doi.org/10.1126/science.1194573.

Spiers, A.G. and Hopcroft, D.H. (1996). Morphological and host range studies of *Melampsora* rusts attacking *Salix* species in New Zealand. *Mycological Research* 100: 1163–1175.

Sweetmore, A., Simons, S.A., and Kenward, M. (1994). Comparison of disease progress curves for yam anthracnose (*Colletotrichum gloeosporioides*). *Plant Pathology* 43: 206–215.

Szabo, L.J. and Bushnell, W.R. (2001). Hidden robbers: the role of fungal haustoria in parasitism of plants. *Proceedings of the National Academy of Sciences USA* 98 (14): 7654–7655. https://doi.org/10.1073/pnas.151262398.

Torto-Alalibo, T., Collmer, C.W., Gwinn-Giglio, M. et al. (2010). Unifying themes in microbial associations with animal and plant hosts described using the gene ontology. *Microbiology and Molecular Biology Reviews* 74 (4): 479–503. https://doi.org/10.1128/MMBR.00017-10.

Turner, J.G. (1984). Role of toxins in plant disease. In: *Plant Diseases: Infection, Damage and Loss* (eds. R.K.S. Wood and G.J. Jellis), 3–12. Oxford: Blackwell Scientific Publications.

Vanderplank, J.E. (1963). *Plant Diseases: Epidemics and Control*. New York: Academic Press.

Wang, S., Welsh, L., Thorpe, P. et al. (2018). The *Phytophthora infestans* haustorium is a site for secretion of diverse classes of infection-associated proteins. *MBio* 9: e01216-18. https://doi.org/10.1128/mBio.01216-18.

Wellings, C.R. and McIntosh, R.A. (1990). *Puccinia striiformis* f.sp. *tritici* in Australasia: pathogenic changes during the first 10 years. *Plant Pathology* 39: 316–325.

Wolfe, M.S. (1993). Can the strategic use of disease resistant hosts protect their inherent durability? In: *Durability of Disease Resistance* (eds. T. Jacobs and J.E. Parlevliet). Dordrecht: Kluwer Academic Publishers.

Woloshuk, C.P. and Kolattukudy, P.E. (1986). Mechanism by which plant cuticle triggers cutinase gene expression in the spores of *Fusarium solani* f.sp. *pisi*. *Proceedings of the National Academy of Sciences USA* 83: 1704–1708. https://doi.org/10.1073/pnas.83.6.1704.

Yun, H.S., Lee, J.H., Park, W.J. et al. (2018). Plant surface receptors recognizing microbe-associated molecular patterns. *Journal of Plant Biology* 61: 111–120.

Zadoks, J.C. and Shein, R.D. (1979). *Epidemiology and Plant Disease Management*. Oxford: Oxford University Press.

Zhu, Q., Maher, E.A., Masoud, S. et al. (1994). Enhanced protection against fungal attack by constitutive co-expression of chitinase and glucanase genes in transgenic tobacco. *Biotechnology* 12: 807–812.

Index

Note: Page references in *italics* refer to Figures; those in **bold** refer to Tables.